METHODS IN MOLECULAR BIOLOGY

Series Editor
John M. Walker
School of Life and Medical Sciences
University of Hertfordshire
Hatfield, Hertfordshire, UK

For further volumes:
http://www.springer.com/series/7651

For over 35 years, biological scientists have come to rely on the research protocols and methodologies in the critically acclaimed *Methods in Molecular Biology* series. The series was the first to introduce the step-by-step protocols approach that has become the standard in all biomedical protocol publishing. Each protocol is provided in readily-reproducible step-by-step fashion, opening with an introductory overview, a list of the materials and reagents needed to complete the experiment, and followed by a detailed procedure that is supported with a helpful notes section offering tips and tricks of the trade as well as troubleshooting advice. These hallmark features were introduced by series editor Dr. John Walker and constitute the key ingredient in each and every volume of the *Methods in Molecular Biology* series. Tested and trusted, comprehensive and reliable, all protocols from the series are indexed in PubMed.

Digital PCR

Methods and Protocols

Edited by

Guillaume Pavlovic

PHENOMIN-Institut Clinique de la Souris, CELPHEDIA, CNRS, INSERM, Université de Strasbourg, Illkirch, France

Humana Press

Editor
Guillaume Pavlovic
PHENOMIN-Institut Clinique de la
Souris
CELPHEDIA, CNRS, INSERM, Université de Strasbourg
Illkirch, France

ISSN 1064-3745 ISSN 1940-6029 (electronic)
Methods in Molecular Biology
ISBN 978-1-0716-4766-0 ISBN 978-1-0716-4767-7 (eBook)
https://doi.org/10.1007/978-1-0716-4767-7

This Humana imprint is published by the registered company Springer Science+Business Media, LLC, part of Springer Nature.
The registered company address is: 1 New York Plaza, New York, NY 10004, U.S.A.

Preface

Polymerase chain reaction (PCR) technologies have revolutionized biology by providing a simple, rapid, and highly sensitive method for amplifying specific DNA sequences.

Standard PCR, developed in the 1980s, enabled DNA amplification but lacked quantification capabilities. This limitation led to the development of the second generation of PCR technologies: real-time PCR, also known as quantitative PCR (qPCR). Developed in the 1990s, qPCR introduced real-time fluorescence monitoring to measure DNA accumulation during the exponential amplification phase, allowing relative quantification against a standard curve.

Digital PCR (dPCR) has emerged as the third generation of PCR technologies and is considered to be as transformative as PCR and qPCR in molecular biology. First conceptualized in 1988 and formally named in 1999, dPCR differs from qPCR by partitioning samples into thousands of microreactions (droplets or nanowells), enabling absolute quantification by counting positive and negative partitions after endpoint amplification, thus eliminating the need for standard curves.

Early commercial dPCR systems (2006–2007) used microfluidic chips, but the introduction of droplet-based systems in 2011 significantly simplified implementation in any laboratory. Today, both droplet digital PCR (ddPCR) and chip digital PCR (cdPCR) are widely available, offering a broad range of applications.

This book aims to provide an overview of the various possibilities currently available, both in terms of instrumentation and applications. The selected protocols cover a wide range of uses, including fundamental research, preclinical studies, human applications, and regulatory contexts.

Additionally, protocols that further extend potential applications, such as protein detection, the development of custom dPCR systems, and non-digital PCR approaches like hpPCR technology, are presented.

The third generation of PCR is now fully mature and will continue to enhance target quantification, notably through its multiplexing capabilities, tolerance to inhibitors, and increased precision.

Illkirch, France *Guillaume Pavlovic*

Contents

Contributors

ALASDAIR J. ALLAN • *The Mary Lyon Centre, MRC Harwell, Didcot, UK*
RIKKE FREDSLUND ANDERSEN • *Department of Biochemistry and Immunology, Vejle Hospital, University Hospital of Southern Denmark, Vejle, Denmark; Department of Oncology, Vejle Hospital, University Hospital of Southern Denmark, Vejle, Denmark*
HENRIK ARONSSON • *Department of Biological and Environmental Sciences, University of Gothenburg, Gothenburg, Sweden*
SAID ASSOU • *IRMB, University of Montpellier, INSERM, CHU Montpellier, Montpellier, France*
JOHAN BANÉR • *Aplex Bio AB, Solna, Sweden*
AARON BIVINS • *Department of Civil & Environmental Engineering, Louisiana State University, Baton Rouge, LA, USA*
ALEXANDRA BOGOŽALEC KOŠIR • *Department of Biotechnology and Systems Biology, National Institute of Biology, Ljubljana, Slovenia*
SARA BRANDT-WINGE • *Department of Molecular Oncology, Institute for Cancer Research, Norwegian Radium Hospital, Oslo University Hospital, Oslo, Norway*
ELISABETTA BROSEGHINI • *IRCCS Azienda Ospedaliero-Universitaria di Bologna, Bologna, Italy*
VICTORIA CANNON • *Translational Science and Therapeutics Division, Fred Hutchinson Cancer Center, Seattle, WA, USA*
TINA BRANDT CHRISTENSEN • *Department of Biochemistry and Immunology, Vejle Hospital, University Hospital of Southern Denmark, Vejle, Denmark*
ANNA MARIA D'ERCHIA • *Department of Biosciences, Biotechnology and Environment, University of Bari Aldo Moro, Bari, Italy*
RÉMI DANGLA • *Stilla Technologies, Villejuif, France*
JOHN DE VOS • *IRMB, University of Montpellier, INSERM, CHU Montpellier, Montpellier, France; Department of Cell and Tissue Engineering, Saint-Eloi Hospital, Montpellier, France*
TINA DEMŠAR • *Department of Biotechnology and Systems Biology, National Institute of Biology, Ljubljana, Slovenia*
DAVID DOBNIK • *Department of Biotechnology and Systems Biology, National Institute of Biology, Ljubljana, Slovenia*
JULIE DONZIER • *Stilla Technologies, Villejuif, France*
GIORGIO DURANTE • *IRCCS Azienda Ospedaliero-Universitaria di Bologna, Bologna, Italy*
MANUELA FERRACIN • *IRCCS Azienda Ospedaliero-Universitaria di Bologna, Bologna, Italy; Department of Medical and Surgical Sciences (DIMEC), University of Bologna, Bologna, Italy*
ERMES FILOMENA • *Department of Biosciences, Biotechnology and Environment, University of Bari Aldo Moro, Bari, Italy*
BEATRICE FONTANA • *Department of Medical and Surgical Sciences (DIMEC), University of Bologna, Bologna, Italy*
ETIENNE FRADET • *Stilla Technologies, Villejuif, France*
JÖRG FUCHS • *Leibniz Institute of Plant Genetics and Crop Plant Research (IPK) OT Gatersleben, Seeland, Germany*

SHIMUL GHOSH • *Department of Civil & Environmental Engineering, Louisiana State University, Baton Rouge, LA, USA*
WILLIAM M. GRADY • *Translational Science and Therapeutics Division, Fred Hutchinson Cancer Center, Seattle, WA, USA; Department of Medicine, University of Washington School of Medicine, Seattle, WA, USA*
CHRISTY J. GREENWOOD • *The Mary Lyon Centre, MRC Harwell, Didcot, UK*
KYUDONG HAN • *Smart Animal Bio Institute, Dankook University, Cheonan, Republic of Korea; Center for Bio-Medical Engineering Core Facility, Dankook University, Cheonan, Republic of Korea; Department of Microbiology, College of Bio-convergence, Dankook University, Cheonan, Republic of Korea; HuNBiome Co., Ltd., R&D Center, Seoul, South Korea*
TORBEN FRØSTRUP HANSEN • *Department of Oncology, Vejle Hospital, University Hospital of Southern Denmark, Vejle, Denmark*
STEFAN HECKMANN • *Leibniz Institute of Plant Genetics and Crop Plant Research (IPK) OT Gatersleben, Seeland, Germany*
AMADEJ JELENČIČ • *Department of Biotechnology and Systems Biology, National Institute of Biology, Ljubljana, Slovenia; Jožef Stefan International Postgraduate School, Ljubljana, Slovenia*
JINUK JEONG • *Center for Bio-Medical Engineering Core Facility, Dankook University, Cheonan, Republic of Korea*
CÉCILE JOVELET • *Stilla Technologies, Villejuif, France*
JANINE KEHRLI • *Université de Strasbourg, CNRS, Architecture et Réactivité de l'ARN, UPR9002, Strasbourg, France*
RADI KHODR • *Université de Strasbourg, CNRS, Architecture et Réactivité de l'ARN, UPR9002, Strasbourg, France*
SONGMI KIM • *Smart Animal Bio Institute, Dankook University, Cheonan, Republic of Korea*
CLAUDIA LEONI • *Institute of Biomembranes, Bioenergetics and Molecular Biotechnologies, National Research Council, Bari, Italy*
GURO E. LIND • *Department of Molecular Oncology, Institute for Cancer Research, Norwegian Radium Hospital, Oslo University Hospital, Oslo, Norway; Department of Biosciences, Faculty of Mathematics and Natural Sciences, University of Oslo, Oslo, Norway*
LAURENT MANCHON • *Stem Genomics, CHU Montpellier, Montpellier, France*
MARINELLA MARZANO • *Institute of Biomembranes, Bioenergetics and Molecular Biotechnologies, National Research Council, Bari, Italy*
KEVIN MATTHAEI • *QIAGEN GmbH, Hilden, Germany*
MOJCA MILAVEC • *Department of Biotechnology and Systems Biology, National Institute of Biology, Ljubljana, Slovenia*
SEYOUNG MUN • *Smart Animal Bio Institute, Dankook University, Cheonan, Republic of Korea; Center for Bio-Medical Engineering Core Facility, Dankook University, Cheonan, Republic of Korea; Department of Cosmedical Materials, College of Bio-convergence, Dankook University, Cheonan, Republic of Korea*
SAMUEL NARMACK • *Aplex Bio AB, Solna, Sweden*
UMEAR NASEEM • *Aplex Bio AB, Solna, Sweden*
AFIF ABDEL NOUR • *QIAGEN GmbH, Hilden, Germany*
RALF PEIST • *QIAGEN GmbH, Hilden, Germany*
HEIDI PHARO • *Department of Molecular Oncology, Institute for Cancer Research, Norwegian Radium Hospital, Oslo University Hospital, Oslo, Norway*

Margarita Psallida • *Aplex Bio AB, Solna, Sweden*
Roberta Roncarati • *Department of Medical and Surgical Sciences (DIMEC), University of Bologna, Bologna, Italy; CNR Institute of Molecular Genetics "Luigi Luca Cavalli-Sforza", Unit of Bologna, Bologna, Italy*
Michael Ryckelynck • *Université de Strasbourg, CNRS, Architecture et Réactivité de l'ARN, UPR9002, Strasbourg, France*
Katharina Schaepe • *QIAGEN GmbH, Hilden, Germany*
Gerald Schock • *QIAGEN GmbH, Hilden, Germany*
Ruben R. G. Soares • *Aplex Bio AB, Solna, Sweden*
Dejan Štebih • *Department of Biotechnology and Systems Biology, National Institute of Biology, Ljubljana, Slovenia*
Suriya Tamilselvan-Nattar-Amutha • *Leibniz Institute of Plant Genetics and Crop Plant Research (IPK) OT Gatersleben, Seeland, Germany*
Lydia Teboul • *The Mary Lyon Centre, MRC Harwell, Didcot, UK*
Evy Marie Thorkildsen • *Department of Molecular Oncology, Institute for Cancer Research, Norwegian Radium Hospital, Oslo University Hospital, Oslo, Norway*
Sylvain Ursuegui • *Stilla Technologies, Villejuif, France*
Hege Marie Vedeld • *Department of Molecular Oncology, Institute for Cancer Research, Norwegian Radium Hospital, Oslo University Hospital, Oslo, Norway*
Esteri Viitanen • *Department of Biological and Environmental Sciences, University of Gothenburg, Gothenburg, Sweden*
Juline Vincent • *Stem Genomics, CHU Montpellier, Montpellier, France*
Jonathan Weiss • *QIAGEN GmbH, Hilden, Germany*
Jocelyn H. Wright • *Translational Science and Therapeutics Division, Fred Hutchinson Cancer Center, Seattle, WA, USA*
Cecilia C. S. Yeung • *Translational Science and Therapeutics Division, Fred Hutchinson Cancer Center, Seattle, WA, USA; Department of Laboratory Medicine and Pathology, University of Washington School of Medicine, Seattle, WA, USA*
Ming Yu • *Translational Science and Therapeutics Division, Fred Hutchinson Cancer Center, Seattle, WA, USA*
Reda Zenagui • *Stem Genomics, CHU Montpellier, Montpellier, France*

Chapter 1

A Beginner Step-by-Step Guide for Gene Expression Analysis Using QIAcuity Digital PCR

Jonathan Weiss, Gerald Schock, Afif Abdel Nour, Katharina Schaepe, Kevin Matthaei, and Ralf Peist

Abstract

Gene expression (GEX) is the process by which information from a gene is used to synthesize functional gene products, typically proteins, with messenger ribonucleic acid (mRNA) playing a crucial role. This process is tightly regulated to ensure genes are expressed at the right time, in the right cell type, and in appropriate amounts, which is essential for cellular function, development, and response to environmental signals. Aberrant gene expression can significantly impact cellular behavior and contribute to various diseases, making it a key area of research. Digital PCR (dPCR) offers a robust method for highly precise gene expression analysis by partitioning the PCR reaction into thousands of individual reactions, allowing for absolute quantification of nucleic acids without the need for standard curves. This method addresses several limitations of quantitative PCR (qPCR), such as the need for high PCR efficiency and susceptibility to inhibitors. dPCR's ability to accurately quantify low-abundance targets minimizes variability, artifacts, and nonreproducibility that may be observed with qPCR at low expression levels. However, qPCR remains advantageous for analyzing targets with widely different expression levels, provided that proper optimization and sample dilution are applied. Overall, dPCR offers substantial advantages for gene expression analysis, providing improved accuracy and greater resistance to inhibitors compared to qPCR. These capabilities empower researchers across diverse fields to explore gene expression hypotheses that were previously inaccessible or difficult to access using qPCR.

Key words Digital PCR (dPCR), Gene expression (Gex), Multiplexing, QIAcuity, RNA Quantification

1 Introduction

Gene expression (GEX) is the process by which information from a gene is used to synthesize functional gene products, such as proteins and noncoding RNAs. For proteins, messenger ribonucleic acid (mRNA) plays a crucial role in this process as it serves as the intermediary molecule that carries genetic information from DNA to protein synthesis. GEX is a tightly regulated process with several levels of regulation. Regulatory mechanisms ensure that genes are

Guillaume Pavlovic (ed.), *Digital PCR: Methods and Protocols*, Methods in Molecular Biology, vol. 2969,
https://doi.org/10.1007/978-1-0716-4767-7_1,

expressed at the right time, in the right cell type, and in the appropriate amount. This regulation is crucial for cellular function, development, and response to environmental signals. The expression of certain genes that differs from the normal state can have an enormous impact on cellular behavior, contributing to various diseases. Both gene expression during normal cellular processes and aberrant expression in diseases are part of the research in a large variety of fields and many different workgroups [1].

Digital PCR is based on a similar principle to end-point PCR but partitions the sample into many individual reactions (or partitions), each containing a limited number of DNA molecules. The template DNAs are randomly distributed across all available partitions. After PCR, the target molecule is detected by measuring the fluorescence of either sequence-specific DNA probes or intercalating dyes. The random distribution of template DNA molecules into the partitions allows the use of Poisson statistics to estimate the number of DNA molecules in each partition. The total number of target DNAs in all partitions of a well is then calculated by multiplying the average amount of target DNA per partition with the number of all valid partitions. Valid partitions are those filled with reaction mix, which is identified by the presence of a fluorescent dye within the mix. The target concentration is then calculated based on the volume of all valid partitions by dividing the total number of target DNAs by the total volume of all valid partitions. Absolute quantification by dPCR eliminates the need for standard curves to determine amounts of target DNA in a given sample [2, 3].

Elucidating the expression of genes using digital PCR solves several problems of quantitative PCR (qPCR)-based gene expression. Quantitative PCR requires a high and similar degree of PCR efficiency across all assays used, which in turn usually requires optimization and makes the PCR reaction susceptible to PCR inhibitors. A disadvantage even more pronounced when setting up a multiplex PCR reaction with several primers and hydrolysis probes. In digital PCR, imaging and result calculations usually only occur post-cycling and therefore PCR efficiency plays only a very minor role and the PCR inhibitor resistance is much higher [3]. Another significant limitation of qPCR-based gene expression analysis is its inherent inaccuracy [4]. In qPCR, one Cq difference between two PCR reactions is the smallest reliably discernable difference. Since Cq values are being determined during the logarithmic phase of the PCR reaction, one Cq difference corresponds to a doubling of template material [5]. In dPCR, due to the nature of absolute quantification, different numbers of mRNA molecules below 20% can be detected, allowing for a more refined gene expression analysis. Quantitative PCR remains advantageous over dPCR when analyzing two targets with significantly different expression levels, for instance, a gene of interest represented with a low number of molecules, alongside a highly expressed reference gene (represented by a large number of templates), such as *ACTB* (ß-Actin) or *GAPDH*. In qPCR, this is typically feasible without requiring sample dilutions. However, in dPCR, care must be taken to avoid overloading the system in such cases. This can be achieved by reducing the input material to a level where the target of interest remains reliably detectable or by selecting a reference gene that is stably expressed at a lower level [6].

Taken together, digital PCR offers tremendous advantages over qPCR when performing gene expression analysis. This protocol presents a basic step-by-step guide, which can be further refined and expanded upon.

2 Materials

1. Quantinova cDNA synthesis kit (QIAGEN).
2. QIAcuity Digital PCR System (QIAGEN).
3. QIAcuity Nanoplate including rubber seal (8.5 K 96-well) (QIAGEN).
4. QIAcuity Nanoplate including rubber seal (26 K 24-well) (QIAGEN).
5. QIAcuity Probe PCR Kit (QIAGEN).
6. QIAcuity Eva Green PCR Kit (QIAGEN).
7. QIAcuity OneStep advanced Probe PCR Kit (QIAGEN).
8. QIAcuity OneStep advanced Eva Green PCR Kit (QIAGEN).
9. QIAcuity Software Suite (QIAGEN).
10. Extracted sample RNA.
11. Prepared Cdna.
12. Primers and probes (*see* Table 1 for reference).
13. Pipettes, sterile filter tips, Eppendorf tubes, 96 Well plates, normal labware.

Table 1
Recommended primer/probe concentrations

QIAcuity PCR Mastermix	Recommended final primer concentration	Recommended final hydrolysis probe concentration
Probe PCR Mastermix	0.8 μM	0.4 μM
EvaGreen PCR Mastermix	0.8 μM	
OneStep Probe PCR Mastermix	0.4 μM	0.2 μM
OneStep Evagreen PCR Mastermix	0.4 μM	

3 Methods

3.1 Assay Development

Assays for digital PCR applications can be obtained from several sources. For example, existing qPCR assays (both intercalating dye-based and hydrolysis probe-based) can be transferred to dPCR. Alternatively, researchers can either purchase ready-made assays or obtain sequences from publications. The latter is particularly relevant when the authors of the publication have followed the dMIQE guidelines [6]. If no established (qPCR) assay is available, de novo designs are also possible. Online tools like Primer3 [7] or commercial tools (e.g., https://eu.idtdna.com/pages/tools/primerquest) can be utilized following the dMIQE guidelines to design specific assays for the targets of interest [6].

3.2 Gene Expression Analysis Workflow

1. Gene expression analysis is fundamentally based on the quantification of messenger RNA (mRNA) or noncoding RNA, as higher RNA levels should correlate with increased gene expression. Unlike DNA, RNA must first undergo a preprocessing step before PCR amplification, known as complementary DNA (cDNA) synthesis. During this process, RNA serves as the template for synthesizing a complementary DNA strand, enabling the subsequent amplification of the gene of interest.
2. There are two primary methods for performing cDNA synthesis: OneStep and two-step reverse transcription PCR (RT-PCR). Each method has its advantages depending on the experimental requirements, such as sample quantity, complexity, and the need for precise control over amplification conditions.
   (a) In the OneStep method, both cDNA synthesis and PCR amplification occur in a single reaction, simplifying the workflow and reducing the time required. This ensures that only the target RNA fragment is transcribed into cDNA and subsequently amplified. For dPCR, the use of either of QIAcuity OneStep Advanced Probe MM or QIAcuity OneStep Advanced EvaGreen MM is required, depending on whether probe-based or dye-based detection is used.
   (b) In contrast, the two-step RT-PCR method separates cDNA synthesis and PCR amplification into distinct stages, offering greater flexibility in optimizing reaction conditions and allowing for the reuse of cDNA in multiple analyses. RNA is initially transcribed into cDNA using universal oligo(dT) primers, random hexamers, or a mix of both. For optimal results, the use of the QuantiTect Reverse Transcription Kit is recommended. The resulting cDNA is then utilized as a template in dPCR. If the

expression level of your gene of interest is unknown, it is recommended to dilute the cDNA to 1:10 and 1:100 before using it in the dPCR reaction. The two-step approach offers the advantage of creating a reusable template for multiple experiments, allowing users to optimize PCR conditions and test different detection chemistries, such as EvaGreen or probe-based methods.

3.3 Preparation of PCR Mixes

1. When using hydrolysis probes, primers-probe (PP) mixes can be prepared for multiple runs. Stock solutions of primers and probes should be mixed according to Table 1 and filled with low-TE buffer. PP mixes should be aliquoted and stored at –20 °C to maintain stability.
2. Prepare the PCR mastermix by combining PCR mix, PP(s), and PCR-grade water according to Tables 2, 3, 4, and 5 (*see* **Note 1**). Mix by vortexing, then briefly centrifuge to collect contents at the bottom of the tube.
3. Depending on the type of nanoplate used (*see* **Note 2**) and added sample volume, transfer the prepared mastermix into either an Eppendorf tube or a 96-well plate.

Table 2
Recommended QIAcuity probe PCR setup

QIAcuity Probe PCR Mastermix	Final concentration	8.5 K plate μL	26 K plate μL
4× Probe PCR Mastermix	1×	3 μL	10 μL
20× Primer/Probe mix	0.4 μM Probe 0.8 μM/Primer	0.6 μL	2 μL
Sample		Variable	Variable
PCR grade water		Variable	Variable
Total reaction volume		12 μL	40 μL

Table 3
Recommended QIAcuity EvaGreen PCR setup

QIAcuity EvaGreen PCR Mastermix	Final concentration	8.5 K plate μL	26 K plate μL
3× EvaGreen PCR Mastermix	1×	4 μL	13.3 μL
20× Primer mix	0.4 μM	0.6 μL	2 μL
Sample		Variable	Variable
PCR grade water		Variable	Variable
Total reaction volume		12 μL	40 μL

Table 4
Recommended QIAcuity OneStep advanced probe RT-dPCR setup

QIAcuity OneStep Probe PCR Mastermix	Final concentration	8.5 K plate μL	26 K plate μL
4× OneStep PCR Mastermix	1×	3 μL	10
100× OneStep Advanced RT Mix	1×	0.12 μL	0.4 μL
20× Primer mix	0.4 μM/Primer 0.2 μM Probe	0.6 μL	2 μL
Enhancer GC[a]	Optional	1.5 μL	5 μL
Sample		Variable	Variable
PCR grade water		Variable	Variable
Total reaction volume		12 μL	40 μL

[a]GC Enhancer is recommended for use with all Applied Biosystems TaqMan assays, amplicons longer than 150 bp, GC-rich targets, and RNA sequences with complex secondary structures

Table 5
Recommended QIAcuity OneStep advanced EvaGreen RT-PCR setup

QIAcuity OneStep Probe PCR Mastermix	Final concentration	8.5 K plate μL	26 K plate μL
4× OneStep advanced EvaGreen PCR Mastermix	1×	3 μL	10
100× OneStep Advanced RT Mix	1×	0.12 μL	0.4 μL
20× Primer mix	0.4 μM/Primer	0.6 μL	2 μL
Q-Solution[a]		1.5 μL	3.3 μL
Sample		Variable	Variable
PCR grade water		Variable	Variable
Total reaction volume		12 μL	40 μL

[a]The use of Q-Solution is highly recommended, though the PCR reaction can still proceed without it. Q-Solution is particularly beneficial for amplicons longer than 150 bp, GC-rich targets, and RNA sequences with complex secondary structures. Adding Q-Solution may cause the reaction mix to appear slightly cloudy; however, this does not affect PCR performance and typically clears once the reaction mix reaches its final 1× concentration

4. The volume of a sample added to the prepared PCR mastermix can be varied (*see* Table 2 and **Note 3**). In addition to the sample(s) of interest, positive and negative controls including non-template controls (NTC) should be included (*see* **Note 4**). Mix the sample and mastermix thoroughly by pipetting up and down. Avoid creating bubbles in the reaction mix.
5. Transfer the sample/mastermix mixture into the wells of the nanoplate. Seal the nanoplate with adhesive rubber foil. Ensure proper alignment and apply firm pressure to seal all the reaction wells. Transfer the plate to the QIAcuity system.

3.4 Plate and Experiment Setup

1. Login to the QIAcuity Software Suite.
2. Navigate to Plates → New plate.
3. In the General Data tab, fill out the basic details such as plate name and type.
4. Go to the dPCR parameters → Priming tab, choose Priming Profile Probe (RT-) PCR.
   (a) For the OneStep RT-PCR method, configure the following conditions in the cycling tab (*see* **Note 5**): 40 min at 50 °C, 2 min at 95 °C, followed by 40 cycles of 15 s at 95 °C and 30 s at the appropriate annealing temperature for your assay (*see* **Note 6**).
   (b) For a two-step RT-PCR method, configure the following conditions in the cycling tab (*see* **Note 5**): 2 min at 95 °C, followed by 40 cycles of 15 s at 95 °C and 30 s at the annealing temperature suitable for your assay (*see* **Note 6**).
5. Enable all optical channels and keep Exposure duration and Gain at their default settings for a first test.
6. Click on Reaction Mixes → New Reaction Mix and fill in the necessary information (mix name; target name and channel (s) for each target) (*see* **Note 7**).
7. Go to Samples and controls → Samples tab, click New samples and label each sample (refer to *see* **Note 8**).
8. Assign the reaction mixes to the wells using the plate layout tab.
9. Assign samples and controls to the wells by clicking the three dots icon at the bottom right of the well selection, then click Add sample/control/NTC, select the appropriate sample/control/NTC and click Assign (*see* **Note 9**).
10. Click Save plate.

3.5 Result Analysis

1. Click on the Analyze button (*see* **Notes 10** and **12**).
2. Select the wells you want to analyze, then go to analyze per target and select all targets.
3. 1D Scatterplot: Scatterplots of the analyzed partitions for each channel. If the threshold is not automatically set correctly (above the negative cluster, *see* Fig. 1 and **Notes 11** and **12**), click in the graph to set the threshold manually and on recalculate to update the results.
4. After setting the thresholds, view the detected target copy concentration and other information for the selected wells under the List, Heatmap, Histogram, and Concentration diagram tables [10].
5. For further analysis, the data can be exported as a CSV file by clicking on Export to CSV → Current results under the List tab.

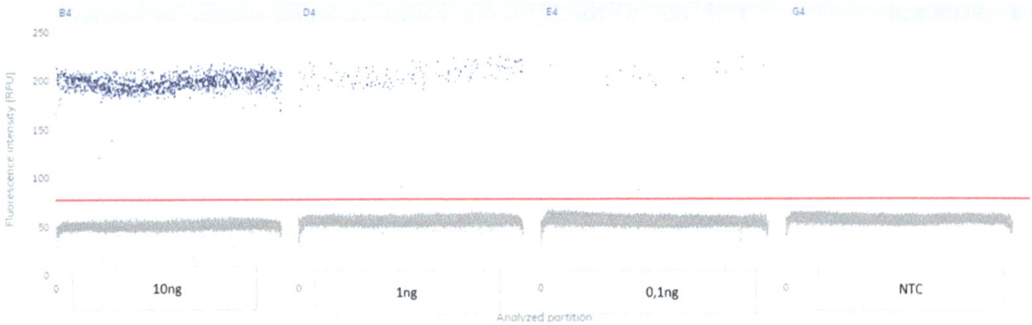

Fig. 1 Example of a GEX assay 1D plot on the QIAcuity system. 1D plot illustrating gene expression data for a single target. The blue cluster represents positive partitions, while the gray cluster represents negative partitions, separated by the red threshold line. From left to right, decreasing amounts of cDNA were added to the PCR reaction. Even at a 0.1 ng input, digital PCR reliably quantifies the number of mRNA molecules

4 Notes

1. Prepare the reaction mixtures with 10% additional volume to account for pipetting errors (e.g., if you need a mixture for 10 reactions, prepare enough for 11). In addition to the listed main components, additives such as dimethylsulfoxide (DMSO), magnesium chloride ($MgCl_2$), bovine serum albumin (BSA), or formamide or substances as Q-Solution (QIAGEN) or High GC Enhancer solution (NEB) can be added to the reaction mix [8]. This is recommended for difficult-to-amplify genomic region, for example, due to high G/C content or the formation of heteroduplexes.
2. The Nanoplate 8.5 K and Nanoplate 26 K contain approximately 8500 and 26,000 partitions per well, respectively, with a 12 μL or 40 μL dPCR reaction. The Nanoplate 8.5 K is available in both 24-well and 96-well formats, while the Nanoplate 26 K comes in 8-well and 24-well formats. Due to its 96-well capacity, the Nanoplate 8.5 K 96-well is ideal for higher throughput applications. In contrast, the Nanoplate 26 K is better suited for applications requiring higher accuracy, rare event detection, and a broader dynamic quantification range.
3. The amount of sample in each reaction is not fixed and is only limited by the maximum amount that can be added. If the expected gene expression is very low and only a few mRNA molecules are anticipated, the maximum amount of sample should be added. For medium and highly expressed targets, it is recommended to test at least a 1:10 dilution of sample and check if the target is detectable or if the system is overloaded. If the target cannot be detected, try switching to the 26 K plate to increase the sample volume or run more replicates to pool

results from multiple wells. If the system is overloaded, further dilute your sample to 1:100 and check again. This is especially important when running a duplex dPCR with (a) target(s) of interest and a reference gene. When selecting a reference gene, care should be taken to choose an appropriate one; highly expressed genes, like ACTB (ß-Actin) or GAPDH, are not well-suited and should be replaced with medium expressed reference genes (for further information refer to the Qiagen Appguide [9]).

4. Controls are crucial for ensuring the validity, reliability, and reproducibility of experimental results. At a minimum, include a positive control (a sample containing the target of interest) to verify assay performance and a non-template control (NTC) to assess the false-positive rate and detect potential contamination.

5. The annealing temperature is assay-dependent. For published or commercial assays, the value should be provided or obtainable from the source of the assay sequence. For new assays, a temperature gradient should be applied across the nanoplate; the temperature that shows the best separation between positive and negative partitions should be used. Good separation includes the highest difference between the positive and negative clusters, the lowest amount of rain (i.e., non-clustered signals that are scattered across the negative partitions, which can make it difficult to distinguish between true positive signals and noise) and no unspecific intermediate clusters.

6. The Nanoplate dPCR technology allows for multiple imaging steps. For example, after 40 cycles, an imaging step can be performed, followed by more cycles and another imaging step. This allows for easy determination of whether additional cycling steps are beneficial for assay performance in dPCR.

7. **Steps 6** through **11** can be carried out once the run is complete.

8. You can enter a concentration factor and the added sample volume into the software to account for any dilution factors and calculate the concentration of the undiluted sample or control. This step is optional and can alternatively be performed using an external spreadsheet program for manual calculations.

9. The same layout can be used across multiple PCR experiments, with different samples, in several runs. To do so, export the experiment layout by, click on CSV import/export → Plate layout → Export to CSV. To import the CSV for the next run, navigate to the Plate layout tab → CSV import/export → Import from CSV. In addition, the export/import function allows for exporting whole plates, including sample names, to program a new PCR run.

10. Data analysis can be performed externally on another PC. To do this, locate the run in the Plates Overview, click on the three dots in the plate header → Export. The resulting compressed file in the Downloads folder can then be transferred to another computer using a USB flash drive or imported from the export location. In the QIAcuity software, installed on the other PC, click on Import plate, locate the exported plate and click Import.
11. Thresholds must be set individually for each channel but can be adjusted simultaneously for all selected wells. To set a threshold for an individual well, click on the header of the scatterplot → select the desired threshold location→ Recalculate. The previously set thresholds for all other wells retained.
12. In some samples, partitions with fluorescence intensities between the negative and positive clusters may appear. In such cases, set the threshold above these partitions to ensure accurate quantification. For example, positive partitions in the red channel may appear as an intermediate fluorescence cluster. To ensure proper thresholding, always include a positive control containing targets for both the red and crimson probes to enable setting thresholds accurately. With Qiagen Software version 3.1, a custom crosstalk matrix function was introduced, allowing for accurate compensation and the elimination intermediate fluorescence artifacts ("medium" positive partitions).

References

1. Latchman D (2020) Gene control. Garland Science
2. Karlin-Neumann G, Bizouarn F (2018) Digital PCR. Methods in molecular biology. Springer
3. Vogelstein B, Kinzler KW (1999) Digital PCR. Proc Natl Acad Sci 96(16):9236–9241
4. Renault C, Bolloré K, Pisoni A, Motto-Ros C, Van de Perre P, Reynes J, Tuaillon E (2022) Accuracy of real-time PCR and digital PCR for the monitoring of total HIV DNA under prolonged antiretroviral therapy. Sci Rep 12(1): 9323
5. Taylor SC, Laperriere G, Germain H (2017) Droplet Digital PCR versus qPCR for gene expression analysis with low abundant targets: from variable nonsense to publication quality data. Sci Rep 7(1):2409
6. Wils G, Hamerlinck L, Trypsteen W, Van DenEeckhaut C, Weiss J, Nour AA, VandesompeleJ (2025) Digital PCR-based gene expressionanalysis using a highly multiplexed assay withuniversal detection probes to study induced pluripotentstem cell differentiation into cranialneural crest cells. Methods Mol Biol (Clifton,NJ) 2880:17–47
7. Huggett JF (2020) The digital MIQE guidelines update: minimum information for publication of quantitative digital PCR experiments for 2020. Clin Chem 66(8):1012–1029
8. Kõressaar T, Lepamets M, Kaplinski L, Raime K, Andreson R, Remm M (2018) Primer3_masker: integrating masking of template sequence with primer design software. Bioinformatics 34(11):1937–1938. The QIAcuity User Manual for Software 3.1. www.Qiagen.com
9. Strien J, Sanft J, Mall G (2013) Enhancement of PCR amplification of moderate GC-containing and highly GC-rich DNA sequences. Mol Biotechnol 54:1048–1054
10. The QIAcuity Appguide V2. www.Qiagen.com

Chapter 2

Multiplex Quantification of 19 GM Soybean Lines Using Digital PCR

Amadej Jelenčič, Dejan Štebih, Tina Demšar, and David Dobnik

Abstract

The European Union (EU) imposes strict regulations on the presence of genetically modified (GM) material in food and feed, requiring thorough testing of samples for various GM lines. Although traditional quantitative real-time PCR (qPCR) methods are sensitive and robust, they are not cost-effective for managing large numbers of GM events due to their limited multiplexing capabilities. Conversely, digital PCR (dPCR) is capable of robust quantitative multiplexing in addition to other benefits such as absolute quantification and better tolerance of PCR inhibitors. In this context, we present a protocol for multiplex quantification of 19 GM soybean lines using dPCR as an improvement over the currently used simplex qPCR approach. This method enables simple and robust quantification of common GM soybean lines with a relatively low number of reactions.

Key words Digital PCR (dPCR), Genetically modified (GM) soybean, Multiplexing, Food and feed, Food regulation

1 Introduction

The European Union (EU) has one of the most stringent regulatory systems regarding the presence of genetically modified (GM) material in food and feed [1]. To comply with the legal requirements, food and feed samples must be rigorously tested for the presence of different GM lines. Precise, accurate, and robust methods for detection and quantification of GM material are required for this purpose. The current gold standard method, quantitative real-time polymerase chain reaction (qPCR), while being sensitive and robust, is not the most labor- and cost-effective approach when dealing with large numbers of GM events due to its limited multiplexing capabilities [2, 3]. Conversely, digital PCR (dPCR) has been shown to possess comparable robustness and sensitivity while being more tolerant to PCR inhibitors and providing better multiplexing capabilities [3–7]. Furthermore, the

Guillaume Pavlovic (ed.), *Digital PCR: Methods and Protocols*, Methods in Molecular Biology, vol. 2969,
https://doi.org/10.1007/978-1-0716-4767-7_2,

lack of requirement of a standard curve, better tolerance of PCR inhibition, and straightforward result analysis make it an excellent choice for quantifying large numbers of GM events [8, 9].

The use of dPCR for multiplex quantification of GM lines has been demonstrated in maize [7] and soybean [4]. These approaches allowed for determination of the quantity of GM material in the sample but did not discriminate between individual GM events. Discriminative quantification of GM events with dPCR has been shown in maize [3], soybean, and canola [10]. As the dPCR systems used in those studies only allowed for simultaneous detection of two fluorescent dyes, higher multiplexing was achieved by varying the concentrations of primers and probes. While this has proven to be a viable approach, the recent introduction of dPCR systems with up to seven optical channels [11–18] has made multiplexing using several dyes possible. This approach significantly simplifies assay design and result analysis, as shown by Bogožalec Košir and colleagues [5] in the context of GM soybean detection and quantification by a 6-plex assay on the Stilla Naica platform.

Described below is a protocol for four 5-plex dPCR assays covering the detection and quantification of the 19 GM soybean lines for which EU reference quantification methods were available at the time of preparation of the protocol [19]. One 5-plex assay includes the assay targeting the soybean reference gene, *Le1*, in addition to four GM soybean events, while the other three 5-plex assays target five individual events each. The inclusion of the reference gene is necessary for the calculation of the relative GM content in a sample. The four 5-plex assays are performed on the same plate, ensuring better comparability of results. In each assay, the five probes are labeled on the 5′ end with the fluorophores FAM, HEX, TAMRA, Texas Red, and Cy5, respectively. This protocol enables accurate quantification of each GM soybean event present in a sample, providing data of the highest quality required by regulatory authorities.

2 Materials

1. QIAcuity Digital PCR System (QIAGEN).
2. QIAcuity Nanoplate including rubber seal (8.5 k 96-well) (QIAGEN).
3. QIAcuity Probe PCR Kit (QIAGEN).
4. QIAcuity Software Suite (QIAGEN).
5. Extracted sample and control DNA.
6. Primers and probes (*see* Tables 1, 2, 3, and 4).
7. Pipettes (e.g., Eppendorf Research plus, different volumes).
8. Filtered pipette tips (e.g., Corning IsoTip).

Table 1
Primers and probes used in soybean 5-plex 1

Event/target name	Oligonucleotide name	DNA sequence of the oligonucleotide (5′-sequence-3′)	Final concentration in the reaction [nmol/L]	Volume of 100 μM stock -solution needed for 1000 μL of reaction PP mix (μL)
DBN9004	DBN-09004-6-F	GCCGTATCCGCAATGTGTTA	600	24.00
	DBN-09004-6-R	GCTCCATAAACGTGTGCTTTCA	600	24.00
	DBN-09004-6-P	FAM-TTGTTTACAACTCTGTGACCC-MGBNFQ	200	8.00
GMB151	GMB151-F	TCAAATCAACATGGGTGACTAGAAA	400	16.00
	GMB151-R	CATTGTGCTGAATAGGTTTATAGCTATGAT	400	16.00
	GMB151-P	HEX-CAGTACTGG/ZEN/GCCCTTGTGGCGCT-IAbFQ[a]	200	8.00
SYHT0H2	SYHT0H2-F	GGGAATTGGGTACCATGCC	600	24.00
	SYHT0H2-R	TGTGTGCCATTGGTTTAGGGT	600	24.00
	SYHT0H2-P	TAMRA-CCAGCATGGCCGTATCCGCAA-BHQ2	200	8.00
MON87751	MON87751-F	CTAAATTGCTCTTTGGAGTTTATTTTGTAG	500	20.00
	MON87751-R	GGCCTAACTTTTGGTGTGATGATG	500	20.00
	MON87751-P	TexRd-TGACTGGAGATCTCCAAAGT GAGGGGAAA-IAbRQ	300	12.00
Le1 (reference gene)	Lectin-F	CCAGCTTCGCCGCTTCCTTC	650	26.00
	Lectin-R	GAAGGCAAGCCCATCTGCAAGCC	650	26.00
	Lectin-P	Cy5-CTTCACCTT/TAO/CTATGCCCCTGACAC-IAbRQ[a]	180	7.20

[a] *see* **Note 4**

Table 2
Primers and probes used in soybean 5-plex 2

Event/target name	Oligonucleotide name	DNA sequence of the oligonucleotide (5′-sequence-3′)	Final concentration in the reaction [nmol/L]	Volume of 100 μM stock solution needed for 1000 μL of reaction PP mix (μL)
DP305423	DP305-F	CGTGTTCTCTTTTTGGCTAGC	800	32.00
	DP305-R	GTGACCAATGAATACATAACACAAACTA	500	20.00
	DP-305-P	FAM-TGACACAAA/ZEN/ TGATTTTCATACAAAAGTCGAGA-IAbFQ[a]	220	8.80
CV127	CV127-F	AACAGAAGTTTCCGTTGAGCTTTAAGAC	400	16.00
	CV127-R	CATTCGTAGCTCGGATCGTGTAC	400	16.00
	CV127-P	HEX-TTTGGGGAA/ZEN/GCTGTCCCATGCCC-IAbFQ[a]	100	4.00
MON87769	MON87769-F	CATACTCATTGCTGATCCATGTAGATT	600	24.00
	MON87769-R	GCAAGTTGCTCGTGAAGTTTTG	600	24.00
	MON87769-P	TAMRA-CCCGGACATGAAGCCATTTACAATTGAC-BHQ2	200	8.00
MON87701	87701-F	TGGTGATATGAAGATACATGCTTAGCAT	600	24.00
	87701-R	CGTTTCCCGCCTTCAGTTTAAA	600	24.00
	87701-P	TexRd-TCAGTGTTTGACACACACACTAAGCGTGCC-IAbRQ	250	10.00
MON87708	87708-F	TCATACTCATTGCTGATCCATGTAG	300	12.00
	87708-R	AGAACAAATTAACGAAAAGACAGAACG	300	12.00
	87708-P	Cy5-TCCCGGACT/TAO/TTAGCTCAAAATGCATGTA-IAbRQ[a]	150	6.00

[a] *see* Note 4

Table 3
Primers and probes used in soybean 5-plex 3

Event/target name	Oligonucleotide name	DNA sequence of the oligonucleotide (5′-sequence-3′)	Final concentration in the reaction [nmol/L]	Volume of 100 μM stock solution needed for 1000 μL of reaction PP mix (μL)
MON40-3-2	40-3-2-AF	TTCATTCAAAATAAGATCATACATACAGGTT	600	24.00
	40-3-2-AR	GGCATTTGTAGGAGCCACCTT	600	24.00
	40-3-2-AP	FAM-CCTTTTCCATTTGGG-MGBNFQ	200	8.00
A5547	A5547-F	GCTATTTGGTGGCATTTTTCCA	400	16.00
	A5547-R	CACTGCGGCCAACTTACTTCT	400	16.00
	A5547-P	HEX-CCGCAATGT/ZEN/ CATACCGTCATCGTTGT-IAbFQ[a]	200	8.00
A2704	A2704-F	GCAAAAAAGCGGTTAGCTCCT	400	16.00
	A2704-R	ATTCAGGCTGCGCAACTGTT	400	16.00
	A2704-P	TAMRA-CGGTCCTCCGATCGCCCTTCC-BHQ2	200	8.00
MON89788	MON-89788-F	TCCCGCTCTAGCGCTTCAAT	600	24.00
	MON-89788-R	TCGAGCAGGACCTGCAGAA	600	24.00
	MON-89788-P	TexRd-CTGAAGGCGGGAAACGACAATCTG-IAbRQ	200	8.00
MON87705	87705-F	TTCCCGGACATGAAGCCATTTAC	450	18.00
	87705-R	ACAACGGTGCCTTGGCCCAAAG	450	18.00
	87705-P	Cy5-AAGAGACTC/TAO/ AGGGTGTTGTTATCACTGCGG-IAbRQ[a]	250	10.00

[a]*see* **Note 4**

Table 4
Primers and probes used in soybean 5-plex 4

Event/target name	Oligonucleotide name	DNA sequence of the oligonucleotide (5′-sequence-3′)	Final concentration in the reaction [nmol/L]	Volume of 100 μM stock solution needed for 1000 μL of reaction PP mix (μL)
FG72	FG72-F	AGATTTGATCGGGCTGCAGG	400	16.00
	FG72-R	GCACGTATTGATGACCGCATTA	400	16.00
	FG72-P	FAM-AATGTGGTTCATCCGTCTT-MGBNFQ	200	8.00
DAS68416	DAS68416-F	GTACATTAAAAACGTCCGCAATGTGT	550	22.00
	DAS68416-R	GTTTAAGAATTAGTTCTTACAGTTTATTGTTAG	550	22.00
	DAS68416-P	HEX-TTAAGTTGT/ZEN/CTAAGCGTCAATA-IAbFQ[a]	150	6.00
DP356423	DP356-F	GTCGAATAGGCTAGGTTTACGAAAAA	750	30.00
	DP356-R	TTTGATATTCTTGGAGTAGACGAGAGTGT	750	30.00
	DP356-P	TAMRA-CTCTAGAGATCCGTCA ACATGGTGGAGCAC-BHQ2	200	8.00
DAS81419	DAS81419-F	TCTAGCTATATTTAGCACTTGATATTCAT	400	16.00
	DAS81419-R	GCTTCAAGATCCCAACTTGCG	400	16.00
	DAS81419-P	TexRd-CGGTCCTCCGATCGCCCTTCC-IAbRQ	120	4.80
DAS44406	DAS44406-F	TTATTGTTCTTGTTGTTTCCTCTTTAGG	300	12.00
	DAS44406-R	CCTCAATTGCGAGCTTTCTAATTT	300	12.00
	DAS44406-P	Cy5-ATTCGGACC/TAO/ TCCATGATGACCTTACCGTT-IAbRQ[a]	180	7.20

[a] *see* Note 4

9. 96-well plates for sample preparation (e.g., Eppendorf twin. tec® PCR Plate).
10. Adhesive Sealing Sheet (e.g., AB-0558, Thermo Fisher Scientific).
11. Vortex.
12. Microcentrifuge.
13. Centrifuge with rotor for 96-well plates.

3 Methods

3.1 Multiplex Assay Development

EU authorized soybean GM event quantification methods were used as a basis for the development of the four 5-plex assays [19]. The primers and probes from the reference methods were combined into 5-plex assays in accordance with data found in literature [5] and in silico analysis. Potential primer and probe dimer formation was excluded based on analysis with the software Autodimer [20]. Furthermore, specificity was predicted using online software MFEPrimer [21]. The multiplex primer combinations were checked for specificity against the reference genomes of soybean, maize, oilseed rape, wheat, and rice. Although these tools are a good starting point for determining potential multiplex primer and probe combinations, specificity should also be assessed experimentally when developing an assay. This is done by testing samples containing DNA from a variety of nontarget GM events in addition to various non-transgenic crops. These should all yield negative results.

3.2 Sample Homogenization and DNA Extraction

Food and feed samples should be properly homogenized prior to DNA extraction to ensure that the analyzed portion is representative of the whole sample. The DNA extraction method utilized must yield DNA of sufficient quality and quantity for analysis. Three methods have been shown to be fit for this purpose in our lab, namely, the cetyltrimethylammonium bromide (CTAB) method with RNase A and Proteinase K solutions (as described in Annex 3 of ISO 21570:2005 [22]), MACHEREY-NAGEL Nucleospin Food kit (as described by manufacturer [23]), and QIAGEN DNeasy Plant Mini Kit (as described by manufacturer [24]).

3.3 Preparation of Reaction Mixtures

1. Prepare the primers-probe (PP) mixes by mixing the stock solutions of primers and probes according to Tables 1, 2, 3, and 4. Add nuclease-free water up to the final volume. PP mixes can be aliquoted and stored at −20 °C (*see* **Note 1**).
2. Prepare the reaction mixtures by mixing 3 μL of each PP mix with 3 μL of QIAcuity Probe Master Mix and 2 μL of deionized

water for one reaction (*see* **Note 2**). Mix by vortexing and collect the contents at the bottom of the tube by short spin in the microcentrifuge.

3. Distribute 9.6 μL of the reaction mixtures into the wells of nuclease-free 96-well plate.
4. Add 4.8 μL of DNA sample in each well with reaction mixture according to the prepared plate scheme. In addition to tested samples, positive control (mixture of DNA containing all targets) and non-template control (NTC; nuclease-free water) should be included. Optionally negative extraction and/or environment controls can be used as well.
5. Seal the plate with clear adhesive sealing sheet, mix by vortexing and centrifuge the plate to collect the liquid at the bottom of wells (*see* **Note 3**).

3.4 Plate and Experiment Setup

1. Start the QIAcuity Software Suite (QIAGEN) on the computer connected to the instrument and log in.
2. Click "Plates," then "New plate."
3. Enter the basic information (plate name, type, and barcode) under the "General Data" tab (*see* **Note 5**).
4. Under the "dPCR parameters" -> "Priming" tab, select "QIAGEN Priming Profile Probe (RT-) PCR."
5. Under the "dPCR parameters" -> "Cycling" tab, set the following conditions:
   (a) 2 min at 50 °C.
   (b) 2 min at 95 °C.
   (c) 40 cycles of a two-step thermal profile comprising of 15 s at 95 °C and 30 s at 60 °C.
6. Under the "dPCR parameters" -> "Imaging" tab, enable all optical channels. Leave the "Exposure duration" and "Gain" on default settings (*see* **Note 6**).
7. Under the "Reaction Mixes" tab, click "New Reaction Mix" and enter the required information (Reaction mix name; target name, and channel for each of the targets in a given assay) (*see* **Note 7**).
8. Under the "Samples and controls" -> "Samples" tab, click "New samples" and enter the name for each sample (*see* **Note 8**).
9. Under the "Samples and controls" -> "Controls" tab, click "Add control" and enter the name for each control (*see* **Note 8**).

10. Under the "Samples and controls" -> "Non-Template Controls," click "Add NTC" and enter the information for each NTC.
11. Open the "Plate layout" tab and assign the reaction mixes to the wells. This is done by selecting specific well (seconds), selecting "Add reaction mix," selecting the appropriate reaction mix and clicking "Assign."
12. Assign the samples and controls to the wells. This is done by clicking the three dots icon on the bottom right of the well selection, clicking "Add sample," "Add control," or "Add NTC," selecting the appropriate sample/control/NTC, and clicking "Assign" (*see* **Note 9**).
13. Click "Save plate" (*see* **Note 10**).

3.5 Plate Setup, Partitioning, PCR, and Detection

1. Remove the adhesive seal from the reaction mix plate.
2. Use a multichannel pipette to transfer 12 μL of each reaction mixture from the 96-well plate to the QIAcuity Nanoplate (QIAGEN) (*see* **Note 11**).
3. Seal the plate with the QIAcuity Nanoplate seal and roll horizontally, vertically, and diagonally with the nanoplate roller to ensure proper sealing.
4. Power on the QIAcuity instrument and log in.
5. Press the button on the instrument or tap "Eject Tray" on the touchscreen.
6. Place the plate in the ejected tray and tap "Close Tray."
7. If the barcode had been entered during plate creation in the Software Suite, "Plate is detected" will appear on the screen. Otherwise, a plate can be selected among the plates that were created in the Software Suite but have no barcode linked.
8. Tap the "Run" icon to start the run. Partitioning, temperature cycling, and fluorescence detection are performed automatically by the QIAcuity instrument according to the set experiment.

3.6 Result Analysis

1. After the run is completed, the results can be analyzed by clicking on "Analyze" (*see* **Note 12**).
2. Select the wells you want to analyze, click on "Analyze per" -> "Target" and select all targets (*see* **Note 13**).
3. Click on "1D scatterplot" to see the scatterplots of the analyzed partitions for each channel. Set the appropriate thresholds by clicking on the graph where you want the threshold and then click "Recalculate." This will recalculate the data about positive and negative partitions according to the newly set

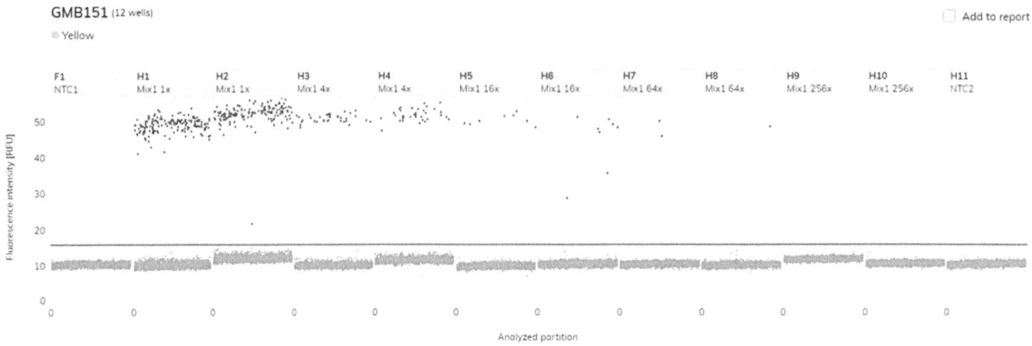

Fig. 1 Amplification plots for multiplex assay on QIAcuity. Results for the detection of the event GMB151 in the yellow channel are presented. Samples are dilutions of a DNA mixture containing events quantified by the 5-plex 1 assay. Similar plots are obtained for all detection channels. The red line represents the set threshold. RFU relative fluorescence units

thresholds. Thresholds should be set above the negative clusters in the NTC wells (*see* Fig. 1 and *see* **Notes 14** and **15**).

4. In some cases, the thresholds might need to be set in the 2D view. To do so, click on "2D scatterplot," select the appropriate targets for the X and Y axes, click on the graph where you want the thresholds to be and click "Recalculate." Alternatively, you can select the partitions belonging to a particular cluster with the lasso tool (*see* Fig. 2 and *see* **Notes 16** and **17**).
5. After setting the thresholds, detected target copy concentration and other information for all the selected wells can be seen under the "List," "Heatmap," "Histogram," and "Concentration diagram" tabs.
6. Images of the wells can be analyzed under the "Signalmap" tab, which can serve as an additional quality control step (*see* Fig. 3).
7. The data can be exported as CSV for further analysis in spreadsheet software. To do this, click on "Export to CSV" -> "Current results" under the "List" tab (*see* **Note 18**).
8. The percentage of a given event in a sample can be calculated using the following equation:

$$\text{GM event content}[\%] = \frac{\text{average concentration of event}}{\text{average concentration of } Le1} \times 100$$

Concentrations in undiluted sample should be used (if dilutions were performed, they should be back-calculated to stock sample). An example of calculation for targets in 5-plex 1 is presented in Table 5.

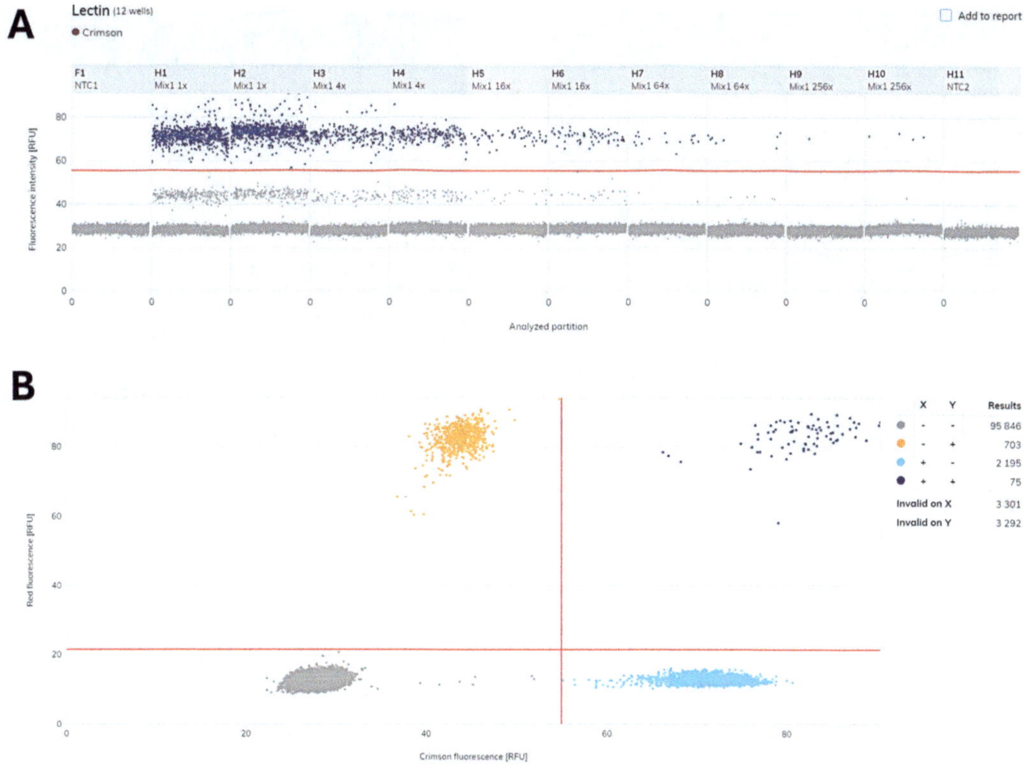

Fig. 2 Example of cross-talk between the red and crimson detection channels. (**a**) 1D scatterplot view of the crimson channel. An additional population of partitions can be seen with fluorescence intensity above the negatives but below the true positives. (**b**) 2D scatterplot view of the crimson and red channels. The cluster positive for red fluorescence is tilted toward the right, indicating higher crimson fluorescence than the double negatives. Setting the threshold in 2D view helps avoid misclassification of partitions. If using QIAcuity Software Suite version 3.0 or greater, a color compensation matrix is automatically applied for new experiments, eliminating the cluster tilt. The red lines represent the set thresholds. RFU relative fluorescence units

4 Notes

1. Preparation of PP mixes is advisable to avoid having to pipet all the primer and probe solutions each time an assay is run. This saves time and helps in avoiding errors, which is especially important in the case of multiplex assays.
2. Prepare the reaction mixtures with 20% excess volume to offset pipetting losses and ensure proper loading of samples onto the nanoplate. Prepare the mixtures for 10% more reactions than needed (e.g., if you need a mixture for 20 reactions, prepare it for 22).
3. Vortex with low speed to avoid bubble formation.

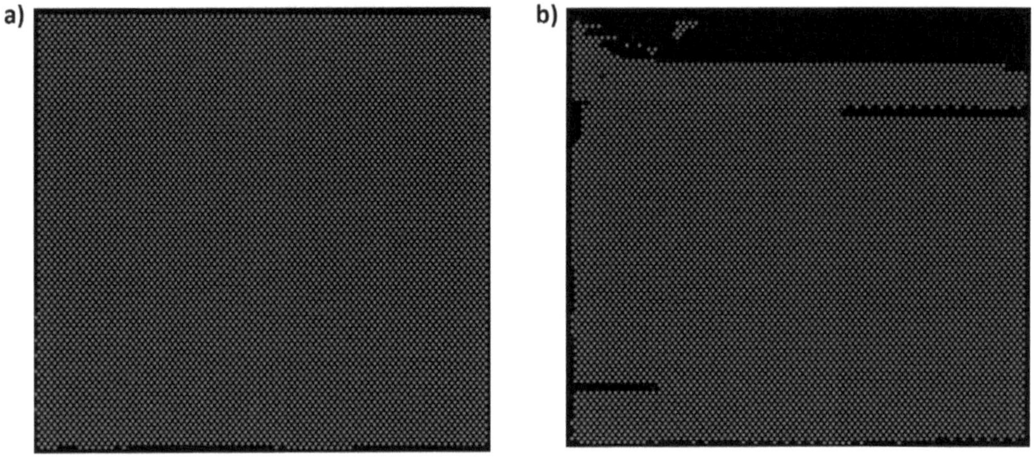

Fig. 3 Signal maps of two nanoplate wells. Reference dye signal is shown for (**a**) a completely filled well and (**b**) an incompletely filled well. Partitions lacking reference dye signal are automatically excluded from the analysis

4. When possible, the probes are double-quenched by addition of ZEN or TAO internal quencher. This helps reduce background fluorescence, which can be problematic in the case of multiplex assays. The use of nonfluorescent quenchers, such as black hole quencher (BHQ) or Iowa Black (IAb), is recommended for the same reason.
5. The barcode on the side of the nanoplate should be entered (or scanned by barcode scanner) in this step to connect the plate settings in the software with the specific nanoplate and avoid having to set up the reaction conditions on the instrument.
6. Additional cycling and imaging steps can be added before or after the run, e.g., to determine if additional cycles improve the separation between the negative and positive clusters, or to check how different exposure times affect detection. This might prove useful when developing new assays.
7. **Steps 6** through **12** of Subheading 3.4 can also be performed after the run has been finished. In that case, click "Save plate" after setting the imaging conditions. It is advisable to save multiple times during the setup to avoid losing the changes made.
8. Concentration factor and volume of added matrix DNA can be entered for software to account for dilution and calculate the concentration for undiluted sample/control. This, however, is not required, as the calculation can easily be performed in spreadsheet software after exporting the results.
9. The plate layout can be exported as a .csv file. This is useful if a similar layout with different samples is used in several runs. To

Table 5
Example of GM event content calculation

Well	Sample	Target	Concentration [cp/μL]	Concentration of undiluted sample [cp/μL]	Average concentration [cp/μL]	GM event content [%]
H1	Mixl 1×	DBN9004	105	105	94	$=94/330 \times 100$ $=28.48$
H2	Mixl 1×	DBN9004	86	86		
H3	Mixl 4×	DBN9004	19	75		
H4	Mixl 4×	DBN9004	28	110		
H1	Mixl 1×	GMB151	62	62	56	16.97
H2	Mixl 1×	GMB151	56	56		
H3	Mixl 4×	GMB151	13	53		
H4	Mixl 4×	GMB151	13	52		
H1	Mixl 1×	SYHT0H2	81	81	83	25.15
H2	Mixl 1×	SYHT0H2	70	70		
H3	Mixl 4×	SYHT0H2	22	87		
H4	Mixl 4×	SYHT0H2	23	93		
H1	Mixl 1×	MON87751	113	113	113	34.24
H2	Mixl 1×	MON87751	106	106		
H3	Mixl 4×	MON87751	29	114		
H4	Mixl 4×	MON87751	30	119		
H1	Mixl 1×	*Le1*	347	347	330	
H2	Mixl 1×	*Le1*	330	330		
H3	Mixl 4×	*Le1*	82	328		
H4	Mixl 4×	*Le1*	79	314		

export, click on “CSV import/export” -> “Export to CSV” under the “Plate layout” tab. The .csv file can be opened in spreadsheet software (e.g., Microsoft Excel), where the find and replace function can be used to change the sample names. To import, go to the “Plate layout” tab in the new plate, click on “CSV import/export” -> “Import from CSV” and locate the relevant .csv file.

10. You can save the plate as a template if you intend to use a similar layout in multiple experiments. This is done by clicking “Plate templates” -> “Save template,” entering the template name and clicking “Save.” To create a new plate based on the

template, click “Plates” -> “New plate” -> “Plate templates” -> “Import template,” select the appropriate template and click “Import.”

11. Be careful to avoid forming bubbles, as those can reduce the number of partitions suitable for analysis and thus negatively affect data interpretation.
12. The plate can be exported to be analyzed on another computer (e.g., in the office). To do so, find the plate in the “Plates Overview” view, click on the three dots in the plate header, then click on “Export.” A compressed file will be created in the “Downloads” folder of the computer. This can then be transferred to another computer using a USB flash drive or network connection. On the receiving computer, the latest QIAcuity software version should be installed from the QIAGEN website. Open the software and login using “admin” for both username and password (unless other users are set). Click on “Import plate,” locate the exported plate on the computer and then click “Import.”
13. The wells can also be analyzed per channels, which is useful, e.g., to get a broad overview of the results before plate layout has been defined.
14. Thresholds are set for each channel separately. The thresholds of selected wells can be set together. It is advisable to set the thresholds of samples of similar type together with the corresponding NTCs. If a threshold for a certain well needs to be set individually (e.g., if the negative cluster is higher than in other wells), this can be done by clicking on the header of the corresponding scatterplot, clicking where you want the threshold to be and clicking “Recalculate.” This will set the threshold for the given well while retaining the previously set thresholds for all the other wells. If the result obtained in such a well differs significantly from other replicates, it should be excluded from analysis.
15. With some samples, some partitions with fluorescence higher than the negative population might appear, but their fluorescence is lower than the true positives. In such case, the threshold should be set above these partitions to avoid false positive results and ensure correct quantification (*see* Fig. 4).
16. In software versions below 3.0, thresholds for the red and crimson channels usually need to be set in this way due to cross-talk between the channels. Positive partitions in the red channel appear in the crimson channel as a cluster with fluorescence higher than the negative cluster but lower than the true positives, which is seen as a leaning of the red-positive cluster into the crimson-positive area in 2D view. A positive control containing targets for both the red and crimson probes should

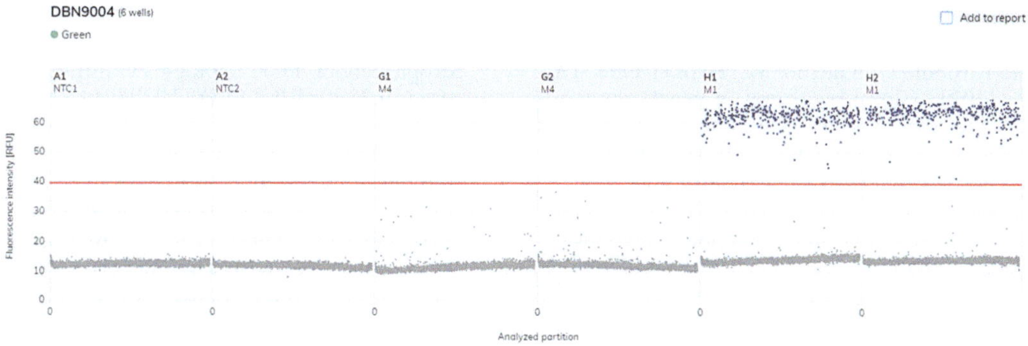

Fig. 4 Example of partitions with intermediate fluorescence above the cluster of negative partitions observed with assay DBN9004. The first two wells contain NTCs, the middle two wells contain samples negative for the target, and the last two wells contain samples positive for the target. The threshold is set somewhere in the middle between positive and negative partitions separating the true positive partitions from the negative ones

always be included to enable setting thresholds in this way. All the channel combinations should be checked for leaning of the clusters to avoid misclassification of partitions.

17. A new version of the QIAcuity Software Suite (3.0) was released while this protocol was in revision phase. The new version allows cross-talk color compensation. A default color compensation matrix is used for all new experiments using the standard detection channels and a custom matrix can be created if needed (e.g., for high-order color combination multiplexing that the new software version allows).

18. If the computer is set to a locale where commas are used for denoting decimal points, data formatting problems can be encountered when opening the .csv file in Excel. For this reason, it is recommended to import the data into Excel using Power Query. Another possible workaround is to use an alternative spreadsheet application (e.g., LibreOffice Calc) to open the file in English (US) locale, use the find and replace function to change the decimal dots to commas, and save as .xlsx file.

Acknowledgments

This work was supported by the Slovenian Research and Innovation Agency (research core funding contract no. P4-0165, including young researcher grant (AJ)). Equipment for dPCR was kindly provided by Mediline as a demo unit and consumables for QIAcuity runs were kindly provided by QIAGEN.

References

1. Turnbull C, Lillemo M, Hvoslef-Eide TAK (2021) Global regulation of genetically modified crops amid the gene edited crop boom – a review. Front Plant Sci 12. https://doi.org/10.3389/FPLS.2021.630396/BIBTEX
2. Jacky L, Yurk D, Alvarado J et al (2021) Robust multichannel encoding for highly multiplexed quantitative PCR. Anal Chem 93:4208–4216. https://doi.org/10.1021/ACS.ANALCHEM.0C04626
3. Dobnik D, Štebih D, Blejec A et al (2016) Multiplex quantification of four DNA targets in one reaction with Bio-Rad droplet digital PCR system for GMO detection. Sci Rep 6:1–9. https://doi.org/10.1038/srep35451
4. Bogožalec Košir A, Spilsberg B, Holst-Jensen A et al (2017) Development and inter-laboratory assessment of droplet digital PCR assays for multiplex quantification of 15 genetically modified soybean lines. Sci Rep 7:1–11. https://doi.org/10.1038/s41598-017-09377-w
5. Bogožalec Košir A, Muller S, Žel J et al (2023) Fast and accurate multiplex identification and quantification of seven genetically modified soybean lines using six-color digital PCR. Foods 12. https://doi.org/10.3390/FOODS12224156
6. Bogožalec Košir A, Demšar T, Štebih D et al (2019) Digital PCR as an effective tool for GMO quantification in complex matrices. Food Chem 294:73–78. https://doi.org/10.1016/J.FOODCHEM.2019.05.029
7. Dobnik D, Spilsberg B, Bogožalec Košir A et al (2015) Multiplex quantification of 12 European Union authorized genetically modified maize lines with droplet digital polymerase chain reaction. Anal Chem 87:8218–8226. https://doi.org/10.1021/acs.analchem.5b01208
8. Pinheiro LB, Coleman VA, Hindson CM et al (2011) Evaluation of a droplet digital polymerase chain reaction format for DNA copy number quantification. Anal Chem 84:1003–1011. https://doi.org/10.1021/AC202578X
9. Demeke T, Dobnik D (2018) Critical assessment of digital PCR for the detection and quantification of genetically modified organisms. Anal Bioanal Chem 410:4039–4050. https://doi.org/10.1007/S00216-018-1010-1/TABLES/3
10. Demeke T, Lee SJ, Eng M (2022) Increasing the efficiency of canola and soybean GMO detection and quantification using multiplex droplet digital PCR. Biology 11. https://doi.org/10.3390/BIOLOGY11020201/S1
11. Stilla Technologies Nio™ + dPCR. https://www.stillatechnologies.com/multiplex-pcr/nio-dpcr/. Accessed 27 Mar 2024
12. QIAGEN QIAcuity Digital PCR System. https://www.qiagen.com/us/products/instruments-and-automation/pcr-instruments/qiacuity-digital-pcr-system. Accessed 13 Oct 2023
13. Roche Digital LightCycler® dPCR System. https://sequencing.roche.com/us/en/products/group/digital-lightcycler-dpcr-system.html. Accessed 13 Oct 2023
14. Thermo Fisher Scientific QuantStudio Absolute Q Digital PCR System. https://www.thermofisher.com/si/en/home/life-science/pcr/digital-pcr/quantstudio-absolute-q-system.html. Accessed 13 Oct 2023
15. Select Science (2022) Introducing the QX600 Droplet Digital PCR System. https://www.selectscience.net/application-articles/introducing-the-qx600-droplet-digital-pcr-system/?artid=58177. Accessed 10 Oct 2023
16. Bio-Rad Laboratories (2019) Bio-Rad launches its QX ONE Droplet Digital PCR System to early access customers at AMP 2019 annual meeting & expo. https://www.bio-rad.com/en-si/life-science-research/news/bio-rad-launches-its-qx-one-droplet-digital-pcr-system-early-access-customers-at-amp-2019-annual-meeting-expo?ID=Bio-Rad-Launches-Its_1573066471. Accessed 10 Oct 2023
17. Stilla Technologies naica® system: flexible multiplex digital PCR solution. https://www.stillatechnologies.com/. Accessed 20 Oct 2023
18. Sniper Technologies Sniper DQ24 Digital PCR System. https://www.sniper-tech.com/DQ24.html. Accessed 24 Sep 2024
19. EURL GMFF GMOMETHODS. https://gmo-crl.jrc.ec.europa.eu/gmomethods/. Accessed 9 Oct 2023
20. Vallone PM, Butler JM (2004) AutoDimer: a screening tool for primer-dimer and hairpin structures. Biotechniques 37:226–231. https://doi.org/10.2144/04372ST03
21. Wang K, Li H, Xu Y et al (2019) MFEprimer-3.0: quality control for PCR primers. Nucleic Acids Res 47:W610–W613. https://doi.org/10.1093/NAR/GKZ351
22. International Organization for Standardization (2005) ISO 21570:2005 – foodstuffs —

methods of analysis for the detection of genetically modified organisms and derived products — quantitative nucleic acid based methods. https://www.iso.org/standard/34615.html. Accessed 16 Oct 2023

23. MACHEREY-NAGEL NucleoSpin Food, Mini kit for DNA from food. https://www.mn-net.com/nucleospin-food-mini-kit-for-dna-from-food-740945.50. Accessed 8 May 2024

24. QIAGEN DNeasy Plant Mini Kit (EN). https://www.qiagen.com/us/resources/resourcedetail?id=6b9bcd96-d7d4-48a1-9838-58dbfb0e57d0&lang=en. Accessed 8 May 2024

Chapter 3

Multiplex Digital PCR Assay Targeting DNA Methylation for Early Detection of Cancer

Victoria Cannon, Jocelyn H. Wright, Cecilia C. S. Yeung, William M. Grady, and Ming Yu

Abstract

Digital PCR has been demonstrated to enable the precise nucleic acid absolute quantification across a wide range of applications. Historically, dPCR technology was limited to two channels for fluorescence detection. Recent advances in digital PCR have made it possible to multiplex and allow for simultaneous analysis of multiple targets within a PCR reaction. Here, we describe a multiplex dPCR assay for the detection and quantification of methylated DNA (also known as DNA-methylation-specific multiplex digital PCR, or DNA-methylation specific dPCR). We also demonstrate the application of this technology for the development of a candidate DNA methylation-based biomarker panel for the early detection of high-grade dysplasia and esophageal adenocarcinoma. Further, we discuss common technical challenges and troubleshooting for performing successful DNA-methylation-specific MS-dPCR assays.

Key words Multiplex digital-PCR, DNA methylation, Biomarkers, Early detection of cancer, Esophageal adenocarcinoma

1 Introduction

DNA methylation is an epigenetic modification that occurs when a methyl group is covalently added to a cytosine base of DNA, typically at cytosine-guanine (CpG) dinucleotides. DNA methylation is one of the most common epigenetic modifications and plays an important role in regulating gene expression. Unsurprisingly, abnormal changes in DNA methylation have been implicated in pathological processes or diseases, such as cancer. Aberrant DNA methylation patterns in cancer tissue, when compared to normal counterparts, can contribute to the entire tumorigenesis process and have been recognized as one of the hallmarks of cancer [1]. Consequently, DNA methylation-based biomarkers hold the promise for early detection of cancer.

Guillaume Pavlovic (ed.), *Digital PCR: Methods and Protocols*, Methods in Molecular Biology, vol. 2969,
https://doi.org/10.1007/978-1-0716-4767-7_3,

Digital PCR (dPCR) technology is an advancement in quantitative PCR (qPCR) for the quantification of nucleic acids. In contrast to traditional qPCRs, where PCR is performed in a bulk reaction, dPCR involves collecting data from a large number of individual reaction chambers, termed partitions. Recent advances in nanofluidic techniques and emulsion chemistries have allowed the generation of tens of thousands of partitions of uniform nanoliter size, thus greatly enhancing the capacity of dPCR technology for the detection and absolute quantification of rare molecular events with unprecedented precision. The quantification depends on an independent distribution of target molecules into equally sized partitions, meaning that the number of targets in each partition follows a standard Poisson distribution [2]. This partitioning system gives dPCR superior sensitivity over other iterations of PCR. It has been demonstrated that the analysis method used in dPCR is less affected by reaction inhibitors because it uses endpoint measurements, whereas qPCR relies on amplification kinetics [3]. Therefore, current dPCR technologies have the potential to have an impact on biomarker applications [4, 5]. While early detection biomarker development has benefited from the increase in sensitivity and precision offered by the dPCR approaches, most digital assays for DNA methylation have been limited to single- or duplex-assays for the detection of one or two targets in one reaction. When it comes to a biomarker panel that is often composed of multiple targets, the assessment of each biomarker by single- or duplex assays in clinical samples can be labor-, time-, and cost-intensive, in addition to the fact that many types of precious clinical samples are difficult to obtain and are only available in small quantities. Therefore, there is an unmet need to develop multiplex PCR assays for DNA methylation-based biomarker development.

Here, we describe the application of dPCR technology for the detection and quantification of methylated DNA (also known as DNA-methylation-specific multiplex digital PCR, or MS-dPCR) using the QIAcuity dPCR system (Qiagen), which achieves partitioning through nanoplates. We present a quadraplex dPCR assay for a candidate DNA methylation-based biomarker panel for the early detection of high-grade dysplasia and esophageal adenocarcinoma. Furthermore, we discuss special considerations in the development of a biomarker panel for the early detection of cancer using MS-dPCR assays, in addition to common technical challenges and troubleshooting needed to perform successful experiments and fully unleash the potential of this technology. *See* Fig. 1 for a diagram of the workflow.

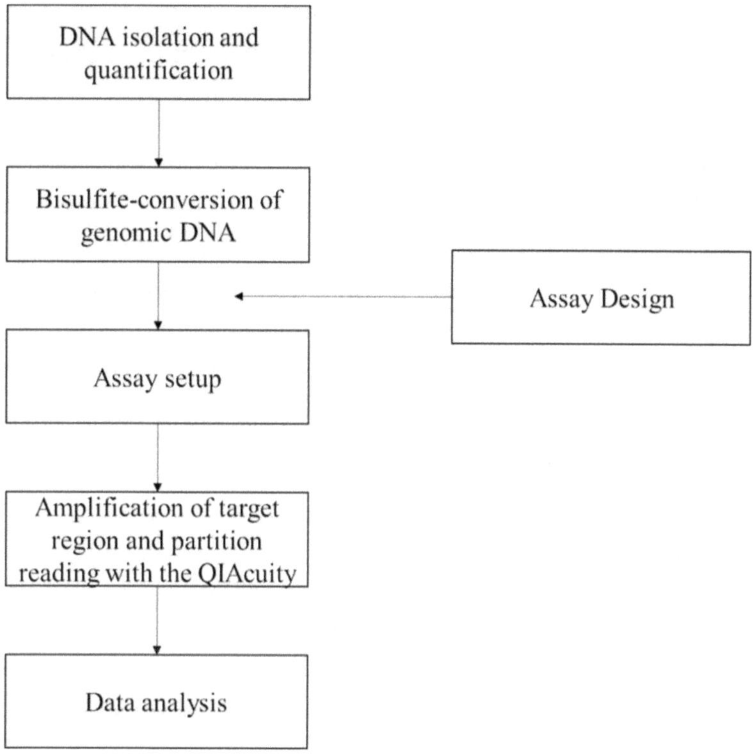

Fig. 1 Workflow. Workflow diagram for the Methylation-specific digital PCR protocol

2 Materials

2.1 Assay Components and Controls

1. Bisulfite converted Methylated and Unmethylated control DNA, from the EpiTect PCR Control DNA Set (catalog #59695) (*see* **Note 1**).
2. Primer Express® V3.0.1 (Life Technologies, catalog #4363991). This software can be used to predict the secondary structure of the assay primers and probes and to visualize their Tm and %GC content.

2.2 Sample Preparation Extraction of gDNA

1. For DNA isolated from fresh frozen tissue, we recommend the use of the DNeasy® Blood and Tissue Kit (QIAGEN, catalog #69504). For FFPE tissue, we recommend the QIAamp® DNA FFPE Tissue Kit (QIAGEN, catalog # 56404).
2. DNAse and RNase free microcentrifuge tubes, 1.5 mL.
3. Vortex and microcentrifuge.
4. Centrifuge.
5. Molecular grade water.

6. 100% ethanol.
7. Xylene (for FFPE paraffin removal).
8. A heat block or water bath.

2.3 Bisulfite Conversion

1. Molecular biology grade water.
2. EZ DNA methylation Kit (Zymo Research, Catalog # D5002).
3. Strip tubes.
4. A thermocycler, incubator, or other device that allows for a constant temperature overnight.
5. Microcentrifuge tubes 1.5 mL (DNase and RNAse free).
6. Microcentrifuge.
7. Centrifuge.

2.4 DNA Quantification

1. QuantiT™ PicoGreen dsDNA Assay Kit (Life Technologies, p7589).
2. 96-well Black/Clear Falcon™ 96-Well, Cell Culture-Treated, Flat-Bottom Microplate (Fisher Scientific, cat#353219).
3. DNAse and RNase-free microcentrifuge tubes, 1.5 mL.
4. A microcentrifuge and vortex.
5. Molecular grade water.

2.5 Qiagen QIAcuity Consumables and Equipment

1. 26 k 24-well, QIAcuity Nanoplate (catalog # 25003).
2. Nanoplate Seals (catalog #250099).
3. QIAcuity Probe PCR Kit containing QIAcuity Probe Mastermix and water (catalog # 250102) (*see* **Note 2**).
4. Forward and reverse primers.
5. Probes to label each target.
6. QIAcuity One 5plex Device (catalog # 911021).
7. QIAcuity Nanoplate Tray (catalog #250098).
8. QIAcuity Software Suite.
9. Nanoplate roller.
10. Electronic multichannel pipette.

3 Methods

3.1 Extraction of gDNA from Clinical Samples

1. Genomic DNA (gDNA) isolation can be performed using either the Qiagen's DNeasy® blood and tissue kit, or the QIAamp® DNA FFPE kit. Select the appropriate kit depending on the starting material and follow the manufacturer's instructions.

2. DNA can be quantified using the Quant-iT™ PicoGreen® dsDNA Assay Kit according to the manufacture instructions (*see* **Note 3**).

3.2 Bisulfite Conversion of gDNA

Bisulfite conversion can be performed using a kit such as the EZ DNA methylation kit (Zymo Research) following the manufacturer's instructions with the modifications included in the protocol below. Note that after bisulfite conversion, the gDNA is transformed in single-stranded fragments, which will be partitioned in dPCR.

1. Prepare the CT conversion reagent by combining 750 μL of water and 210 μL of the kits M-Dilution Buffer to the tube of CT Conversion Reagent. Mix by vortexing for 10 min.
2. Add 5 μL of M-Dilution Buffer to the DNA sample and use water to adjust the total volume to 50 μL.
3. Incubate the sample for 15 min at 37 °C in a thermocycler or heat block.
4. Once the incubation is over, add the 100 μL of previously prepared CT Conversion Reagent to the sample and mix by pipetting.
5. Incubate for 12–16 h at 50 °C in the dark then hold at 4 °C until ready to proceed, using a water bath, thermocycler, or heat block. Sample can remain at 4 °C for up to 20 h.
6. Next add 400ul of M-Binding Buffer the Zymo-Spin™ IC Column and place it into a collection tube.
7. Load your sample from **step 5** into the Zymo-Spin™ IC Column with buffer prepared in **step 6**.
8. Centrifuge at >10,000 g for 30 s, then discard the flow through.
9. Wash the column using 100 μL of M-Wash Buffer. Centrifuge at >10,000 g for 30 s.
10. Add 200 μL of M-Desulfonation Buffer to the column and incubate at room temperature for 15–20 min. After the incubation centrifuge at >10,000 g for 30 s.
11. Add 200 μL of M-Wash Buffer to the column. Centrifuge at >10,000 g for 30 s. Add 200 μL of M-Wash Buffer and centrifuge at >10,000 g for 30 s.
12. After the final wash, spin-dry the column for an additional 1 min (total of 2 min).
13. Place the column into a 1.5 mL microcentrifuge tube. Add 10 μL of M-Elution Buffer directly to the column. Centrifuge for 30 s at >10,000 g to elute the DNA. To optimize the DNA yield, perform two elutions of 10 μL into the same collection tube.

14. Bisulfite converted DNA can be used immediately or stored at −80 °C for long-term storage. Single-use aliquots are recommended to avoid repeated freeze-thawing of the converted DNA.

3.3 Methylation-Specific dPCR Assay Design Guidelines

1. The maximum length of the amplicon is 160 bp for optimal performance, with a preferred size of 80–100 bp.
2. The bisulfite-converted sequence of either the top or the bottom strand can be used as the input sequence in the PyroMark Assay Design Tool.
3. As a rule, each primer and probe sequence should contain more than two CG oligodeoxynucleotides (CpGs). For primer design, CpGs should be located near 3′ end of the primers but try to avoid repeating more than 3 C or G mononucleotide at the end of the sequence. For probe design, the CG of interest should be located in the middle or near 3′ end of the probe. If only one CpG can be found (e.g., in a CpG sparse region), the CpG should be positioned near the 3′ end in the primers or in the middle of the probe to improve the specificity toward methylated alleles. Additional specificity can be achieved by using minor groove binding (MGB) probes. Multiple target analysis is possible using multiplexing (*see* **Note 4**).

3.4 QIAcuity Digital PCR Assay Setup and PCR Cycling

1. All procedures are to take place at room temperature unless otherwise noted. Pre-and post-PCR procedures need to occur in designated rooms, such that the pre-PCR steps can occur in a clean room free of post-PCR products (*see* **Note 5**).
2. Assemble the PCR reaction using the correct proportion as described in Table 1, which contains enough of the PCR reaction mix for two technical replicates.
3. Use an electronic multichannel pipette to transfer 38 μL of the reaction mixture from the strip tubes to the QIAcuity plate. Dispense the volume onto the wall of the well in the QIAcuity

Table 1
Master mix composition and calculations in dPCR reactions

PCR reaction mix component	Starting	Final	Volume (μL) for 1 well
QIAcuity Probe Mastermix	4×	1×	10
Methylated target primer probe mix	10×	1×	4
C-LESS-C1 primer probe mix	10×	1×	4
RNase-free water	–	–	20
Total volume	–	–	38

Table 2
Thermocycling conditions for dPCR

Step	Time	Temperature	Number of cycles
Heat activation	2 min	95 °C	1×
Denaturation	15 s	95 °C	40×
Annealing/extension	1 min	60 °C[a, b]	

[a]Annealing temperature varies for each assay. Use a temperature gradient to establish optimal annealing temperature
[b]The annealing temperature for C-LESS-C1 is 60 °C

plate, being careful not to introduce bubbles into the well. Bubbles, as well as defects in the plate or damage to the plate, can cause dead space in the partition data.

4. Once the QIAcuity plate has been loaded, it needs to be sealed properly. Follow manufacturer instructions to seal the plate and then to transfer it to the QIAcuity machine for cycling and reading following conditions described in Table 2. Be careful not to tilt, bump, or drop the plate during transfer.

3.5 Data Analysis and Interpretation

1. After the run and image transfer have been completed, the QIAcuity Software Suite can be used to analyze the data (*see* **Note 6**). First, check for any error messages, such as one that notes some wells are oversaturated. If this message appears, follow the manufacturer's guide to adjust the gain and exposure appropriately and reimage the plate.
2. Check partition images in the signal map tab for the reference channel. Note any areas of dead space. Any large regions of dead space in the data may impact the results and should be documented (Fig. 2).
3. Look at the partition data in the signal map for the target channels; the image should show a random distribution of positive partitions (Fig. 3).
4. A threshold is used to determine positive and negative partitions. Rely on non-template control (NTC) and negative control samples to set the proper threshold level (Fig. 4).
5. After setting the threshold for each assay, the data can be exported into Excel software for further analysis.
6. First, calculate the RM% for each individual sample replicate separately using the formula below.

$$\mathrm{RM\%} = \left(\frac{\text{Methylation Assay}\left(\frac{\text{copies}}{\mu\text{L}}\right)}{C - \text{LESS} - C1\left(\frac{\text{copies}}{\mu\text{L}}\right)} \right) \times 100$$

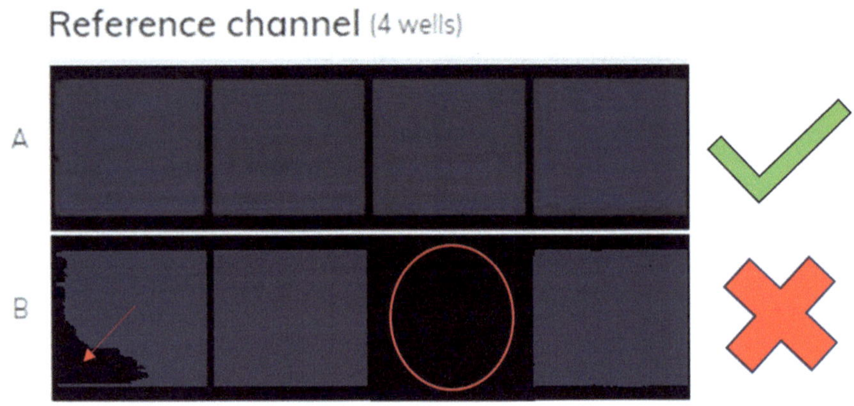

Fig. 2 QIAcuity Reference channel partition data. Representative images of the QIAcuity reference channel from a MS-dPCR assay. Each Row (A and B) represents a data from and individual QIAcuity nanowell. The blue in each row represents the reference dye that flows into the partitions when the plate is rolled. Row A shows the ideal result; each square is blue indicating the reference dye was distributed into every partition evenly. Row B is a representative depiction of incomplete filling of the reference dye leading to areas of dead space, which are indicated by the red arrow and circle. Dead space partitions will not be analyzable, and a large amount of dead space will impact the quality and reproducibility of the final data

Fig. 3 QIAcuity fluorescence channel partition data. A representative image of a QIAcuity fluorescence channel, with the random distribution of positive partitions shown in green

7. Average the RM% of the replicates to get the average RM% of the sample.
8. Calculate the RM% standard deviation between replicates (Fig. 5).
9. Next, use the average (μ) and the standard deviation (σ) of the RM% to calculate the coefficient of variation (%CV) for the replicates.

$$\%CV = \frac{\sigma}{\mu} \times 100$$

10. Finally, screen the data for any samples with a %CV of greater than 30; consider rerunning these samples if DNA is available.

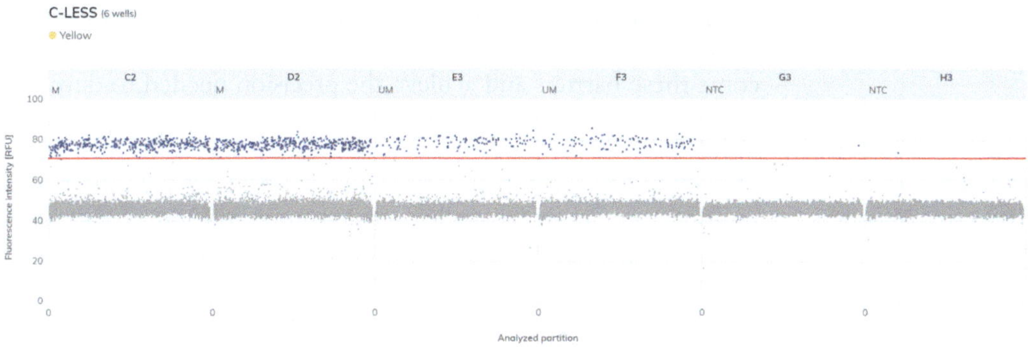

Fig. 4 QIAcuity 1D plot. 1D plot of C-LESS-C1 in the EpiTect controls run in duplicate. The controls appear in the following order; 100% EpiTect methylated (M) (wells C2 and D2), 100% Epitect unmethylated (UM) (wells E3 and F3), and a NTC (wells G3 and H3). The 1D plot shows the separation of positive partitions (navy) and negative partitions (gray). The red line is the threshold that has been set, between the positive and negative partitions. The threshold setting is done based on the non-template control (NTC). The threshold is set so that the baseline noise of negative partitions in the NTC well is negative

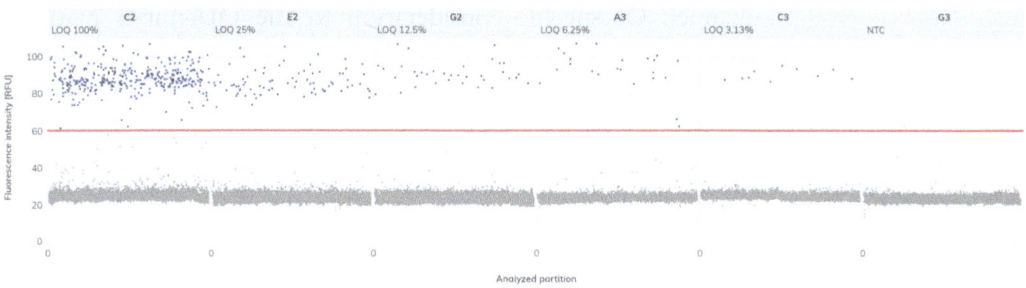

Fig. 5 MS-dPCR LOQ 1D-plot. 1D plot for a MS-dPCR assay showing representative results of a limit of quantification (LOQ) experiment, in which the 100% methylated Epitect control was diluted down into a background of the 100% unmethylated Epitect control for a final total DNA input of 8 ng. The positive signal (blue) is determined by the threshold (red line), which is set above the negative partition baseline (gray) and the baseline noise from the non-template control (NTC) well (G3). For each fourfold dilution of the 100% Epitect methylated control, a corresponding decrease in positives signal detected as positive partitions for the MS-dPCR assay is observed

3.6 Consideration on MS-dPCR Panel for Early Cancer Detection

3.6.1 Integrity and Amount of Input DNA

Working with archived clinical specimens poses unique challenges for the development of DNA-methylation-based biomarker assays. First, the integrity of input DNA from FFPE tissue blocks is suboptimal due to the chemical modification from formalin cross-linking, denaturation, and fragmentation during the bisulfite conversion process, in addition to the potential PCR inhibitors present in clinical samples. Therefore, even though a dPCR assay may be both precise and reproducible when measuring methylated genes in reference samples, its performance may become suboptimal in DNA extracted from FFPE blocks. Second, due to the precious nature of the clinical samples, the availability of input DNA can become the major limiting factor, thus requiring a highly sensitive assay. In this regard, dPCR offers advantages over qPCR to overcome these barriers and achieve the precision needed to develop a robust biomarker assay for clinical samples.

3.6.2 Choice of Fluorescence Channels on the dPCR Platform

Designing an assay for the detection and absolute quantification of multiple rare, methylated alleles from tumor DNA in a background of unmethylated DNA from non-tumor DNA is technically constraining. A robust and specific assay for each target is the most crucial of the criteria that must first be met before a multiplex panel of assays can be considered. Once the assays for individual targets have been worked up, there are further technical considerations that need to be considered. Probe synthesis itself can be another design constraint when the user considers assigning a fluorophore to an assay. Manufacturers have specifications that may limit what length of probe can be synthesized for a specific fluorophore, and MGB's which give added specificity to an assay are not available for all fluorophores. One strategy is to place the most robust assay in the darkest channel (Cy5), and the least robust one in the brightest channel. Of specific consideration to the QIAcuity's platform, among the five available channels, it is likely that only four (green, yellow, red, and crimson) can be successfully multiplexed as the fifth channel (orange) is known to leak out into other channels.

3.6.3 Plate Design

1. To ensure a successful experiment using methylation-specific dPCR assays, it is important to carefully perform plate design to include proper controls for data analysis and interpretation. Both positive and negative controls are crucial for all PCR assays, including dPCR. Three types of control samples must be used along with the experimental samples.
   (a) A positive control for methylation detection assays. This positive control can be commercially available, e.g., 100% methylated EpiTect control DNA (Qiagen), or from reference tumor DNA that is known to be methylated in the CpGs of interest. The use of the positive control is needed for initial assay development and performance evaluation (LOD and LOQ). A positive control with a defined amount can also serve as a calibrator to include on each plate, a valuable tool to assess run-to-run variability and to perform data analysis.
   (b) A negative control for methylation detection assays ensures that the assay is methylation-specific. This negative control can be commercially available, e.g., 100% unmethylated (UM) EpiTect control DNA (Qiagen), or from reference samples that are known to be unmethylated in the CpGs of interest.
   (c) No-template control (NTC) detects any contamination in the reaction mix that might lead to false positive reactions.

Table 3
Primers and probe sequences for the C-LESS-C1 assay

C-LESS-C1	Primer/probe sequence (5′–3′)	Reference
Forward primer	TTGTATGTATGTGAGTGTGGGAGAGAGA	[6]
Reverse primer	TTTCTTCCACCCCTTCTCTTCC	
Probe	VIC-CCTCCCCCTCTAACTCTAT-MGBNFQ[a]	

[a]MGBNFQ refers to a Minor Groove Binder (MGB) nonfluorescent quencher at the 3′ end of the probe. This quencher is essential for the use of a 60 °C annealing temperature with C-LESS-C1 (*see* **Note 7**)

The NTC can be used to set the threshold to distinguish positive from negative partitions.

(d) Inclusion of a control reaction (C-LESS-C1) in a multiplex assay to simultaneously measure the amount of amplifiable DNA in a sample, in addition to the target of interest (Table 3), is highly recommended.

3.6.4 Reaction Optimization

Since dPCR relies on data collection from a large number of independent partitions, in which a single reaction occurs, reaction optimization is particularly important for a successful dPCR run. In dPCR reactions using the QIAcuity instrument, an error message from wells reported with oversaturated signals is usually an indicator that too much probe is being used for the assay. On the other hand, if the concentration of either the primers or the probe is too low, the dynamic range of a dPCR assay is suppressed. Therefore, it's crucial to conduct a preliminary assessment for primers/probe concentration, after a careful primers/probe design, to achieve the maximum signal from a PCR reaction without saturating the signal with too much probe. For multiplex assays, the whole reaction optimization process should be performed, and the performance of a multiplex assay should be compared to single-plex reactions. For trouble shooting issues with assay performance and optimization please *see* Table 4.

3.6.5 Evaluation of Assay Performance

To ensure the performance of MS-dPCR, it's important to understand and appropriately determine both the limit of detection (LOD) and the limit of quantification (LOQ). While the LOD is defined as the lowest copies of methylated DNA that can be accurately and reliably quantified by the assay using water as a diluent, the LOQ is a more comprehensive parameter to consider since it indicates an assay's capacity to detect abnormal methylation from tumor cells in a background of unmethylated DNA from normal cells, with acceptable accuracy and precision. While an assay's LOD can be used to guide determination of LOQ, extrinsic factors such

Table 4
Troubleshooting

Observation	Explanation	Suggestion
No signal	Incomplete reaction mix assembly	Check the input volumes of each component and the concentration added to the master mix
Large areas of dead space	Improper plate handling	Take care not to drop or bump the plate during transfer and loading
Bisulfite converted DNA has a loss of signal (copies/μL) over time	Bisulfite converted DNA degradation due to freeze-thaw cycles between uses	Aliquot the bisulfite-converted DNA for single, double, or triple use, and discard after three freeze-thaw cycles
Loss of probe amplitude over time	Probe degradation due to light exposure and/or freeze-thaw cycles	Aliquot the stock probe into single, double, or triple use aliquots
The negative baseline is below 30 relative fluorescence units (rfu)	The PCR reaction is being limited by probe availability	Increase the probe concentration
Data comes back and is reported to be oversaturated	Excess probe is being used in the reaction	Test a gradient of probe concentrations to optimize probe input

as the availability of input DNA from clinical samples can influence an LOQ of a biomarker assay; i.e., LOQ can be improved with increasing input DNA. Both LOD and LOQ should be determined using the dilution series with 100% methylated, and 100% unmethylated EpiTect controls as background diluents for LOQ (Fig. 5). For all multiplex assays, we utilize the signal from C-LESS-C1 to account for the total amount of DNA in the sample regardless of methylation status. This allows us to use the control DNA to create ratios of mixtures of methylated and unmethylated DNA and report the RM% for each mixture, and to see at what point the expected ratio of methylation diverges from the calculated RM%.

3.6.6 Experimental Replicates

To increase the chance of detecting rare methylated DNA from tumor cells, a common strategy is to pool multiple wells (hyperwell) and generate larger numbers of partitions than single wells. For clinical samples with limited availability, we suggest a minimum of duplicate wells, but triplicates or quadruplicates could be used to achieve desirable LOQ.

4 Notes

1. The control DNA used in the experimental design is a DNA kit from QIAGEN. The 100% unmethylated/methylated EpiTect

DNA is provided in the bisulfite converted state, and needs to be stored at −80 °C. The DNA should be stored in aliquoted to prevent degradation by freeze thaw.

2. The QIAcuity Probe Mastermix is light sensitive and needs to be stored at −20 °C for long-term storage.
3. Other DNA quantification methods may also be successfully used for this protocol; however, the Pico Green DS assay kit is what we have adopted in our lab. If looking to use an alternative method, we suggest using a fluorometry-based method for measuring DNA, as this method is insensitive to dust and other biomolecules that may impact the accuracy of DNA quantification by other methods.
4. To have multiple assays run concurrently, each primer/probe set should have a similar annealing temperature, and the probe should be labeled with fluorophores whose spectra do not overlap. The combinations of primer/probe sets should be evaluated to check for any interference of each assay by non-cognate primers/probes.
5. Handle the DNA and PCR reaction assembly steps in a designated PCR workstation decontaminated with UV light (254 nm) for 15 min. All microcentrifuge tubes destined to be used in the PCR reaction assembly can be autoclaved and then exposed to UV during the experimental preparation. Everything that enters the PCR workstation should be treated with similar caution, to guarantee consistency and confidence in the absolute quantification data. For this purpose, PCR workstations such as an Air Clean 600 system are recommended for all PCR reaction setup.
6. After a run is completed, the QIAcuity software counts the number of positive and negative partitions for each fluorophore, depending on a manually set threshold level. The resulting data is fed into a Poisson statistical model to report the concentration (copies/μL) of specific methylated genes in the sample. This method allows for absolute quantification without requiring a standard curve.
7. For C-LESS-C1, a minor grove-binding (MGB) domain quencher is required for this assay to work at an annealing temperature of 60 °C. Without the MGB domain, the annealing temperature of the assay will be lower. This methodology can also be employed for assay designs with lower annealing temperature than other assays you desire to multiplex to bring them to the same annealing temperatures as the other assays. MGB in shorter probes can help to achieve greater specificity by increasing the melting temperature.

Acknowledgments

This work was supported by National Cancer Institute (NCI) RO1CA220004, UO1AG077920, U2CCA271902; U54CA274374, Listwin Family Foundation; and Cottrell Family Fund; (WMG); R50CA233042 (MY).

References

1. Nishiyama A, Nakanishi M (2021) Navigating the DNA methylation landscape of cancer. Trends Genet 37(11):1012–1027
2. Taylor SC, Laperriere G, Germain H (2017) Droplet Digital PCR versus qPCR for gene expression analysis with low abundant targets: from variable nonsense to publication quality data. Sci Rep 7(1):2409
3. Sidstedt M, Rådström P, Hedman J (2020) PCR inhibition in qPCR, dPCR and MPS-mechanisms and solutions. Anal Bioanal Chem 412(9):2009–2023
4. Yu M et al (2015) MethyLight droplet digital PCR for detection and absolute quantification of infrequently methylated alleles. Epigenetics 10(9):803–809
5. Yu M et al (2022) Novel DNA methylation biomarker panel for detection of esophageal adenocarcinoma and high-grade dysplasia. Clin Cancer Res 28(17):3761–3769
6. Weisenberger DJ et al. (2008) DNA methylation analysis by digital bisulfite genomic sequencing and digital MethyLight. Nucleic Acids Res 36 (14):4689–4698

Chapter 4

Multiplexed Digital PCR for Evaluation of Karyotype Quality of Mouse Embryonic Stem Cells

Christy J. Greenwood, Alasdair J. Allan, and Lydia Teboul

Abstract

Digital PCR enables the high precision, absolute quantification of nucleic acid target sequences through volumetrically defined partitioning. This precision, coupled with the removal of PCR efficiency biases and the need for a standard curve, has instituted its widespread adoption over other quantitative methods. Digital PCR is particularly well suited for precise copy counting of chromosome numbers in cells, making it a viable alternative to traditional karyotyping in many instances. Here we describe a multicolor multiplex digital PCR method of screening ES cell clones for aneuploidy for cheaper and swifter validation. This method can also be easily applied to other in vitro cellular models that require accurate analysis for copy number variations.

Key words Digital PCR, Karyotype, Aneuploidy, Embryonic stem cells, Gene targeting, Multiplex assay

1 Introduction

While genome editing enables efficient animal model modification using donors up to a few kilobases, modifying or integrating large genomic segments in the mouse genome still requires gene targeting in embryonic stem (ES) cells [1, 2]. This process requires maintaining cell pluripotency for germline transmission throughout the multistep in vitro culture, including electroporation and cloning of ES cells. After careful molecular characterization of the modified locus, cells that contain the required allele are microinjected into host blastocysts, which they will colonize to yield a chimeric mouse. The ability of the mouse to transmit alleles belonging to the targeted ES cells, an event called germline transmission (GLT), is associated with these cells maintaining genomic integrity [3–6]. Studies have shown that the chromosomal composition of mouse ES cells is a predictor of the efficiency of GLT [6]. The exclusion of mutant ES cell clones with abnormal

Guillaume Pavlovic (ed.), *Digital PCR: Methods and Protocols*, Methods in Molecular Biology, vol. 2969,
https://doi.org/10.1007/978-1-0716-4767-7_4,

karyotypes prior to microinjection is thus essential to prevent foreseeable microinjection failures and ensure the ethical use of animals. We and others have empirically determined that ES cell colonies isolated from some broadly employed parental ES cell lines conserve their ability to produce GLT provided that at least 50% of the cell population has remained euploid [6]. The evaluation of the karyotype of ES cell colonies was classically performed by metaphase spread analysis, a time-consuming exercise that requires specific microscopes and rare expertise in chromosome morphology evaluation [7]. This type of analysis became increasingly challenging to implement as the scale of animal engineering increased [2] and in the absence of staff skilled in karyotype evaluation using metaphase spreads.

A review of a large screen for aneuploidy in ES cell clones, using chromosome spreads, showed that in the widely used mouse ES cell line JM8 and in subclones derived from this parental line, the number of at least one of a group of four chromosomes (1, 8, 11, and Y) is altered in most of the aneuploid cells (with either loss or acquisition of chromosomes, and sometimes in combination with abnormalities in the number of other chromosomes) [6]. Copy counting, that use qPCR [8] or droplet digital PCR (ddPCR) [6], of a locus on each (or some) of these four chromosomes was developed as an alternative to direct chromosome examination and yielded a more convenient method for karyotype validation.

Here, we have taken advantage of the availability of new modalities for multicolor multiplex digital PCR (dPCR) as an alternative to our ddPCR karyotype screen. By employing multiplex dPCR, all four chromosomes are counted within a single reaction. We have identified assays that are compatible when run in a multiplex format and have optimized DNA input and annealing temperature to achieve screening outcomes comparable to those of previous protocols that were based on duplex reactions [6]. This protocol is based on the use of the QIAcuity (Qiagen) system, for cheaper and faster implementation (*see* **Note 1**) of a chromosome count-based validation of ES cell karyotype (*see* **Note 2**). This principle of copy counting can also be applied to karyotype validation in other in vitro cellular models [9].

2 Materials

2.1 Sample Lysis

1. Heat cabinet set to 55 °C.
2. Plate shaker.
3. Plate/tube centrifuge.
4. ES clones amplified in 96-well round-bottom plates.

Table 1
Details of assays employed for dPCR

Genomic location (GRCm39)	Gene target	Probe *Channel*	Primer/ probe	Sequence and chemical modifications (5′–3′)
Ch1: 153629544–153629620	*RnaseL*	TAM RA *Orange*	Forward	GGACCGTTGTCGAAGATGATT
			Reverse	TAACAATTGCTGGACCCTGAC
			Probe	/56-TAMN/CGTTGATCAAAGCTGTTCA GAAGGGAGA/3IAbRQSp/
Ch8: 34324389–34324478	*Rbpms*	FAM *Green*	Forward	TCCGCTTCGATCCTGAAATC
			Reverse	GAGTCCCTACGAGTTTGTTCTT
			Probe	/56-FAM/TGCTAAGGC/ZEN/ AAACACGAAGATGGC/3IABkFQ/
Ch11: 5084869–5084947	*Emid1* [6]	ROX *Red*	Forward	GCCAGGACTGGGTAGCAC
			Reverse	AGGAGGCTCCTGAATTTGTGACAAG
			Probe	/56-ROXN/ CCTGGGTCATCTGAGCTGAGTCC/ 3IAbRQSp/
Ch17: 13232436–13232535	*Sod2* [6]	HEX *Yellow*	Forward	AAGGAGCAAGGTCGCTTACA
			Reverse	GGCACTGACATGGGAAGAGT
			Probe	/5HEX/CAGGACCCA/Zen/ TTGCAAGGAACA/3IABkFQ/
ChY: 1010549–1010676	*Eif2s3y*	Cy5 *Crimson*	Forward	CCCATAGGCTTGTGCGTT
			Reverse	AAGATCCTGACGAGAAAGATGC
			Probe	/5Cy5/CTTTCGTGA/TAO/ GCAAGATGGCGGG/3IAbRQSp/

Primers and probes are sourced from IDT and resuspended to a 100 μM stock concentration upon receipt, ready to be diluted further when in use. Stock primers and probes are kept at −20 °C to reduce degradation

5. Microcentrifuge tube.
6. DNA Extract All Reagents Kit (Applied Biosystems™).
7. Pipette filter tips.
8. Parafilm (Merck).

2.2 dPCR

1. TaqMan assays (*see* Table 1).
2. Bench-top plate centrifuge.
3. QIAcuity Nanoplate 26 K 24-well plate (QIAgen).
4. Nanoplate plate holder (QIAgen).
5. QIAcuity rubber plate seal (QIAgen).
6. Nanoplate roller (QIAgen).
7. QIAcuity One 5plex system (QIAgen).
8. QIAcuity Probe PCR 4× mastermix (QIAgen).
9. Microcentrifuge tubes.

10. Pipette filter tips.
11. Nuclease-free water.
12. A standard 96-well PCR plate.

3 Methods

3.1 DNA Extraction

Crude genomic DNA (gDNA) is extracted from ES cells using the DNA Extract All Reagents Kit (*see* **Note 3**).

1. Seed ES cells in exponential growth phase on a 2% coated gelatinized well of a 96-well round-bottomed plate for 3–5 days to allow for some growth to confluence and differentiation to occur. Remove all media. Alternatively, if more convenient, move cells into individual microcentrifuge tubes and pellet.
2. If needed, the plate or tubes can be stored at −20 °C after removal of media and until ready for DNA extraction.
3. If needed, defrost the samples and add 50 μL Lysis Solution to each sample (*see* **Note 4**).
4. Seal the plate and lid together with Parafilm. If the samples are in microcentrifuge tubes, ensure that the lid is tightly closed.
5. Incubate the plate or tubes at 55 °C for 1 h.
6. Briefly centrifuge the plate or tubes to pull down any condensate. Add 50 μL DNA Stabilizing Solution. Mix thoroughly. To prevent sample degradation/evaporation, aliquot a required volume into a separate plate to be used for testing (*see* **Note 5**).
7. Undiluted crude lysate is used as the template for dPCR.
8. Store the crude lysate at 4 °C if testing is being carried out the same day. Crude lysate should be stored long term at −20 °C.

3.2 Assay Design

3.2.1 General Considerations for Assay Design

New assay primers are designed using the IDT SciTools OligoAnalyzer tool [10] with the following generic parameters: Tm: 60–64 °C, Length: 18–24 nt, GC%: 40–60%, Amplicon length: 70–150 nt. Probes are designed to have a Tm of 66–70 °C, 4–10 °C above the Tm of the primers.

Depending on the attached fluorophore, additional principles are applied to the design of probes: For probes with a 5′ FAM label, a guanine on the 5′ end is avoided as this has previously been shown to decrease fluorescence intensity. The preferential order for the choice of 5′ end nucleotide is $T > A \approx C \gg G$ [11]. For probes with a 5′ Cy5 label, the order of preference for the 5′ end nucleotide is $A > G > T > C$, as previous studies have shown that Cy5 probe fluorescence intensity is positively affected by an increased purine presence [12] (*see* **Note 6**).

The NCBI Primer-BLAST tool (https://www.ncbi.nlm.nih.gov/tools/primer-blast/index.cgi) is used to check primer specificity [13, 14]. Primer-BLAST is also used to check for potential amplification that may occur between primers from different assays (*see* **Note 7**).

3.2.2 Design of Multiplexed Assays

1. Several specific assays are designed for genomic loci on mouse chromosomes 1, 8, 11, 17, and Y (*see* Subheading 3.2.1) (*see* **Note 8**) [13].

   Due to the increase in the number of primer and probe sequences in the multiplex mix, the potential for sequence interactions is greatly increased. Potential dimer or hairpin formation could have a significant impact on the assay efficiency during thermal cycling, leading to uneven amplification, which could affect fluorescence intensity and copy number evaluation accuracy. Care should be taken to check for these potential dimers and hairpins during the design process, including checking the amplicon for inter-assay dimers (*see* **Note 9**) (Fig. 1).

   Check for primer-dimers and hairpins using IDT OligoAnalyzer (https://eu.idtdna.com/pages/tools/oligoanalyzer) [10]. This tool can be used for gathering the details on potential dimers and hairpins but when designing an assay panel with many primer/probe sequences within the same mix it can

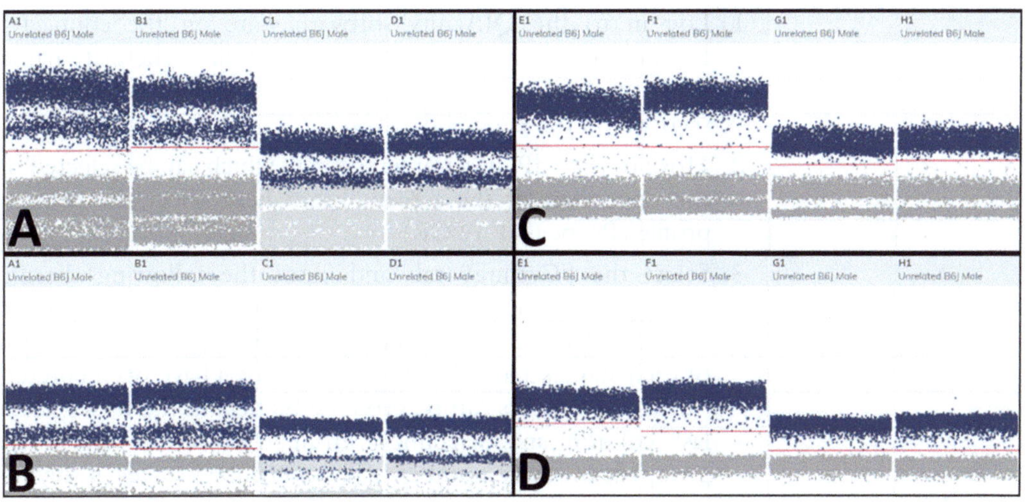

Fig. 1 Examples of orange channel imaging issues with different causes. (**a**, **c**) show examples of four sample replicates with multiple banding present within the well. Samples (**a**, **b**) are two images taken of the same samples being run with the *Dot1l* (VIC) assay [6] present within the mix and a variety of orange channel assays. Samples (**c**, **d**) are run with the *Sod2* (HEX) assay instead of *Dot1l* and the same variety of orange channel assays. When reimaged with increased exposure time, the assays with the *Sod2* mix (**d**) resolve to just two bands present, whereas the assays with the *Dot1l* mix (**b**) resolve slightly, but remain with multiple bands present in each well

become slow to run (*see* **Note 10**). In this case, to speed up this process, MFEprimer v4.0 (https://m4.igenetech.com) [14] can be used to check for potential dimers/hairpins and any that are flagged can be double-checked with IDT OligoAnalyzer. MFEprimer v4.0 is well suited to checking multiplex mixes, as all sequences can be checked at once and the sensitivity of the output can be adjusted (*see* **Note 11**).

2. Once an assay panel has been assembled in silico, set up the multiplex assay conditions: run with an annealing temperature gradient and choose an optimal annealing temperature. Compare results from the multiplex assay with those from the duplex assay, using known samples to look for consistency and to test for assay interactions. In the case of assay interaction, alternative assays can be swapped in and the new multiplex combination evaluated. Check for sensitivity of multiplex assays and set cutoffs for aneuploidy using euploid samples spiked with increasing proportions of aneuploid samples.

   The final assays used in the multiplex panel are shown in Table 1.

3.3 QIAcuity Template Setup

Before running the karyotyping panel on the QIAcuity system, a run template must be made in the software suite that describes the priming, thermal cycling, and imaging conditions alongside details of which reaction mixes and samples are being run, as well as the plate layout.

1. Log in to the QIAcuity software suite on the connected PC. Select the "New Plate" button and fill in the plate name, plate type (Nanoplate 26 K 24-well), and plate ownership sections on the general data tab.
2. Move to the "dPCR parameters" tab, which can be selected on the upper left of the screen. From here select the priming profile (Probe (RT-) PCR).
3. Select the "Cycling" tab and input the following thermal cycling conditions: 1× 95 °C 2 min, 40× (95 °C 15 s, 58 °C 30 s) (*see* **Note 12**) by entering the temperature and duration of each stage separately. Once these stages have been entered, the last two stages can be grouped by ticking the boxes in the last column and selecting "group." Once the stages are grouped, the cycle number of these grouped stages in the first column can be increased.
4. Select the "Imaging" tab and input the imaging settings as described in Table 2.
5. Enter the sample details and reaction mix details into the plate layout. This process is facilitated by using the CSV import function. The import CSV template can be found on the "Plate layout" tab by selecting "CSV import/export --> Export to CSV."

Table 2
The imaging settings for the channels used in the multiplex assay

Color (Fluorophore)	Exposure time (ms)	Gain
Green (FAM)	300	8
Yellow (HEX)	300	6
Orange (TAMRA)	400	4
Red (ROX)	300	4
Crimson (Cy5)	400	8

6. Fill out the information in the CSV file with "Sample name," "Type (Sample/NTC/Control)," and "Reaction mix name" for each well.
7. The next columns correspond to each channel and are numbered accordingly:
   (a) The "Channel #" column should be filled out with the channel color of the respective assay in all capitals, e.g., GREEN for FAM, YELLOW for HEX.
   (b) The "Target #" column should be filled out with the name of the assay for that channel, e.g., *Rbpms* for FAM, *Sod2* for HEX.
   (c) The "Dye #" column should be filled out with the name of the dye being used in each assay. These are largely in all capitals (e.g., FAM, HEX); however, Cy5 must have the lower case "y."
   (d) The final column to complete is "Internal control #," which is either "TRUE" for the internal control (*Sod2*) or "FALSE" for any target assays (e.g., *Rbpms*).
8. Import the CSV file in the "Plate layout" tab by selecting "CSV import/export --> Import to CSV," choosing the relevant CSV file in the popup, and selecting "Import." All the information should be placed into the plate layout and the plate can then be set up.

3.4 Digital PCR Assay Setup

1. Collect the relevant samples, shake briefly, and spin down using a bench-top plate centrifuge.
2. Make the mastermixes for each assay according to the values in Table 3 (*see* **Notes 13** and **14**), ensuring to vortex and spin reagents first.
3. Take out a new 96-well pre-plate and load the required sample and mastermix into the wells in the order that will be loaded into the Nanoplate.

Table 3
PCR reaction assembly

Component	Volume (μL) per 40 μL well	Final concentration (μM)
QIAcuity Probe PCR 4× Mastermix	10	1
Nuclease-free water	18.5	N/A
Forward primer (20 μM)—per assay	0.4 (×5)	0.2
Reverse primer (20 μM)—per assay	0.4 (×5)	0.2
Probe (5 μM)—per assay	0.8 (×5)	0.1
Template DNA (crude lysed)	3.5	N/A

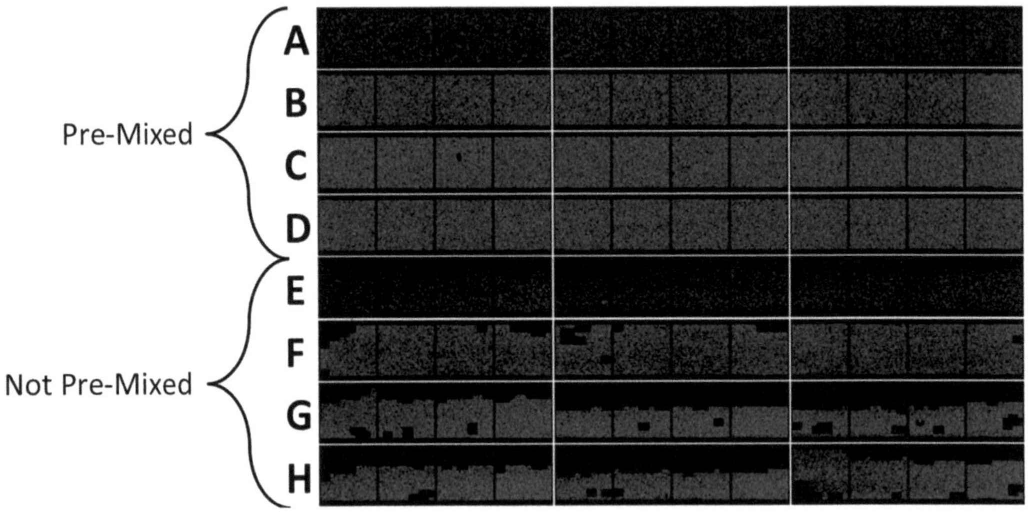

Fig. 2 Improvement in number of accepted partitions with mixing of reactions. Four control samples were run twice in triplicate, the first three repeats (rows A–D) were thoroughly mixed in a pre-plate by pipetting up and down at least ten times before loading into the Nanoplate. The next three repeats (rows E–H) had sample and mastermix consecutively loaded into the Nanoplate without mixing. Rows E–H show many partitions that have not reached the threshold for fluorescence intensity and have been discounted from the analysis by the QIAcuity software (blacked out areas especially prevalent at the top of each well). The exclusion from analysis of regions of low fluorescence by the software typically results in an overestimation of copy number

4. Pipette up and down at least ten times to ensure thorough mixing of the sample and mastermix (*see* **Note 15**) (Fig. 2).
5. Seal the 96-well plate and centrifuge it in a plate centrifuge to bring the sample/mastermix mixture to the bottom of the well.
6. Take a new 26 K 24-well Nanoplate out and examine the underside. If there are significant scratches or damage, discard the plate as these would affect the results.
7. When handling the Nanoplate, ensure to only handle the sides or top of the plate and that the underside is protected. Place the

Nanoplate in a white plate holder that is provided with the system to ensure that the underside is protected during loading.

8. Within the QIAcuity software suite select the "General Data" tab in the previously made run template and input or scan the barcode number, found on the short edge of the plate nearest column 3, into the "Barcode" section. Ensure that the run template is prepared before loading the mastermix into the Nanoplate, as the results can be affected if the plate is left for a long period of time after loading.
9. Pipette 39 μL (26 K 24-well plate) of sample/mastermix mixture from the 96-well pre-plate into the relevant Nanoplate wells, ensuring that no bubbles are present in the pipette tip or well.
10. Apply the plate seal that comes with the Nanoplate, taking care to ensure that the foil is undamaged and overlaid with no more than 1 mm overlapping on any side of the plate (*see* **Note 16**).
11. Use the roller to press the plate seal down into the plate in both horizontal and vertical directions.
12. Carefully remove the top clear protective film and use the roller to firmly seal the edges of the plate.
13. Log in to the QIAcuity One system using the screen on the front of the machine, press the "Eject plate" button and place your prepared plate in the drawer with the barcode facing away from the operator (*see* **Note 17**) (a diagram inside the drawer shows the correct orientation, depicting the cut corner present at the top-left corner of the Nanoplate). Press the "Eject plate" button again once the plate is loaded.
14. When the machine has read the barcode on the plate from the template setup, the "Run" button will light up and the run can be started. Once started, the process will run for approximately 2 h.
15. Once the run has finished, log in to the QIAcuity One system and select "Eject plate" to remove the plate from the machine. Store the completed plate at 4 °C until the results are confirmed (*see* **Note 18**).

    Once the run has been completed the data will automatically be imported into the software suite for analysis.

3.5 Data Analysis

The analysis and extraction of copy number data from the QIAcuity system is a simple process; however, there are several essential aspects that should be checked to ensure data quality and to identify results that may need to be disregarded:

1. Open the QIAcuity software suite after the run is completed, choose the plate from the home screen, and select "Analyse" from the options list on the right of the screen.
2. The command will open to the "Absolute Quantification" tab, which must be used to set the thresholds for each assay before data export.
3. The QIAcuity software suite will display any warnings or errors encountered during the running of the plate at the top of the analysis window. A common warning is oversaturation of a particular color channel. Any wells identified by this warning will need to be reimaged with a lower exposure duration (*see* **Note 19**).
4. Wells can be selected from the plate layout and analysis can be performed according to the color channel or by the assay targets. Once all wells have been selected the "Select targets" drop-down menu will appear. Select all assay targets that need to be analyzed, choose whether the wells should be ordered horizontally or vertically, and select "Show results."
5. To begin analysis, select the "1D Scatterplot" lower tab. The QIAcuity software suite will have automatically set the thresholds for each assay target; however, these will often need to be adjusted manually. Common thresholds for each assay target should be set at a distance that is as consistent as possible between the negative and positive partitions (*see* **Note 20**).
6. Check for any imaging issues by selecting the "2D scatterplot" lower tab. Select an X-axis and Y-axis target from the two drop-down menus and check for any instances in which the partition clusters are significantly leaning to one side (an ideal result will have four populations that are in an orthogonal grid; see Fig. 3 for examples). Run through each target by progressively selecting them in the X-axis and Y-axis drop-down menus. Any targets displaying this lean should be reimaged with an adjusted exposure duration (*see* **Note 21**).
7. Select the "List" lower tab to see the results in a tabular view; there are several elements to pay attention to in this screen. The number of valid partitions should be more than half of the number of possible partitions for the plate (e.g., if there are ~26,000 partitions per plate, >13,000 should be deemed valid) and should be roughly consistent for each target across each sample.
8. If the number of partitions varies between targets or is not close to the number of possible partitions, the "Signalmap" lower tab should be selected. Two types of issues that can impact the improper filling of the partitions include damage to the underside of the plate or insufficient mixing of the sample and mastermix (*see* **Note 22**).

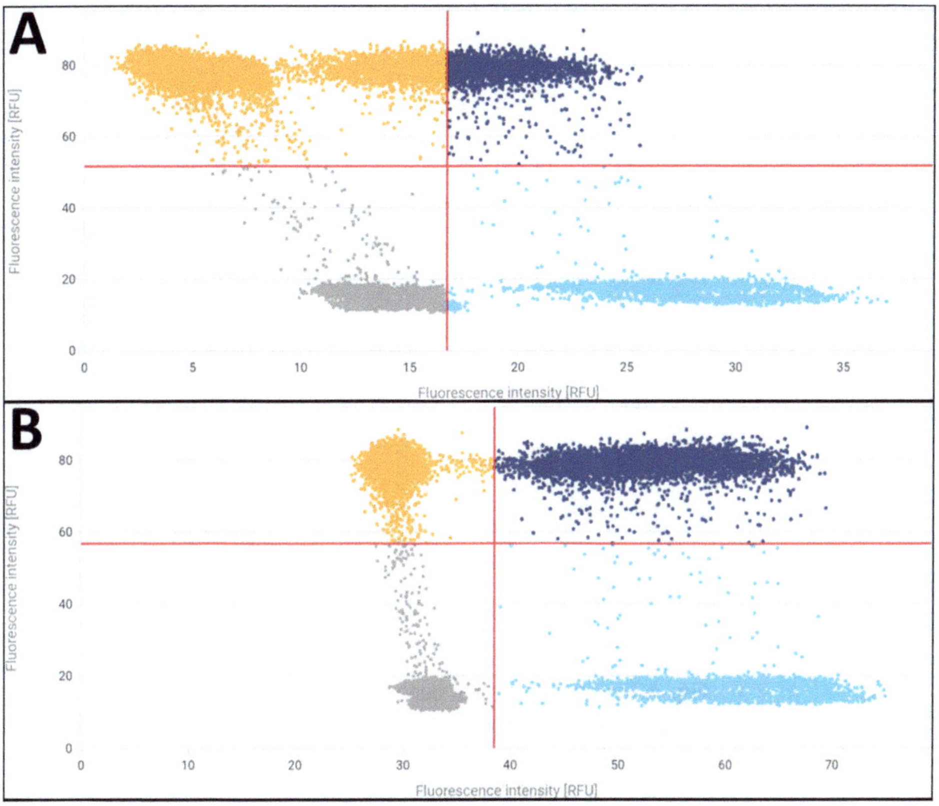

Fig. 3 Example of imaging issues between closely fluorescing probe colors Cy5 (X-axis, crimson channel) and ROX (Y-axis, red channel). (**a**) Shows an example of fluorescence lean and multiple populations, which is resolved by an increase of Cy5 (red channel) exposure duration (**b**)

9. Pay close attention to the number of positive partitions. If this number is lower than 400, then the confidence interval is greatly impacted, and the sample will probably need to be repeated. This number can be improved upon repeat by increasing the DNA input volume on the next run.

10. Once these stages have been completed, select the "Copy Number Variation" tab at the top of the page. Select "CNV with reference target" and choose the *Sod2* reference assay from the reference target drop-down menu. Input "2" into the Copies/Genome box. Select all targets from the "Target of interest" drop-down options and press "Show results."

    The analysis will open in the "List" lower tab. We have used a 95% CI of 5% or higher to identify samples that may need to be repeated with a higher DNA input volume. From this screen, the CNV results can be exported to a comma separated values (CSV) file by selecting "Export to CSV..." (*see* **Notes 23** and **24**).

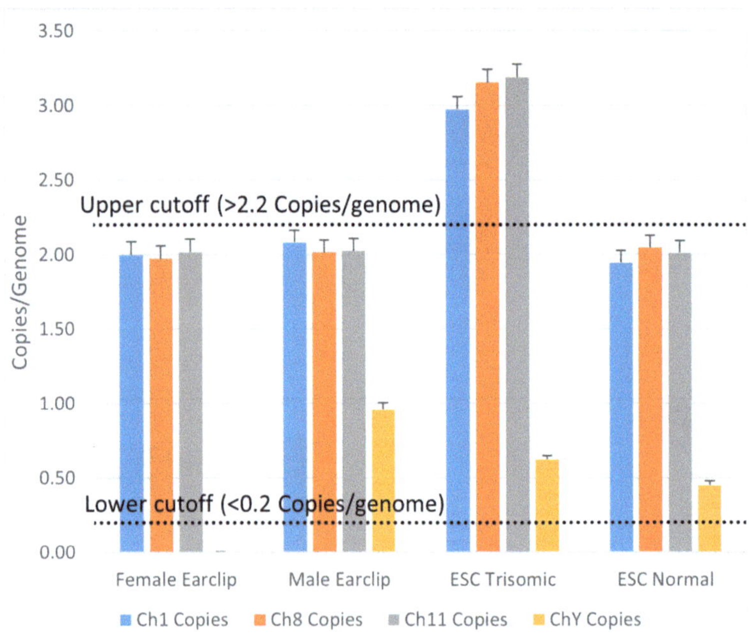

Fig. 4 Example of results generated from a range of samples. The first two samples are female and male mouse ear biopsies. The third sample is an ES cell sample with aneuploidy in chromosome 1, 8, and 11. The fourth sample is a normal ES cell sample, which can be taken forward for microinjection. As previously reported (6), ES cell samples under-report the copy numbers for ChY. The copy number for ChY is often between 0.5 and 0.7 copies/genome in cultured ES cells. Error bars shown denote a 95% confidence interval (CI) for the result

11. Figure 4 shows an example of results obtained with an ear biopsy, a euploid clone, and an aneuploid clone. We are considering any samples that have a result of over 2.2 copies/genome for chromosomes 1, 8, or 11 or under 0.2 copies/genome for chromosome Y to be aneuploid and are not taken forward for microinjection (*see* **Note 25**) (Fig. 5).

4 Notes

1. The setting up of a multiplex protocol represents a significant investment in time and reagent purchase. Therefore, running dPCR in multiplex yields savings in both time and consumable expenditure only when sufficiently large numbers of samples are to be assayed.
2. We present a protocol that employs one of several platforms for multiplex dPCR implementation available on the market. The QIAcuity platform was chosen as its characteristics suited the

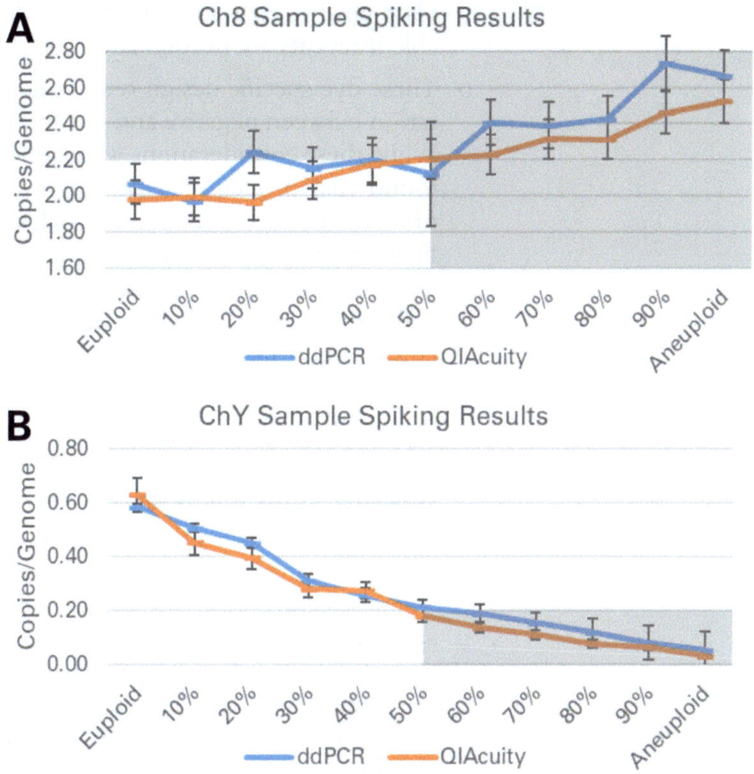

Fig. 5 Results for chromosome 8 (**a**) and chromosome Y (**b**) copy number counting from the sample spiking on the Bio-Rad ddPCR (blue) and the QIAcuity (red) systems. The samples chosen had previously been identified as euploid or aneuploid on both systems. Samples were brought to a similar total DNA concentration then a serial dilution was performed to achieve a range of spiked aneuploid/euploid samples. The QIAcuity cutoff for aneuploidy is set by the copy number result when there are 50% aneuploid cells within the sample (Presence of aneuploidy above 50% of cells has previously been shown to confer a decreased GLT rate in chimeric animals [6]), the previously used ddPCR cutoff of 2.2 for chromosome 8 and 0.2 for chromosome Y are highlighted by gray areas. Error bars denote a 95% CI for the result

needs of our group, but we did not carry out a side-by-side comparison with other equipment with similar technical capability. Other platforms will be suitable to run similar protocols.

3. A crude DNA preparation method is suitable, as it is compatible with dPCR, which produces short amplicons.
4. If the sample has been frozen in a 96-well plate it may be difficult to get the cells back in solution. We found that scraping the bottom of each well with a pipette tip was sufficient to detach the cells and bring them into solution.
5. Crude lysate degrades rapidly when put through multiple freeze/thaw cycles. Keeping back some sample in the original

plate ensures that some non-degraded sample is always available if repeats are needed later in the process.

6. These dye-specific design requirements are beneficial for separation between negative and positive partitions during analysis. While these considerations were adhered to where possible, we found that probes that did not follow these rules were still functional and other factors, such as annealing temperature, final primer/probe concentration, interactions with other assays within the multiplex, and the relative amplicon length and copy number between assays in the multiplex mix, were equally important for optimizing partition separation and fluorescence intensity.
7. Primer-BLAST remains essential to exclude primers that amplify additional genomic segments, as this may impact the efficiency of the multiplex reaction.
8. Like other platforms that detect fluorescent wavelengths from multiple sources (i.e., Flow cytometry), care must be taken when selecting assay-fluorophore combinations to take into account cross-talk between channels. This was most prominent with the orange channel in our hands. It is likely that on setting up a panel from scratch, experimentation will be required to settle on the optimal assay-fluorophore combinations for your experiments.
9. The relative amplicon length of each assay within the multiplex mix was possibly an important factor in improving amplification.
10. IDT offers a batch analysis tool, but we found that it did not provide the same depth of analysis as could be obtained by running through the list manually or using MFEprimer v4.0. Several dimers/hairpins that were identified in the manual version did not appear when the same sequences were run through the batch analysis tool.
11. The initial settings on the MFEprimer v4.0 software are set to exclude dimers that do not affect the 3′ end of each primer and have a delta-G number of higher than −3. We adjust these sensitivity settings, so that all potential dimers are listed, allowing us to review each one individually. While dimers on the 3′ end of the primer are most important to avoid, a strong 5′ dimer can equally negatively affect the efficiency of the reaction.
12. Thermal cycling conditions were optimized by running a series of multiplex assays with an annealing temperature gradient ranging from 64 to 52 °C. The annealing temperature selected was the one that produced the best separation between positive and negative partitions and resulted in consistent DNA concentration between multiplex and singleplex mixes. Note that

temperature gradients are only possible with v2.5 or subsequent versions of the QIAcuity Software Suite.

13. QIAgen recommends using final concentrations of 0.8 μM for primers and 0.4 μM for probes (0.8/0.4 μM). However, we have found that concentrations as low as 0.2/0.1 μM still produced consistent results for the proposed application.
14. The recommended DNA input volume was optimized to yield a lower average error with the ES cell cultures typically produced in our cloning experiments. This volume should be adjusted according to samples (including culture confluence) and DNA extraction methods, to yield an amount of template input that results in a proportion of positive partitions compatible with Poisson's law-based statistical analysis.
15. The thorough mixing of samples and mastermix in a pre-plate is essential for accurate results (Fig. 2).
16. Line up the bottom right corner of the film with the plate edge to ensure that the angle of the film will line up across the whole plate. Accuracy is essential at this step to achieve an appropriate seal.
17. A diagram in the drawer shows the orientation of the cut corner present on the top-left corner of the Nanoplate.
18. Completed plates can be reimaged for up to 1 week after the thermal cycling process has completed.
19. Cross-talk was observed between assays employing fluorophores with close emission wavelength during the validation of the multiplex protocol. Most instances were resolved when the plate was reimaged with adjustment to exposure lengths (Fig. 3). Qiagen have also released an updated software from that used in this analysis (V3.0) that allows the user to perform compensation adjustments of specific assay-fluorophore combinations that have not been covered in this article.
20. To gain a sense of where thresholds should be placed, look at the CNV results of known samples (e.g., ear clips from male wild-type mice should always call at two copies for chromosomes 1, 8, and 11 and 1 copy for chromosome Y).
21. Once imaging settings have been validated for the panel being run, this stage becomes less important but should still be reviewed at regular intervals or if unexpected results are being produced.
22. Damage to the plate will look like long streaky regions of empty wells. Insufficient mixing will look like blocky regions of the plate that are blacked out.
23. The QIAcuity can be networked so that multiple computers can perform analysis/template setup from the software suite at the same time if needed. Utilizing the run template CSV files

and setup of plate templates within the QIAcuity software suite allows for quick run setup leaving the computer available for data analysis.

24. The CNV data export from the QIAcuity system can be difficult to interpret from within the exported CSV file, as there are multiple results per well for each assay that was run. We found utilizing multiple VLOOKUP functions in Excel to be the quickest way of extracting the desired results from the output. As VLOOKUP will only return the first result of a sample, it can be combined with a COUNTIF function along with the well value to extract all values into a more convenient format to analyze.

25. We chose the same stringent cutoffs to exclude ES cell colonies from microinjection (>2.2 copies/genome for chromosomes 1, 8, or 11, <0.2 copies/genome for chromosome Y) with the data obtained from the multiplex assay as with the previous, ddPCR-based, process. We found that this cutoff gave similar calls when testing older clones that yielded germline transmission after microinjection (Fig. 5). This stringent cutoff can be used as gene targeting experiments typically yield an excess number of correctly targeted colonies. Less stringent thresholds were successfully set for IMPC mouse production [2, 6]. It is important to note that these choices for chromosomes assayed and cutoffs will only apply to JM8 cells-derived and R1 parental lines. Assay panels and cutoffs will need to be established individually for other parental ES cell lines [6].

References

1. Capecchi MR (1989) The new mouse genetics: altering the genome by gene targeting. Trends Genet 5:70–76
2. Birling M-C, Yoshiki A, Adams DJ, Ayabe S, Beaudet AL, Bottomley J et al (2021) A resource of targeted mutant mouse lines for 5,061 genes. Nat Genet 53:416–419
3. Nagy A, Gocza E, Diaz EM, Prideaux VR, Ivanyi E, Markkula M et al (1990) Embryonic stem cells alone are able to support fetal development in the mouse. Development 110:815–821. https://doi.org/10.1242/dev.110.3.815
4. Liu X, Wu H, Loring J, Hormuzdi S, Disteche CM, Bornstein P et al (1997) Trisomy eight in ES cells is a common potential problem in gene targeting and interferes with germ line transmission. Dev Dyn 209:85–91
5. Longo L, Bygrave A, Grosveld FG, Pandolfi PP (1997) The chromosome make-up of mouse embryonic stem cells is predictive of somatic and germ cell chimaerism. Transgenic Res 6:321–328. https://doi.org/10.1023/A:1018418914106
6. Codner GF, Lindner L, Caulder A, Wattenhofer-Donzé M, Radage A, Mertz A et al (2016) Aneuploidy screening of embryonic stem cell clones by metaphase karyotyping and droplet digital polymerase chain reaction. BMC Cell Biol 17:1–13
7. Evans E (1996) Standard idiogram. In: Genetic variants and strains of the laboratory mouse, vol 2. Oxford University Press, Oxford; New York, pp 1446–1448
8. D'Hulst C, Parvanova I, Tomoiaga D, Sapar ML, Feinstein P (2013) Fast quantitative real-time PCR-based screening for common chromosomal aneuploidies in mouse embryonic stem cells. Stem Cell Rep 1:350–359

9. Mianné J, Bourguignon C, Nguyen Van C, Fieldès M, Nasri A, Assou S et al (2020) Pipeline for the generation and characterization of transgenic human pluripotent stem cells using the CRISPR/Cas9 technology. Cells 9. https://doi.org/10.3390/cells9051312
10. Owczarzy R, Tataurov AV, Wu Y, Manthey JA, McQuisten KA, Almabrazi HG et al (2008) IDT SciTools: a suite for analysis and design of nucleic acid oligomers. Nucleic Acids Res 36:W163–W169. https://doi.org/10.1093/nar/gkn198
11. Lietard J, Ameur D, Somoza MM (2022) Sequence-dependent quenching of fluorescein fluorescence on single-stranded and double-stranded DNA. RSC Adv 12:5629–5637. https://doi.org/10.1039/D2RA00534D
12. Agbavwe C, Somoza MM (2011) Sequence-dependent fluorescence of cyanine dyes on microarrays. PLoS One 6:1–10. https://doi.org/10.1371/journal.pone.0022177
13. Ye J, Coulouris G, Zaretskaya I, Cutcutache I, Rozen S, Madden TL (2012) Primer-BLAST: a tool to design target-specific primers for polymerase chain reaction. BMC Bioinformatics 13:134. https://doi.org/10.1186/1471-2105-13-134
14. Wang K, Li H, Xu Y, Shao Q, Yi J, Wang R et al (2019) MFEprimer-3.0: quality control for PCR primers. Nucleic Acids Res 47:W610–W613. https://doi.org/10.1093/nar/gkz351

Chapter 5

Optimizing Digital PCR Assays for Multiplexed Microbial Source Tracking in Surface Waters

Shimul Ghosh and Aaron Bivins

Abstract

Digital PCR (dPCR) shows great promise for precise, sensitive, and inhibition-resilient measurement of nucleic acids in surface water, providing an advantage for applications such as microbial source tracking (MST). Herein, we describe our empirical optimization of two triplex format dPCR reactions for MST (Triplex 1—Cow M2, Rum2Bac, Cow M3 and Triplex 2—HF183/BacR287, Pig2Bac, Entero1a) on the QIAcuity One system. Each triplex provides gene copy measurements similar to those obtained from single-plex format assays for a standard reference material (SRM 2917) at a fixed concentration and along a concentration gradient. On water samples previously tested by single-plex assays, each triplex also produces MST marker measurements comparable to the single-plex results. The triplexes we describe here can be directly adopted for MST on the QIAcuity, or the optimization protocol we demonstrate can be used to develop additional multiplex assays on the QIAcuity system.

Key words QIAcuity, Digital PCR, Microbial source tracking, Multiplex, Cow M2, Rum2Bac, Cow M3, HF183/BacR287, Pig2Bac, Entero1a

1 Introduction

Digital PCR (dPCR) makes use of PCR reaction partitioning and the Poisson distribution to quantify nucleic acid templates using a most probable number (MPN) approach. Unlike quantitative PCR (qPCR), dPCR does not require the preparation and analysis of calibration curves and standards to produce a quantitative result. Instead, dPCR estimates the number of template copies by measuring the presence or absence of the target in a large number of independent PCR partitions and then estimating the copy number via a Poisson model [1]. Although digital PCR was conceptualized before qPCR, the nanoscale partitioning required for precise quantification was not operationalized in a reliable format until 1998 [2, 3]. The first commercially available dPCR system was brought to market by Fluidigm in 2006 [4]. In 2011, BioRad Laboratories

Guillaume Pavlovic (ed.), *Digital PCR: Methods and Protocols*, Methods in Molecular Biology, vol. 2969,
https://doi.org/10.1007/978-1-0716-4767-7_5,

introduced the QX100, which revolutionized dPCR via a reliable water–oil emulsion process that made the dPCR format more accessible to research laboratories [5].

Digital PCR saw immediate applications in clinical cell biology, particularly for the sensitive detection and quantification of rare mutations and pathogens in the fields of oncology and pathology. However, the sensitive and calibration-curve-free dPCR format also offers improved resilience to inhibitors, which is advantageous for environmental microbiology, particularly water microbiology, where molecular measurements are becoming increasingly common [6, 7]. Recently, the uptake of dPCR for water microbiology has been accelerated by the widespread use of wastewater testing for public health surveillance during the COVID-19 pandemic, where reliability is crucial to inform decision-making [8, 9]. Another compelling use case for dPCR is microbial source tracking (MST). During MST, molecular assays are designed to be specific to particular sources of fecal contamination and are used to test environmental water samples to inform risk management in recreational waters [10]. The US EPA has developed qPCR-based methods to measure human-specific fecal contamination (HF183/BacR287) and fecal indicator bacteria *Enterococcus* spp. (Entero1a) [11, 12].

Here, we adapt these two TaqMan probe assays from the US EPA, along with four additional MST assays, to dPCR format and optimize two triplex reactions for simultaneous measurement of fecal contamination from ruminants, pigs, cows, and humans in surface waters (Table 1). Because dPCR multiplexing development is not thoroughly described in the peer-reviewed literature, here we describe our multistep and multiphase protocol to optimize and validate triplex reactions in dPCR format. Our experiences indicate crucial variables in the optimization include forward and reverse primer compatibilities (*i.e.*, primer dimers) and concentrations, probe compatibilities and concentration, annealing temperature compatibility, and reliable quantitative standard materials.

Each optimized triplex MST dPCR reaction provides precise quantification of a standard reference material that is statistically similar to single-plex measurements for both a fixed concentration (Fig. 4) and a concentration gradient down to 25 gene copies (GC) per μL of sample material (Fig. 5). Each triplex also produces qualitative and quantitative MST results that are comparable to previous single-plex testing of DNA extracts from surface water samples (Figs. 8 and 9). Although our protocol is specific to the assays, standard reference materials, and sample materials we use, our approach could easily be extended and amended to develop other multiplex assays for dPCR—up to 5-plex on the QIAcuity One system for other MST applications, testing for antibiotic resistance genes, wastewater surveillance testing or myriad other dPCR-based analyses.

Table 1
Primer (FWD, REV) sequences, probe (P) sequences, and fluorophore and quenchers for the MST assays as adapted to dPCR (*see* Note 8)

Assay	Triplex	MST target		Primer and probe sequences ($5' \rightarrow 3'$)	5′ End probe fluorophore (QIAcuity channel)	3′ End quencher	Reference
Cow M2	1	Cattle fecal waste	FWD REV	CGGCCAAATACTCCTGATCGT GCTTGTTGCGTTCCTTGAGATAAT	FAM (Green)	BHQ-1 Nova	[13]
			P	AGGCACCTATGTCCTTTACCTCATCAAC TACAGACA			
Rum2Bac		Ruminant fecal waste	FWD REV	ACAGCCCGCGATTGATACTGGTAA CAATCGGAGTTCTTCGTGAT	TAMRA (Orange)	BHQ-2	[14]
			P	ATGAGGTGGATGGAATTCGTGGTGT			
Cow M3		Cattle fecal waste	FWD REV	CCTCTAATGGAAAATGGATGGTATCT CCATACTTCGCCTGCTAATACCTT	Quasar 670 (Crimson)	BHQ-2	[13]
			P	TTATGCATTGAGCATCGAGGCC			
HF183/ BacR287	2	Human fecal waste	FWD REV	ATCATGAGTTCACATGTCCG CTTCCTCTCAGAACCCCTATCC	FAM (Green)	MGB-EDQ	[11]
			P	CTAATGGAACGCATCCC			
Pig2Bac		Swine fecal waste	FWD REV	GCATGAATTTAGCTTGCTAAATTTGAT ACCTCATACGGTATTAATCCGC	CIV-550 (Yellow)	MGB-EDQ	[15]
			P	TCCACGGGATAGCC			
Entero1a		*Enterococci*	FWD REV	GAGAAATTCCAAACGAACTTG CAGTGCTCTACCTCCATCATT	Quasar 670 (Crimson)	BHQ-2	[12]
			P	TGGTTCTCTCCGAAATAGCTTTAGGGCTA			

2 Materials

Prepare all dPCR reaction solutions, primer, probe, and control material dilutions with nuclease-free molecular-grade water and/or TE (Tris-EDTA) buffer at pH 8.0. To maximize precision and comparability between dPCR experiments, use calibrated pipettes with proper technique for all liquid handling. To minimize the opportunity for cross-contamination, all dPCR pipetting should be performed in a nucleic acid workstation (laminar flow cabinet) using barrier tip pipettes with aseptic techniques appropriate for the relevant biosafety level of the laboratory (BSL 2 in our case). All surface water samples should be handled in a biosafety cabinet per the approved biosafety protocols to minimize the probability of incidental infection among laboratory personnel. Diligently follow all waste disposal regulations and procedures when disposing of waste materials.

2.1 Diluents, Specialty Plasticware, and General Lab Equipment

1. TE (Tris-EDTA) Buffer, pH 8.0 (VWR, E112-100ML).
2. Molecular-grade nuclease-free water (VWR, VWRL0201-0500).
3. 1.5 mL DNA LoBind tubes (Eppendorf, 022431021).
4. Snap-Cap 5 mL centrifuge tubes (VWR, 10002-731).
5. 1.7 mL microcentrifuge tubes (VWR, 87003-294).
6. 0.2 mL PCR strip tubes (Axygen, PCR-0208-AF-C).
7. Microcentrifuge (VWR, 76269-064).
8. Refrigerated benchtop centrifuge 5810R (Eppendorf, 022625501).
9. Centrifuge rotor A-4-62 (Eppendorf, 022638009).
10. Analog Vortex Mixer (VWR, 10153-838).

2.2 Primers and Probes for dPCR

1. 10 μM working solutions of each forward and reverse primer for each assay (Table 1) (*see* **Note 1**).
2. 25 μM working solutions of TaqMan probes for each assay (Table 1) with the fluorophore and quenchers specified (*see* **Note 2**).

2.3 Control Plasmid and Surface Water Samples

1. Standard Reference Material 2917 Plasmid DNA for Fecal Indicator Detection and Identification (National Institute for Standards and Technology, SRM 2917) (*see* **Note 3**).
2. Purified nucleic acids from surface water samples (*see* **Note 4**).

2.4 dPCR System and Reagents

1. QIAcuity One, 5plex dPCR System (Qiagen, 911022).
2. QIAcuity Probe 5 mL PCR Kit (Qiagen, 250102).

3. QIAcuity Nanoplate 26 k 24-well (Qiagen, 250001).
4. QIAcuity Nanoplate 26 k 8-well (Qiagen, 250031).
5. Nanoplate Tray (Qiagen, 250098).
6. Nanoplate Seals (Qiagen, 250099).
7. QIAcuity Suite Software v. 2.5.0.1.
8. QIAcuity Control Software v. 2.5.0.24.
9. QIAcuity Volume Precision Factor v. 5.0 (VPFV50SOW891005).

3 Methods

Digital PCR assays for each MST marker are adapted from published qPCR TaqMan probe assays (Table 1). All dPCR experiments, including software setup, reaction setup, plate sealing, plate loading, partitioning, thermal cycling, and imaging, are performed following the general protocols and procedures detailed in the QIAcuity User Manual, QIAcuity Application Guide, and QIAcuity Probe PCR Kit [16–18]. Prior to beginning the reaction setup and plate loading, as described in the sections below, the QIAcuity One dPCR system should be started, and the QIAcuity Control and Suite software should be initiated to ensure the instrument is in good working order. It is also best to configure the experimental run in the Suite Software prior to dispensing reagents. This will ensure any hardware and/or software issues are identified and resolved prior to expending reagents or consumables. For the sake of brevity, we have not detailed every standard operating procedure for the QIAcuity system. However, we have noted and detailed any deviation from the standard procedures in the sections to follow.

3.1 Annealing Temperature Optimization

We begin our optimization protocol by examining the performance of each MST assay in single-plex format for combinations of annealing temperature (55, 57, 60 °C) and primer/probe concentration (S1, S2, S3). These parameters have been previously identified as important for optimizing dPCR performance [19]. The goal of these single-plex experiments is to ensure reliable separation between the positive and negative dPCR partitions at an annealing temperature that is compatible with the other assays in the proposed triplex and examine the consistency of the separation with different primer/probe concentrations (*see* **Note 5**).

3.1.1 SRM 2917 Working Solution Preparation

1. Based on the anticipated sample size for the single-plex and duplex experiments, prepare a 2500 μL SRM 2917 working solution at a final concentration of approximately 10,000 GC/μL by pipetting 2453.8 μL of molecular-grade water

into a Snap-Cap 5 mL centrifuge tube followed by 46.2 μL of SRM 2917 Level 6 stock solution (certified by NIST at 541,286 ± 7700 GC/μL).

2. Store the SRM 2917 working solution at 4 °C when not in use for the duration of the single-plex and duplex experiments (9 experimental days performed over 22 calendar days).
3. Prior to any subsampling for dPCR experiments, vortex the tube for 30 s and briefly spin down in a centrifuge.

3.1.2 dPCR Super Mix Preparation

Single-plex dPCR optimization experiments are performed for one assay each day by running three different dPCR super mixes (S1, S2, S3) on a single QIAcuity Nanoplate 26 k 24-well plate (one reaction mixture per 8-well column) with a single annealing temperature during thermal cycling. Three separate plates are run each day to assess annealing temperatures of 55, 57, and 60 °C for each assay. The super mixes are formulated for a single dPCR assay per day; for example, on day one of the experiments, S1, S2, and S3 are prepared using the Cow M2 primers and probe. The super mixes required for a single day of experiments are prepared as detailed below. All reagents are thawed from frozen storage (−20 °C) and kept on ice during super mix preparation.

1. *Super Mix 1 (S1)—800 nM Forward primer (F), 800 nM Reverse primer (R), 400 nM Probe (P), 24 reactions plus overage.* Combine the following in a 1.7 mL microcentrifuge tube, vortex for 30 s, briefly spin down in a microcentrifuge, and then store on ice or at 4 °C until use: 449 μL molecular-grade water, 250 μL 4× Probe PCR Master Mix from the QIAcuity Probe PCR Kit, 80 μL of 10 μM forward primer solution, 80 μL of 10 μM reverse primer solution, and 16 μL of 25 μM probe solution for a total volume of 875 μL of S1.
2. *Super Mix 2 (S2)—800 nM F, 800 nM R, 250 nM P, 24 reactions plus overage.* Combine the following in a 1.7 mL microcentrifuge tube, vortex for 30 s, briefly spin down in a microcentrifuge, and then store on ice or at 4 °C until use: 455 μL molecular-grade water, 250 μL 4× Probe PCR Master Mix from the QIAcuity Probe PCR Kit, 80 μL of 10 μM forward primer solution, 80 μL of 10 μM reverse primer solution, 10 μL of 25 μM probe solution for a total volume of 875 μL of S2.
3. *Super Mix 3 (S3) preparation—400 nM F, 400 nM R, 250 nM P for 24 reactions plus overage.* Combine the following in a 1.7 mL microcentrifuge tube, vortex for 30 s, briefly spin down in a microcentrifuge, and then store on ice or at 4 °C until use: 535 μL molecular-grade water, 250 μL 4× Probe PCR Master Mix from QIAcuity Probe PCR Kit, 40 μL of 10 μM forward primer solution, 40 μL of 10 μM reverse primer solution, 10 μL of 25 μM probe solution for a total volume of 875 μL of S3.

3.1.3 dPCR Reaction Setup

1. As previously mentioned, during the single-plex assay optimization, three QIAcuity Nanoplate 26 k 24-well plates are run each day. Each column of the plate is loaded with a different super mix per the following layout: column 1—S1 (8 wells), column 2—S2 (8 wells), column 3—S3 (8 wells).
2. Prepare the dPCR reactions by pipetting 35 μL of the appropriate super mix into each of eight 0.2 mL tubes in a single PCR strip, then add 5 μL of the 10,000 GC/μL SRM 2917 working solution prepared in **step 1** into each one.
3. Prepare three individual strips of eight PCR tubes, each one containing the appropriate super mix (S1, S2, or S3) for each experimental run (i.e., single Nanoplate).
4. After the addition of the SRM 2917 working solution, cap the strip tubes, vortex for 10 s, and then briefly spin down in a microcentrifuge.

3.1.4 Nanoplate 26 k 24-Well Plate Loading and Sealing

1. Transfer the entire 40 μL volume from each 0.2 mL PCR tube into the appropriate well on the Nanoplate, being careful to avoid the introduction of air bubbles (i.e., first-stop pipetting with the pipette tip in contact with the well walls).
2. Once all 24 wells are loaded with reaction mix, seal the plate using a Nanoplate Tray, Nanoplate Seal, and the roller provided with the QIAcuity system.

3.1.5 dPCR Experiment Run 1 (55 °C Annealing Temperature)

1. Once sealed, carefully transport the Nanoplate into the molecular lab, where the experimental run parameters have already been entered into the QIAcuity Suite Software.
2. For the first experiment, an annealing temperature of 55 °C is used with the thermal cycling input as follows: PCR initial heat activation at 95 °C for 2 min followed by 40 2-step cycles of denaturation at 95 °C for 15 s and annealing/extension at 55 °C for 30 s.
3. Scan the Nanoplate barcode to associate the plate with the correct experimental run in the Suite Software and then carefully load the Nanoplate into the instrument tray.
4. Initiate the run in the Suite Software per the default settings for partitioning and imaging. Each single-plex dPCR run requires approximately 2 h to complete on the QIAcuity One system.
5. Upon run completion, export the dPCR results from the instrument as a CSV file for further analysis.

3.1.6 dPCR Experiment Run 2 (57 °C Annealing Temperature)

1. For the second single-plex experiment of the day, prepare and seal another Nanoplate 26 k 24-well plate as described in Subheadings 3.1.3 and 3.1.4 above.

2. For the second experiment, use an annealing temperature of 57 °C with the thermal cycling input as follows: PCR initial heat activation at 95 °C for 2 min followed by 40 2-step cycles of denaturation at 95 °C for 15 s and annealing/extension at 57 °C for 30 s. All other aspects of the run are performed as described in Subheading 3.1.5.

3.1.7 dPCR Experiment Run 3 (60 °C Annealing Temperature)

1. For the third single-plex experiment of the day, prepare and seal another Nanoplate 26 k 24-well plate as described in Subheadings 3.1.3 and 3.1.4 above.
2. Handle and load the plate into the QIAcuity One as previously described, except in this experiment, specify an annealing temperature of 60 °C with the thermal cycling input as follows: PCR initial heat activation at 95 °C for 2 min followed by 40 2-step cycles of denaturation at 95 °C for 15 s and annealing/extension at 60 °C for 30 s. Upon run completion, export the digital PCR results from the instrument as a CSV file for analysis.

3.1.8 Repeat Experiments

1. Repeat the procedures in Subheadings 3.1.2, 3.1.3, 3.1.4, 3.1.5, 3.1.6, and 3.1.7 for a single MST assay daily over 6 days, running three 26 k 24-well Nanoplates for a single dPCR assay each day for a total of 18 plates (*see* **Note 6**).

As mentioned previously, the goal of the single-plex experiments is to assess the reliability of the separation between positive and negative dPCR partitions with various primer/probe concentrations and annealing temperatures. In Fig. 1, we provide examples of the QIAcuity 1D scatterplot output for HF183/BacR287 (1A), Pig2Bac (1B), and Entero1a (1C) for nine unique primer/probe concentration and annealing temperature combinations. We observe similar results for the Cow M2, Rum2Bac, and Cow M3 assays (not shown). The separation between negative partitions (gray cluster) and positive partitions (blue cluster) remains reliable for each single-plex assay across all combinations with equivalent thresholding values (red line) yielding comparable results.

Based on the results for the single-plex optimization experiments, proceed with super mix formulation S3 (400 nM F, 400 nM R, and 250 nM P) to minimize the likelihood of cross talk between channels while returning reliable results during the duplex and triplex experiments. With no difference in cluster separation between annealing temperatures, we also proceed with an annealing temperature of 60 °C (Qiagen's recommendation) for experiments during duplex and triplex development.

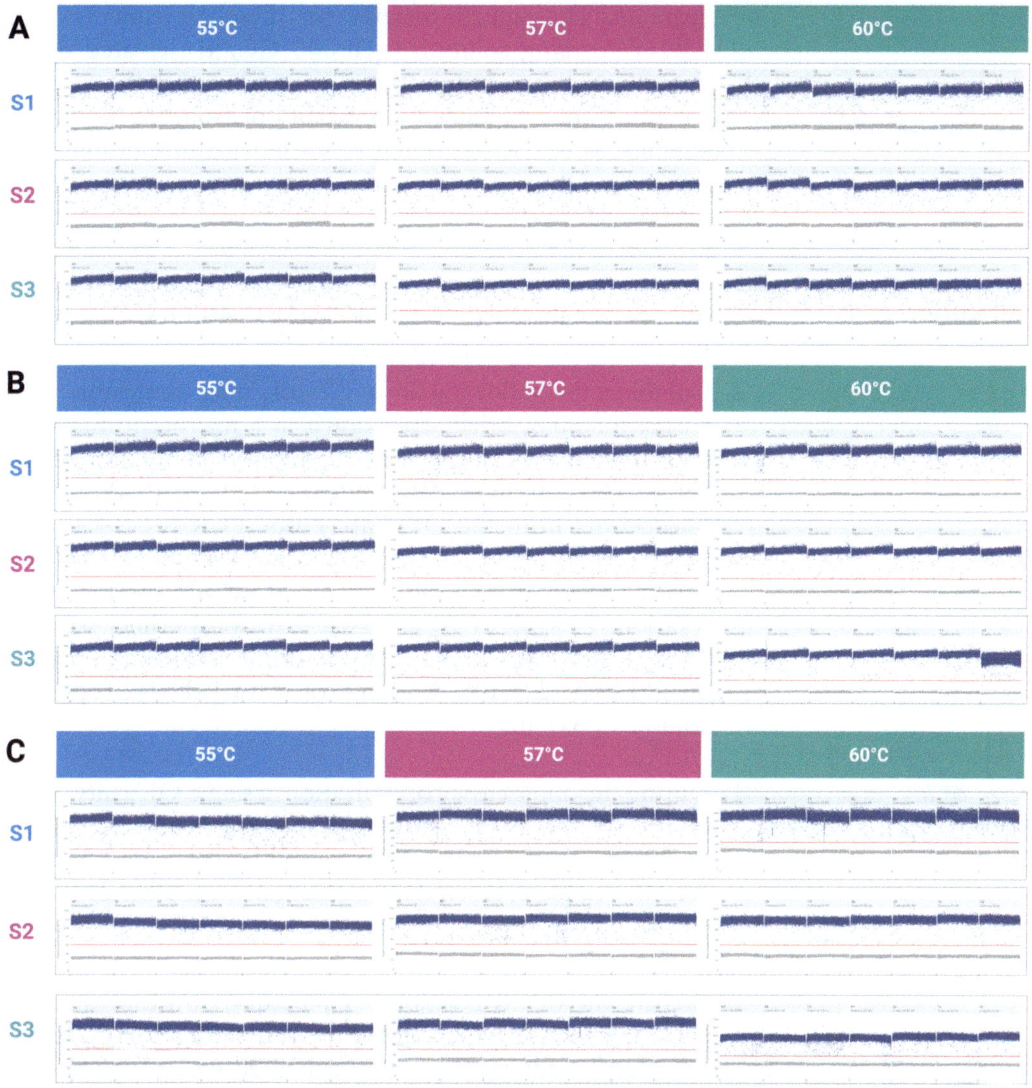

Fig. 1 QIAcuity 1D Scatterplots for HF183/BacR287 (**a**), Pig2Bac (**b**), and Entero1a (**c**) assays at nine combinations of annealing temperature (55, 57, 60 °C) and primer/probe concentrations (S1, S2, S3)

3.2 Duplex Assay Optimization

The goal of the duplex optimization experiments is to begin combining MST assays from the anticipated triplex reactions and ensure reliable separation between positive and negative partitions and minimal cross talk between dye channels in duplex format. For this purpose, we create Duplex 1 (D-1) consisting of Cow M2 (green channel) and Cow M3 (crimson channel) and Duplex 2 (D-2) consisting of Pig2Bac (yellow channel) and Entero1a (crimson channel). The two remaining assays, HF183/BacR287 (green channel) and Rum2Bac (orange channel) are combined into Duplex 3 (D-3) even though in the final triplex reactions, they will be separated. The control material for these experiments is the SRM 2917 working solution (~10,000 GC/μL) created during the single-plex assay optimization (Subheading 3.1.1). As previously mentioned, based on the observations from the single-plex experiments, all dPCR experiments are run with an annealing temperature of 60 °C and with each assay's reagents per formulation S3.

3.2.1 dPCR Duplex Super Mix Preparation

1. Prepare enough duplex super mixes for eight reactions plus overage per formulation S3 (400 nM F, 400 nM R, 250 nM P) for both assays in the duplex.
2. Combine the following in a 1.7 mL microcentrifuge tube, vortex for 30 s, briefly spin down in a microcentrifuge, and then immediately dispense as described in Subheading 3.2.2 below: 160.2 μL molecular-grade water, 90 μL 4× Probe PCR Master Mix from the QIAcuity Probe PCR Kit, 14.4 μL of 10 μM assay 1 forward primer solution, 14.4 μL of 10 μM assay 1 reverse primer solution, 3.6 μL of 25 μM assay 1 probe solution, 14.4 μL of 10 μM assay 2 forward primer solution, 14.4 μL of 10 μM assay 2 reverse primer solution, 3.6 μL of 25 μM assay 2 probe working for a total duplex super mix volume of 315 μL.

3.2.2 dPCR Reaction Setup

1. During the duplex assay optimization, a single QIAcuity Nanoplate 26 k 8-well plate is run each day. For example, D-1 plate on 29 April, D-2 plate on 30 April, and D-3 plate on 1 May. On each 8-well plate, seven duplex reactions with SRM 2917 control material and one no-template control (NTC) duplex reaction are run.
2. Prepare the dPCR reactions by pipetting 35 μL of the day's super mix into each of eight 0.2 mL tubes in a single PCR strip, followed by adding 5 μL of the 10,000 GC/μL SRM 2917 working solution into wells A to G and 5 μL of molecular-grade water into well H (NTC).
3. After the addition of the SRM 2917 working solution (or molecular-grade water for NTCs), cap the strip tubes, vortex for 10 s, and then briefly spin down in a microcentrifuge.

3.2.3 Nanoplate 26 k 8-Well Plate Loading and Sealing

1. Transfer the entire 40 μL volume from each 0.2 mL PCR tube into the appropriate well on the Nanoplate, being careful to avoid the introduction of air bubbles.
2. Once all 8 wells are loaded with the appropriate reaction mix, seal the plate using a Nanoplate Tray, Nanoplate Seal, and the roller provided with the QIAcuity system.

3.2.4 QIAcuity dPCR Experiment Run

1. Once sealed, carefully transport the 8-well Nanoplate into the molecular lab, where the experimental run parameters have already been entered into the QIAcuity Suite Software.
2. Use an annealing temperature of 60 °C with the thermal cycling input as follows: PCR initial heat activation at 95 °C for 2 min followed by 40 2-step cycles of denaturation at 95 °C for 15 s and annealing/extension at 60 °C for 30 s.
3. To avoid oversaturation during plate imaging, adjust the gain and exposure from the default values to achieve satisfactory results (see **Note 7**). Otherwise, each duplex dPCR experiment run was performed as previously described in Subheading 3.1.7.

Just as the single-plex experiments, the overall goal of the duplex experiments is to determine if the separation between positive and negative dPCR partitions is reliable for both assays in each duplex while cross talk between channels is minimal. In Fig. 2, we provide examples of the QIAcuity 1D scatterplot output for single-plex (left column) versus duplex (right column) format for each

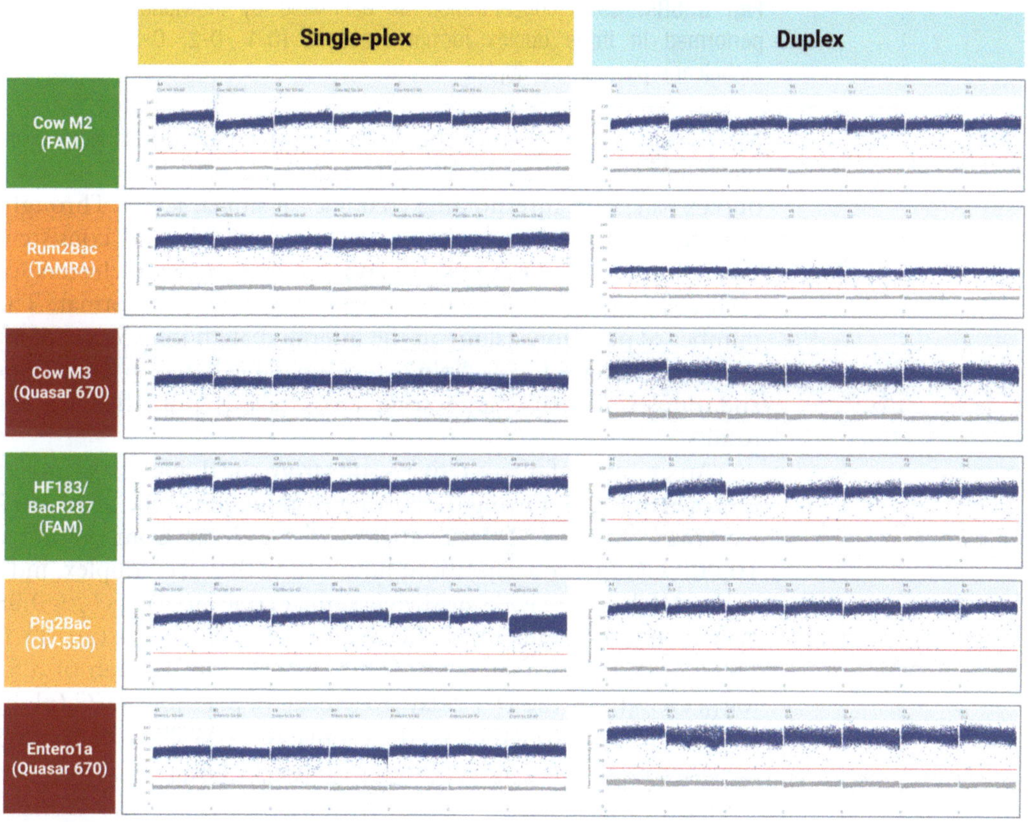

Fig. 2 QIAcuity 1D scatterplots for each dPCR MST assay in single-plex (left column) versus duplex (right column) format

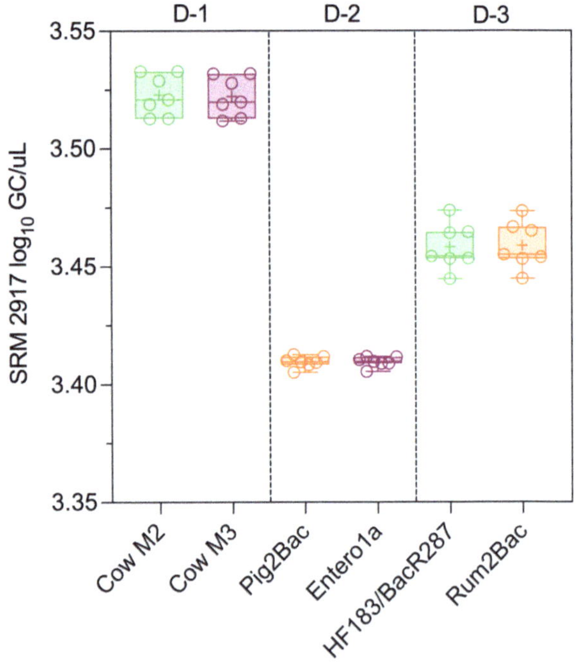

Fig. 3 SRM 2917 concentration as estimated by individual MST assays performed in three duplex format reactions (D-1, D-2, D-3) with seven technical replicates per duplex

MST assay. The separation between negative partitions (gray cluster) and positive partitions (blue cluster) is consistent for each duplex pair, with thresholding (red line) at similar levels. Throughout the duplex experiments, we observe reliable and consistent separation between the positive and negative partition clusters for each assay in duplex format compared to single-plex format. The number of partitions falling in the intermediate fluorescence intensity range (referred to as "rain") is comparable between single-plex and duplex formats, suggesting cross talk is not affecting the results.

An example of quantitative results from the duplex experiments is shown in Fig. 3. Owing to the long working life of the SRM 2917 working solution (22 days), there is some decay from the initial 10,000 GC/μL concentration; however, within each duplex, individual MST assays yield statistically similar concentrations ($p > 0.05$ Wilcoxon signed rank tests). Between experimental runs (i.e., D-1, D-2, D-3), the estimated SRM 2917 varies from a mean of 3.52 $\log_{10}$ GC/μL by the D-1 assays to a mean of 3.41 $\log_{10}$ GC/μL by the D-2 assays. Nonetheless, the estimated SRM 2917 concentrations across all three duplexes are within 0.11 $\log_{10}$ GC/μL of one another.

3.3 Triplex Assay Optimization Fixed Concentration

Based on the successful performance of the MST assays in duplex format, we proceed to the final triplex format by adding Rum2Bac to D-1 to create Triplex 1 (Cow M2, Rum2Bac, Cow M3) and adding HF183/BacR287 to D-2 to create Triplex 2 (HF183/Rum2Bac, Pig2Bac, Entero1a). For the dPCR triplex reaction super mixes, we continue to use formulation S3 for each individual assay, and we maintain an annealing temperature of 60 °C. To assess the performance of the triplex dPCR reactions, we create a fresh batch of SRM 2917 solution at a fixed concentration.

3.3.1 SRM 2917 Working Solution Preparation

1. Based on the anticipated sample size for the triplex experiments, prepare a 300 μL SRM 2917 working solution at a final concentration of approximately 10,000 GC/μL by pipetting 294.5 μL of molecular-grade water into a 1.5 mL LoBind centrifuge tube followed by 5.5 μL of SRM 2917 Level 6 stock solution (certified by NIST at 541,286 ± 7700 GC/μL).
2. Store the SRM 2917 working solution at 4 °C when not in use. Prior to any subsampling for dPCR experiments, vortex the tube for 30 s and briefly spin down in a microcentrifuge.

3.3.2 dPCR Triplex Super Mix Preparation

1. Prepare enough triplex super mix for six reactions plus overage per formulation S3 (400 nM F, 400 nM R, 250 nM P) for each assay in the triplex.
2. Combine the following in a 1.7 mL microcentrifuge tube, vortex for 30 s, briefly spin down in a microcentrifuge, and then immediately dispense as described in Subheading 3.3.3 below: 99.4 μL molecular-grade water, 70 μL 4× Probe PCR Master Mix from the QIAcuity Probe PCR Kit, 11.2 μL of 10 μM assay 1 forward primer solution, 11.2 μL of 10 μM assay 1 reverse primer solution, 2.8 μL of 25 μM assay 1 probe solution, 11.2 μL of 10 μM assay 2 forward primer solution, 11.2 μL of 10 μM assay 2 reverse primer solution, 2.8 μL of 25 μM assay 2 probe solution and 11.2 μL of 10 μM assay 3 forward primer solution, 11.2 μL of 10 μM assay 3 reverse primer solution, 2.8 μL of 25 μM assay 3 probe solution for a total triplex super mix volume of 245 μL.
3. Prepare single-plex reaction mixes for each MST assay per formulation S3 for six reactions plus overage. Combine the following in a 1.7 mL microcentrifuge tube, vortex for 30 s, briefly spin down in a microcentrifuge, and then immediately dispense as described in Subheading 3.3.3 below: 149.8 μL molecular-grade water, 70 μL 4× Probe PCR Master Mix from the QIAcuity Probe PCR Kit, 11.2 μL of 10 μM forward primer solution, 11.2 μL of 10 μM reverse primer solution, 2.8 μL of 25 μM probe solution for a total single-plex super mix volume of 245 μL.

3.3.3 dPCR Reaction Setup

1. During the triplex SRM 2917 fixed concentration experiments, a single QIAcuity Nanoplate 26 k 24-well plate is run each day. On each 24-well plate there are five triplex technical replicates with SRM 2917 ($n = 5$), five single-plex technical replicates with SRM 2917 per individual assay ($n = 15$), and one NTC per triplex and single-plex reaction ($n = 4$) for a total of 24 wells. Running both triplex and single-plex reactions on a single plate allows for the most reliable comparison of performance between the two formats.
2. Prepare the dPCR reactions by pipetting 35 μL of the appropriate super mix into each of six 0.2 mL tubes in a single PCR strip. Add 5 μL of the 10,000 GC/μL SRM 2917 working solution into five tubes and 5 μL of molecular-grade water into a single tube for the five technical replicates and one NTC for each assay.
3. After the addition of the SRM 2917 working solution (or molecular-grade water for NTCs), cap the strip tubes, vortex for 10 s, and then briefly spin down in a microcentrifuge.

3.3.4 Nanoplate 26 k 24-Well Plate Loading and Sealing

1. Transfer the entire 40 μL volume from each 0.2 mL PCR tube into the appropriate well on the Nanoplate, being careful to avoid the introduction of air bubbles.
2. Once all 24 wells are loaded with the appropriate reaction mix, seal the plate using a Nanoplate Tray, Nanoplate Seal, and the roller provided with the QIAcuity system.

3.3.5 QIAcuity dPCR Experiment Run

1. Once sealed, carefully transport the 24-well Nanoplate into the molecular lab, where the experimental run parameters have already been entered into the QIAcuity Suite Software.
2. Using an annealing temperature of 60 °C enter the thermal cycling input as follows: PCR initial heat activation 95 °C for 2 min followed by 40 2-step cycles of denaturation 95 °C for 15 s and annealing/extension 60 °C for 30 s.
3. Perform the dPCR experimental run for each triplex plate as per Subheading 3.2.4.

During both the Triplex 1 and Triplex 2 experiments, the performance of the triplex (T) format versus single-plex (S) format for quantifying SRM 2917 is statistically comparable ($p > 0.05$, Fig. 4). The range of SRM 2917 concentrations for the five technical replicates of each MST assay is within 0.03 $\log_{10}$ of the expected value (4.00 $\log_{10}$ GC/μL) for Triplex 1 (left panel) and 0.02 $\log_{10}$ for Triplex 2 (right panel).

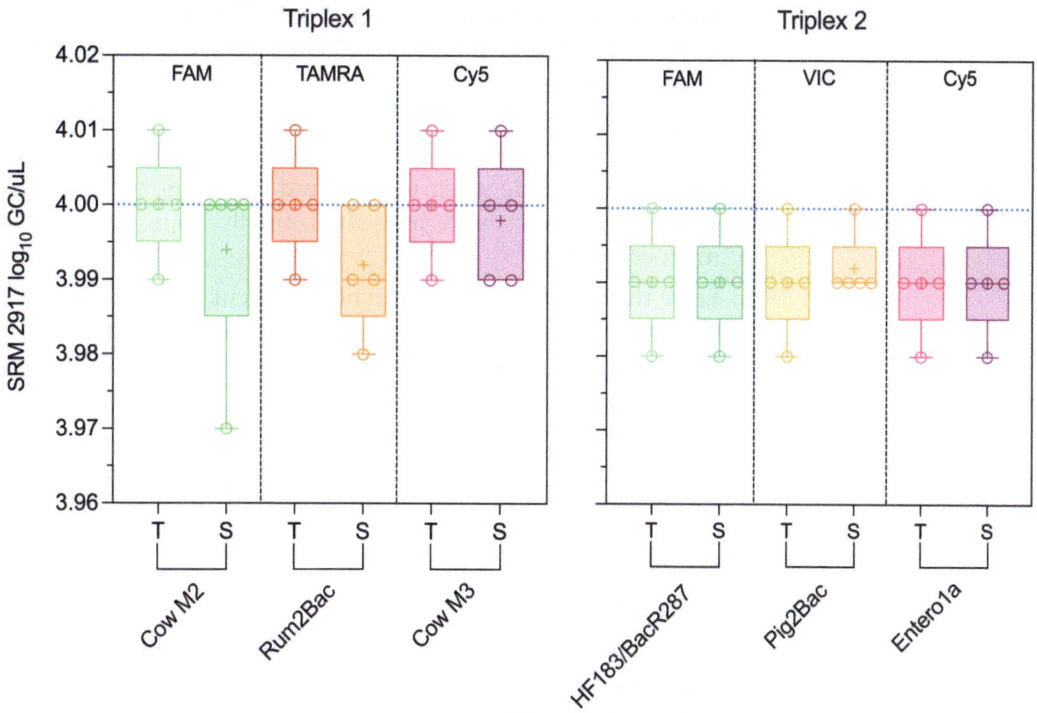

Fig. 4 Quantification of the SRM 2917 control material by two separate triplex reactions (Triplex 1, Triplex 2) yielded statistically similar ($p > 0.05$) quantities of SRM 2917 by each assay in triplex format (T) compared to the same assay in single-plex (S) format

3.4 Triplex Assay Optimization Concentration Gradient

To further assess the performance of the triplex assays, perform triplex versus single-plex comparison experiments with a concentration gradient of SRM 2917. Using 26 k Nanoplate 24-well plates, we perform experiments in a similar format to Subheading 3.3, except during this experimental round, each assay in each format (T, S) is tested in two technical replicates against three dilution levels of SRM 2917. For dPCR triplex and single-plex reaction super mixes, we continue to use formulation S3, and we maintain an annealing temperature of 60 °C.

3.4.1 SRM 2917 Gradient Working Solution Preparation

1. Based on the anticipated sample size for the triplex gradient experiments, prepare a concentration gradient of SRM 2917 working solutions.
2. Make the first-level dilution of 200 μL at 250 GC/μL by mixing 9.40 μL SRM 2917 Level 4 (NIST certified value 5314 ± 71 GC/μL) and 190.60 μL molecular grade water in a 1.5 mL LoBind centrifuge tube. Vortex the tube to mix well and then briefly spin down in a microcentrifuge.
3. Prepare the second-level dilution of 250 μL at 50 GC/μL by mixing 50 uL of the first-level working solution and 200 μL of

molecular-grade water in a 1.5 mL LoBind centrifuge tube. Vortex the tube to mix well and then briefly spin down in a microcentrifuge.

4. Prepare the third-level dilution of 150 μL at 25 GC/μL by mixing 75 μL of the second-level working solution and 75 μL of molecular-grade water in a LoBind centrifuge tube. Vortex the tube to mix well and then briefly spin down in a microcentrifuge.
5. Store the SRM 2917 concentration gradient working solutions at 4 °C when not in use. Prior to any subsampling for dPCR experiments, vortex the tube for 30 s and briefly spin down in a microcentrifuge.

3.4.2 dPCR Triplex Super Mix Preparation

1. Prepare triplex super mixes for six reactions plus overage per formulation S3 (400 nM F, 400 nM R, 250 nM P) for each assay in the triplex.
2. Prepare each triplex super mix as described in Subheading 3.3.2.
3. Prepare single-plex reaction mixes for each MST assay (six reactions plus overage) per formulation S3 as detailed in Subheading 3.3.2, **step 3**.

3.4.3 dPCR Reaction Setup

1. During the triplex SRM 2917 concentration gradient experiments, a single QIAcuity Nanoplate 26 k 24-well plate is run each day. Each 24-well plate includes two triplex technical replicates for each SRM 2917 dilution level ($n = 6$) and two single-plex technical replicates for each SRM 2917 dilution level per individual assay ($n = 18$, 6 per assay by 3 assays) for a total of 24 wells.
2. Prepare the dPCR reactions by pipetting 35 μL of the appropriate super mix into each of six 0.2 mL tubes in a single PCR strip, followed by adding 5 μL of the appropriate SRM 2917 dilution level.
3. After the addition of the SRM 2917 working solution, cap the PCR strip tubes, vortex for 10 s, and then briefly spin down in a microcentrifuge.

3.4.4 Nanoplate 26 k 24-Well Plate Loading and Sealing

1. Transfer the entire 40 μL volume from each 0.2 mL PCR tube into the appropriate well on the Nanoplate, and seal the plate as previously described.

3.4.5 QIAcuity dPCR Experiment Run

1. Once sealed, transfer the 24-well Nanoplate into the molecular lab, where the experimental run parameters have already been entered into the QIAcuity Suite Software.

2. An annealing temperature of 60 °C is used with the thermal cycling as follows: PCR initial heat activation 95 °C for 2 min followed by 40 2-step cycles of denaturation 95 °C for 15 s and annealing/extension 60 °C for 30 s.
3. Perform the dPCR experimental run for each triplex plate as per Subheading 3.2.4.

As shown in Fig. 5, quantification of the SRM 2917 control material along a concentration gradient (2.40, 1.70, 1.40 $\log_{10}$ GC/μL) by Triplex 1 (panel A) and Triplex 2 (panel B) yields similar quantities of SRM 2917 compared to the same assay in

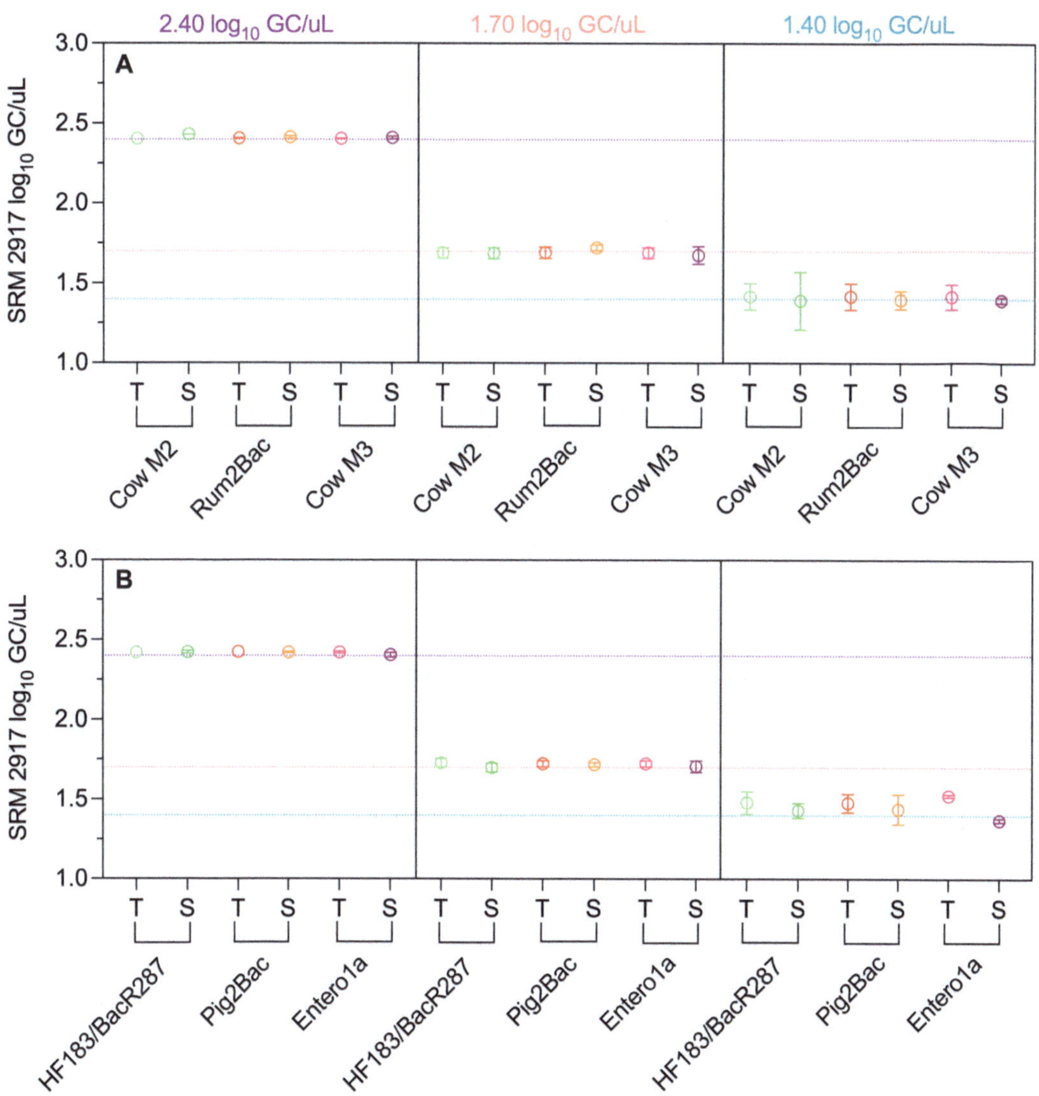

Fig. 5 Quantification of the SRM 2917 control material along a concentration gradient (2.40, 1.70, 1.40 $\log_{10}$ GC/μL) by Triplex (T) 1 (panel **a**) and Triplex (T) 2 (panel **b**) compared to single-plex (S) format results for each MST assay

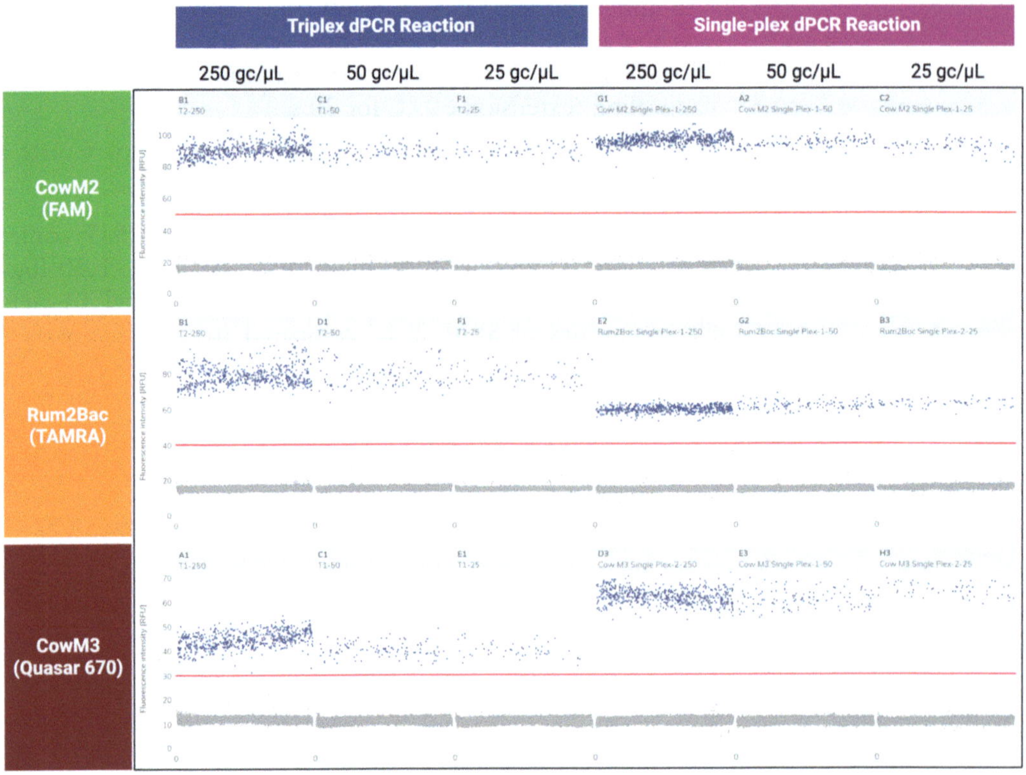

Fig. 6 QIAcuity 1D scatterplots for triplex versus single-plex format dPCR reactions for Triplex 1 (CowM2, Rum2Bac, CowM3) along a concentration gradient of SRM 2917 control material (250 GC/μL, 50 GC/μL, 25 GC/μL) as viewed on the QIAcuity software

single-plex (S) (maximum difference between T/S pairs <0.15 $\log_{10}$ GC/μL). The maximum difference between the mean concentration estimated from two technical replicates for each triplex assay and the expected value across the entire concentration gradient is 0.13 $\log_{10}$ GC/μL. When considering the QIAcuity 1D scatterplots, Triplex 1 (Fig. 6) and Triplex 2 (Fig. 7) provide reasonable separation between the positive (blue) and negative (gray) partitions such that equivalent thresholding (red line) for triplex (left column) and single-plex (right column) reactions yields similar quantitative results.

3.5 Triplex Testing of Archived Surface Water Extracts

Having established the performance of the triplex assays for quantifying an ideal standard reference material, it is also advisable to challenge them against purified nucleic acids derived from actual water samples. In this case, the triplex assays tested against six nucleic acid samples were produced during a previous water quality study (*see* **Note 4**) and had been tested in single-plex format by QIAcuity using some of the MST assays before being stored at

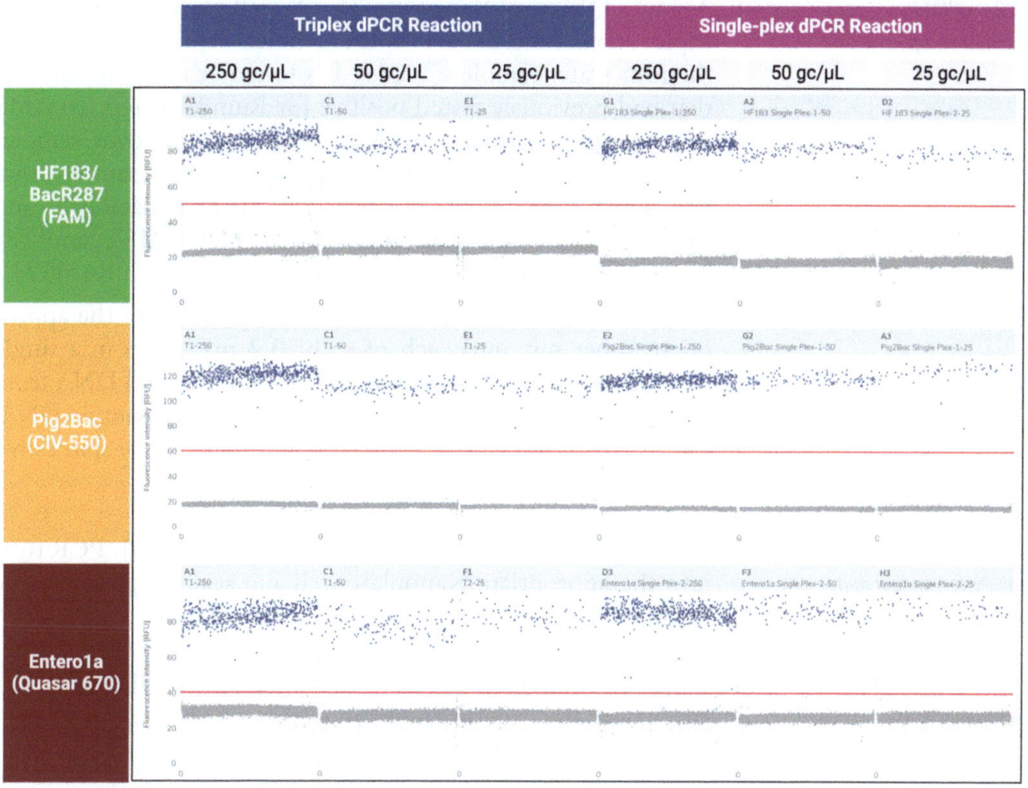

Fig. 7 QIAcuity 1D scatterplots for triplex versus single-plex dPCR reactions for Triplex 2 (HF183/BacR287, Pig2Bac, Entero1a) along a concentration gradient of SRM 2917 control material (250 GC/μL, 50 GC/μL, 25 GC/μL) as viewed on the QIAcuity software

−80 °C. We also include two additional surface water samples that had not been previously tested.

3.5.1 dPCR Triplex Super Mix Preparation

1. Prepare triplex reaction mixes for eight reactions plus overage in a 1.7 mL microcentrifuge tube per formulation S3: 127.8 μL molecular-grade water, 90 μL 4× Probe PCR Master Mix from the QIAcuity Probe PCR Kit, 14.4 μL of 10 μM assay 1 forward primer solution, 14.4 μL of 10 μM assay 1 reverse primer solution, 3.6 μL of 25 μM assay 1 probe solution, 14.4 μL of 10 μM assay 2 forward primer solution, 14.4 μL of 10 μM assay 2 reverse primer solution, 3.6 μL of 25 μM assay 2 probe solution and 14.4 μL of 10 μM assay 3 forward primer solution, 14.4 μL of 10 μM assay 3 reverse primer solution, 3.6 μL of 25 μM assay 3 probe solution for a total triplex super mix volume of 315 μL. After combining the reagents, vortex for 30 s, briefly spin down in a microcentrifuge, and then immediately dispense as described in Subheading 3.5.2 below.

3.5.2 dPCR Reaction Setup

1. A single QIAcuity Nanoplate 26 k 8-well plate is run for each triplex, consisting of two triplex technical replicates for each of the four water samples. Triplex 1 is tested against two samples that had previously tested positive for Rum2Bac and Cow M3 in single-plex format (EA, WA) and two unknown samples (NR4, YW4). Triplex 2 was tested against two samples that had previously tested positive for HF183/BacR287 and Entero1a by single-plex assays (F1021, COM3), and the same two unknown samples as used for Triplex 1 (NR4, YW4).
2. Prepare the dPCR reactions by pipetting 35 μL of the appropriate super mix into each of eight 0.2 mL tubes in a single PCR strip, followed by adding 5 μL of the purified DNA from the appropriate sample. After the addition of the sample DNA, cap the strip tubes, vortex for 10 s, and then briefly spin down in a microcentrifuge.

3.5.3 Nanoplate 26 k 8-Well Plate Loading and Sealing

1. Transfer the entire 40 μL volume from each 0.2 mL PCR tube into the appropriate Nanoplate well and seal the plate per the manufacturer's protocol as previously described.

3.5.4 QIAcuity dPCR Experiment Run

1. Once sealed, transfer the 24-well Nanoplate into the molecular lab, where the experimental run parameters have already been entered into the QIAcuity Suite Software.
2. An annealing temperature of 60 °C is used with the thermal cycling as follows: PCR initial heat activation 95 °C for 2 min followed by 40 2-step cycles of denaturation 95 °C for 15 s and annealing/extension 60 °C for 30 s.
3. Perform the dPCR experimental run for each triplex plate as per Subheading 3.2.4.

As shown in Fig. 8 for Triplex 1, samples that had previously tested positive (EA, WA) for Rum2Bac and CowM3 in single-plex format (S) are also positive for both markers in the triplex format (T). The concentrations estimated by Triplex 1 are comparable to those estimated in single-plex for both Rum2Bac (~4.5 $\log_{10}$ GC/100 mL) and CowM3 (~4 $\log_{10}$ GC/100 mL). The water samples, which were seeded with cow manure, also tested positive for CowM2 in one of two technical replicates for each sample, although previous single-plex results were not available for CowM2. None of the MST targets, which are associated with cows and ruminants, were detected in the two unknown water samples (NR4 and YW4). While there are no single-plex results available for direct comparison, the suspected source of fecal contamination in these two waterbodies is human sewage from failing sewer systems, so negativity for ruminant MST markers is plausible.

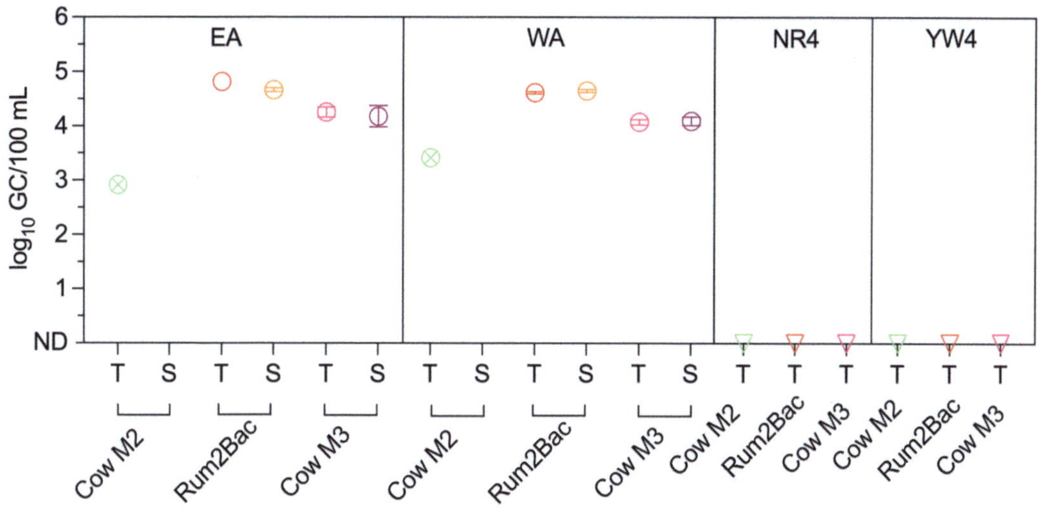

Fig. 8 MST dPCR Triplex 1 (T) results from dPCR duplicates for two archived samples (EA, WA) compared to previously generated single-plex (S) results for Rum2Bac and Cow M3. MST dPCR triplex results from dPCR duplicates for two archived samples (NR4, YW4) not previously tested in single-plex format. Inverted triangles indicate non-detection in both replicates; circles with x's denote one detection among two technical replicates at the concentration estimated from that single detection; open circles with error bars denote the mean and standard deviation for samples testing positive in both replicates; if open circles are shown without error bars then the standard deviation is too small to be displayed

For Triplex 2, samples that had previously tested positive for HF183/BacR287 and Entero1a by single-plex format are also positive in triplex format, as seen in Fig. 9. For sample F1021, the concentration of HF183/BacR287 by Triplex 2 is about 0.6 $\log_{10}$ GC/100 mL higher than by single-plex format. While for COM3, HF183/BacR287 is only detected in a single technical replicate in the triplex format, both single-plex technical replicates are positive. Nonetheless, the estimated concentration (~1.5–2.0 $\log_{10}$ GC/100 mL) is comparable. For Entero1a all technical replicates in both triplex and single-plex format are positive with similar concentration estimates (3.5–4.0 $\log_{10}$ GC/100 mL) in both formats. Pig2Bac, an MST marker for pigs, is detected in a single technical replicate for COM3, but single-plex results are not available for comparison. For the unknown samples (NR4, YW4), Entero1a is detected by triplex format in all technical replicates with an estimated concentration of 3.0–3.5 $\log_{10}$ GC/100 mL. Although single-plex results from these samples are not available, this abundance is consistent with other Entero1a testing of samples from these waterbodies (data not shown). Pig2Bac is not detected in either sample. HF183/BacR287 is detected in all technical replicates from NR4 and YW4 at concentrations of approximately 3 $\log_{10}$ GC/100 mL across both samples. These results are

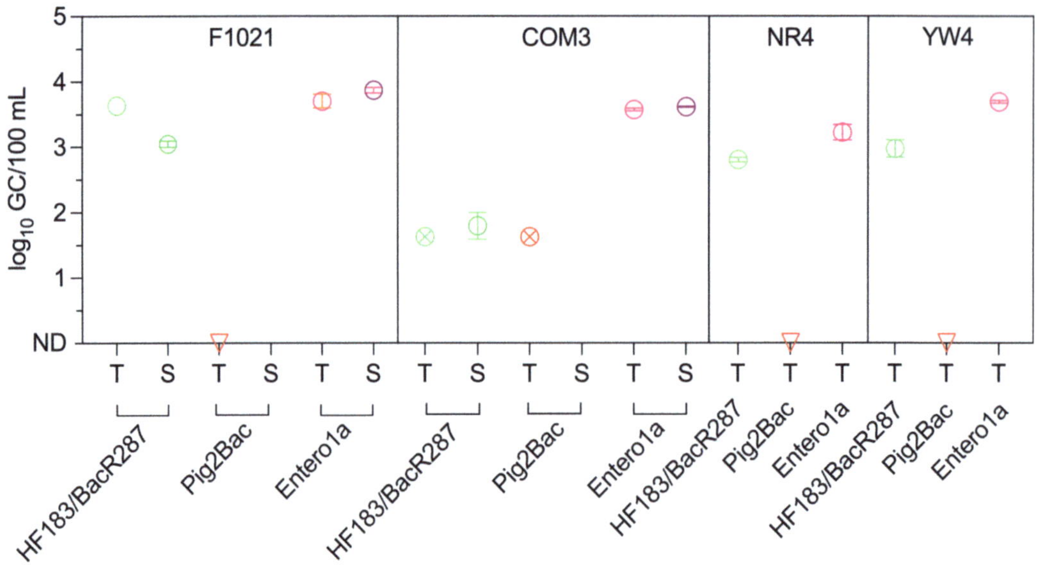

Fig. 9 MST dPCR triplex (T) results from dPCR technical replicates for two archived samples (F1021, COM3) compared to previously generated single-plex (S) results for HF183/BacR287 and Entero1a. MST dPCR triplex results from dPCR duplicates for two archived samples (NR4, YW4) not previously tested in single-plex format. Inverted triangles indicate non-detection in both replicates; circles with x's denote one detection among two technical replicates at the concentration estimated from that single detection; open circles with error bars denote the mean and standard deviation for samples testing positive in both replicates; if open circles are shown without error bars then the standard deviation is too small to be displayed

consistent with the suspected source of fecal contamination for these waterbodies—human sewage from failing on-site sewer systems.

4 Notes

1. We use "LabReady" primers (100 μM stock solution suspended in IDTE buffer at pH 8.0) from Integrated DNA Technologies (IDT).
2. Multiplexing dPCR assays requires careful consideration of probe fluorophore selection to minimize cross talk between dye channels. The probes for each dPCR assay were purchased as custom DNA probes from Biosearch Technologies. Based on conversations with other QIAcuity users, we opted to skip dye channels to minimize the likelihood of cross talk during triplex dPCR reactions. We consulted the Biosearch Technologies Flourophore and BHQ selection chart to select probe and quencher combinations based on the five dye channels on the QIAcuity system [20]. The anticipated spectral overlay for probe selections can be visualized using Biosearch Technologies' Spectral Overlay Tool [21]. For assays requiring a minor

groove binder (MGB) (HF183/BacR287, Pig2Bac), we used Biosearch's MGB probes, which feature an MGB moiety conjugated to an Eclipse Dark Quencher (EDQ). For the Cow M2 assay, which has a probe length greater than 25 bases, we used Biosearch's BHQnova™ probe that uses both a Black Hole Quencher (BHQ™) and an internal Nova quencher to increase quenching efficiency. For the remaining three assays (Entero1a, Cow M3, Rum2Bac), we used Biosearch's BHQ™ probes with fluorophores and quenchers selected per their recommendations. Biosearch Technologies has very useful resources to consult while developing both single-plex and multiplex dPCR assays, including information on: Considerations for multiplexing [22], Fluorophore/Quencher combinations [23], and Quenching Mechanisms [24]. We selected the fluorophores for each probe based on the anticipated assay groups for Triplex 1 and Triplex 2, as summarized in Table 1. Upon receipt, all probes were resuspended to 100 μM in TE (Tris-EDTA) buffer at pH 8.0 per the manufacturer's recommendation. We then created a 25 μM working solution for each probe in a 1.7 mL microcentrifuge tube using an aliquot of the 100 μM stock solution and molecular-grade water. All stock and working solutions for primers and probes were stored in a reagent-specific refrigerator at −20 °C until use.

3. A quantitative standard is extremely useful for dPCR assay validation and optimization. In this case, SRM2917 is provided by NIST in a 6-level dilution series from 5 to 500,000 copies per μL [25]. Upon receipt, we quantified the lowest dilution level (Level 1) via dPCR (QIAcuity) using the HF183/BacR287 assay within one copy per μL of the certified value prior to proceeding with the experiments we describe here. SRM 2917 was stored at 4 °C when not in use per NIST's specification. If no quantitative standards are used, we recommend using dPCR rather than mass-based estimates to establish the control stock solution titer. In our experience, mass-based estimates of the copy number frequently exceed the copy number that can be verified by dPCR.
4. To test the performance of the triplex reactions on environmental material, we made use of six archived nucleic acid extracts derived from surface water samples collected and tested during previous MST projects. Briefly, these extracts were produced by filtering 50 to 100 mL of surface water through a polycarbonate membrane (Millipore, HTTP04700). The membrane was then aseptically rolled and placed in a PowerWater DNA Bead Tube (Qiagen, 14900-50-NF-BT). DNA was then extracted from the filter using the DNeasy PowerWater Kit (Qiagen, 4900-100-NF) following the manufacturer's protocol. Following the initial testing for

a set of MST markers, the purified DNA was stored at −80 °C until interrogation via dPCR using the triplex assays described herein.

5. As previously mentioned, minimizing cross talk between dye channels is one of the key considerations for developing reliable multiplex dPCR assays. In our experience and the experience of others, the manufacturer recommended concentrations for primers and probes on the QIAcuity system are higher than necessary and may create the opportunity for cross talk. Therefore, we experimented with primer and probe concentrations that were lower than the manufacturer's recommendations.
6. These experiments could be planned and executed in a variety of ways, but as an example, our experimental schedule is outlined below:

   9 April—Prepare SRM 2917 working solution.

   11 April—Cow M2: Nanoplate 1: S1, S2, S3 at 55 °C annealing; Nanoplate 2: S1, S2, S3 at 57 °C annealing; Nanoplate 3: S1, S2, S3 at 60 °C annealing.

   12 April—Cow M3 as above.

   15 April—Rum2Bac.

   16 April—Pig2Bac.

   17 April—Entero1a.

   25 April—HF183/BacR287.
7. Optimize imaging parameters for each dye channel by progressively lowering exposure duration and gain settings. Begin with the brightest channel and decrease these values for subsequent, less intense channels. For example: FAM (brightest): 500 ms exposure, gain of 6; TAMRA (medium): 400 ms exposure, gain of 5; Cy5 (dimmest): 325 ms exposure, gain of 4. This stepwise reduction in settings ensures appropriate signal capture across channels with varying fluorescence intensities.
8. It should be noted that each of the molecular assays used in this work were previously optimized and developed for MST purposes and that such considerations have been beyond the scope of our current consideration. While in silico analysis is always advisable for new dPCR/qPCR assay development, ultimately, in our experience, optimization must be performed in an empirical fashion with our protocols described here as a useful example for others.

References

1. Vogelstein B, Kinzler KW (1999) Digital PCR. Proc Natl Acad Sci USA 96:9236–9241
2. Saiki RK, Gelfand DH, Stoffel S et al (1988) Primer-directed enzymatic amplification of DNA with a thermostable DNA polymerase. Science 239:487–491
3. Burns MA, Johnson BN, Brahmasandra SN et al (1998) An integrated nanoliter DNA analysis device. Science 282:484–487

4. Baker M (2012) Digital PCR hits its stride. Nat Methods 9:541–544
5. Hindson BJ, Ness KD, Masquelier DA et al (2011) High-throughput droplet digital PCR system for absolute quantitation of DNA copy number. Anal Chem 83:8604–8610
6. Haas CN (2020) Quantitative microbial risk assessment and molecular biology: paths to integration. Environ Sci Technol 54:8539–8546
7. Tiwari A, Ahmed W, Oikarinen S et al (2022) Application of digital PCR for public health-related water quality monitoring. Sci Total Environ 837:155663
8. Ahmed W, Smith WJ, Metcalfe S et al (2022) Comparison of RT-qPCR and RT-dPCR platforms for the trace detection of SARS-CoV-2 RNA in wastewater. Acs Es&T Water 2: 1871–1880
9. Bivins A, Kaya D, Bibby K et al (2021) Variability in RT-qPCR assay parameters indicates unreliable SARS-CoV-2 RNA quantification for wastewater surveillance. Water Res 203: 117516
10. Rock C, Rivera B, Gerba CP (2015) Chapter 14 – Microbial source tracking. In: Pepper IL, Gerba CP, Gentry TJ (eds) Environmental microbiology, 3rd edn. Academic Press, San Diego, pp 309–317
11. US EPA (2022) Method 1696.1: characterization of human fecal pollution in water by HF183/BacR287 TaqMan® quantitative polymerase chain reaction (qPCR) assay. Office of Water
12. US EPA (2015) Method 1609.1: Enterococci in water by TaqMan quantitative polymerase chain reaction (qPCR) with internal amplification control (IAC) assay. Office of Water, Washington, DC
13. Shanks OC, Atikovic E, Blackwood AD et al (2008) Quantitative PCR for detection and enumeration of genetic markers of bovine fecal pollution. Appl Environ Microbiol 74: 745–752
14. Mieszkin S, Yala J-F, Joubrel R et al (2010) Phylogenetic analysis of Bacteroidales 16S rRNA gene sequences from human and animal effluents and assessment of ruminant faecal pollution by real-time PCR. J Appl Microbiol 108: 974–984
15. Mieszkin S, Furet J-P, Corthier G et al (2009) Estimation of pig fecal contamination in a river catchment by real-time PCR using two pig-specific Bacteroidales 16S rRNA genetic markers. Appl Environ Microbiol 75:3045–3054
16. Qiagen (2023) User manual for QIAcuity instruments and QIAcuity Software 3.1 is available at https://www.qiagen.com/us/resources/resourcedetail?id=948d868f-81ed-4f51-9692-149df3557f73&lang=en
17. Qiagen (2023) QIAcuity Application Guide is available at https://www.qiagen.com/us/resources/resourcedetail?id=5d19083d-fa10-4ed2-88a0-2953d9947e0c&lang=en
18. Qiagen (2021) QIAcuity Probe PCR Kit is available at https://www.qiagen.com/us/resources/resourcedetail?id=1f29fe9f-8983-4a27-af2a-202a0cc94058&lang=en
19. The dMIQE Group, Huggett JF (2020) The digital MIQE guidelines update: minimum information for publication of quantitative digital PCR experiments for 2020. Clin Chem 66: 1012–1029
20. Multiplexing Recommendations for BHQ Probes and primers is available at https://biosearchassets.blob.core.windows.net/assets/bti_bhq_selectionchart.pdf
21. Spectral Overlay Tool for Multiplexed qPCR – Biosearch Technologies. https://www.biosearchtech.com/qpcr-multiplex-spectral-overlay-tool
22. Multiplexing qPCR. https://www.biosearchtech.com/support/education/multiplex-qpcr
23. Fluorophores and quenchers. https://www.biosearchtech.com/support/education/fluorophores-and-quenchers
24. Quenching mechanisms in oligonucleotide probes. https://www.biosearchtech.com/support/education/quenching-mechanisms
25. Kralj J, Servetas S, Hunter M et al (2021) NIST Special Publication 260-221: certification of Standard Reference Material® 2917 plasmid DNA for fecal indicator detection and identification. National Institute of Standards and Technology

Chapter 6

Crystal Digital PCR™-Based Single Nucleus Pollen Genotyping in Barley (*Hordeum vulgare* L.) to Measure Meiotic Crossover Rates in Defined Chromosomal Intervals

Suriya Tamilselvan-Nattar-Amutha, Jörg Fuchs, and Stefan Heckmann

Abstract

Genotyping-based meiotic crossover (CO) rate measurement within defined chromosomal regions typically relies on the segregation analysis of linked markers (polymorphism) in hybrid offspring populations. Pollen (male gametes) nucleus genotyping can be an alternative as pollen is abundant even from single individuals. Hence, CO rates can be measured without segregation distortion and growing offspring populations. Here, we present a protocol for meiotic recombination and CO interference rate measurements within two-linked chromosomal regions in hybrid barley pollen nuclei using multiplex Crystal digital PCR (dPCR)-based single pollen nucleus genotyping on the Naica™ Crystal Digital PCR platform. It consists of hybrid pollen isolation, flow sorting of single pollen nuclei, and their genotyping through dPCR.

Key words Digital PCR, Pollen nuclei isolation, Single pollen nucleus genotyping, Meiotic recombination, Crossover, Meiosis, Barley

1 Introduction

Meiotic homologous recombination (HR) can result in reciprocal DNA exchanges between homologous chromosomes called crossovers (CO) assuring genetic variation in offspring that is harnessed during plant breeding [1, 2]. However, in plants such as barley or wheat, CO events are heterogeneously distributed along chromosomes with large interstitial chromosome regions being nearly devoid of CO and thus untapped during breeding [3]. Hence, manipulation of the number and distribution of meiotic recombination events could accelerate plant breeding programs [4].

Analysis of meiotic recombination events within specific chromosomal regions in crop plants such as barley (*Hordeum vulgare* L.), wheat (*Triticum aestivum* L.), or rice (*Oryza sativa* L.) generally involves analysis of segregating markers (polymorphism) based

Guillaume Pavlovic (ed.), *Digital PCR: Methods and Protocols*, Methods in Molecular Biology, vol. 2969,
https://doi.org/10.1007/978-1-0716-4767-7_6,

on qPCR assays such as KASP (Competitive allele-specific PCR) or TaqMan™, PCR coupled with agarose gel electrophoresis, or genotyping-by-sequencing in hybrid offspring populations [5–7]. However, all these methods require the growth of offspring populations. As an alternative, microspores (pollen nuclei; direct derivate of male meiosis), offering large sample numbers from single individuals and CO detection without segregation distortion, can be genotyped or sequenced without growing offspring [8–11].

Here, we describe a multiplex Crystal Digital PCR-based single pollen nucleus genotyping protocol. It offers rapid measurements of male meiotic recombination rates and CO interference strength within two genetically linked chromosomal intervals in barley hybrid pollen nuclei (cultivars Golden Promise x Barke) without growing segregating offspring populations and without the need for expensive whole-genome amplification procedures even by employing stored barley anthers.

2 Materials

2.1 Plant Material and Cultivation

1. *Hordeum vulgare* cv. Barke (B), cv. Golden Promise (GP), or F_1 B × GP hybrid grains.
2. Whatmann filter paper.
3. Home-made compost.
4. Container substrate 2 (Klassmann-Deilmann GmbH).
5. Sand.
6. Plantacote depot: 4M granulate (Hauert Manna).
7. Fertilizer (1% Hakaphos blau; COMPO EXPERT).
8. Fertilizer (1% Hakaphos rot; COMPO EXPERT).
9. Plant germination pots: 16 cm Ø.
10. Controlled plant cultivation facility.

2.2 Pollen Nuclei Isolation

1. Nuclei isolation buffer [12]: 45 mM $MgCl_2$, 20 mM MOPS, 30 mM sodium citrate, 0.1% (v/v) Triton X-100. Mix well, adjust pH with 1 M NaOH to 7.0, and filter through a 0.22 mm filter. Store at −20 °C in 10 ml aliquots.
2. 1 mg/ml propidium iodide (PI).
3. Disposable filters with a mesh size of 100 μm and 20 μm (Celltrix, Sysmex-Partec).
4. Vortex mixer.
5. Metal beads: 3.5 mm Ø (Askubal).

6. Tissue grinder.
7. Biology tweezers.
8. Polypropylene microcentrifuge tubes.
9. Liquid nitrogen.
10. Micro pestle RNase/DNase free (Carl Roth).
11. Pipette.
12. Sterile pipette tips.
13. 5 mL polystyrene sample tubes (suitable for the cell sorter used).

2.3 Flow Sorting

1. Cell sorter (e.g., BD Influx™ Cell Sorter; BD Biosciences): the instrument should be equipped with either a blue (488 nm) or a green (532 nm) laser to excite PI efficiently. Suitable filters and detectors for PI emission (e.g., 610/40) as well as forward (FSC) and/or side scatter (SSC) signals are required to ensure an optimal gating strategy.
2. Software for controlling the cell sorter (usually provided by the manufacturer of the cell sorter).
3. 10× PBS: 1.37 M NaCl, 27 mM KCl, 80 mM Na_2HPO_4, 15 mM KH_2PO_4, pH 7.4
4. Mini centrifuge.

2.4 Crystal Digital PCR

1. Naica™ Prism6 Instrument (Stilla Technologies).
2. Naica™ Geode Instrument (Stilla Technologies).
3. Naica™ Sapphire chips (Stilla Technologies).
4. Oligonucleotides.
5. Dual-labeled fluorescent probes (TaqMan™).
6. 5× Naica® multiplex PCR Mix (Buffer A, Buffer B) (Stilla Technologies).
7. Restriction enzymes (if required, *see* **Note 13**).
8. Naica™ Crystal Reader Software (Version 3.0.7.3) (Stilla Technologies).
9. Naica™ Crystal Miner Software (Version 3.1.6.3) (Stilla Technologies).
10. Anti-static spray (Screen clean; AF international).
11. Petri dish.
12. Precision wipes™ Tissue wipes (Kimtech Science).
13. Primers and Probes.

3 Methods

3.1 Design of Oligonucleotides and TaqMan Probes

1. Select three SNP/InDel markers along a single chromosome (*see* Fig. 1a) (*see* **Note 1**).
2. Design suitable primers to amplify a 100–200 bp region flanking each selected marker.

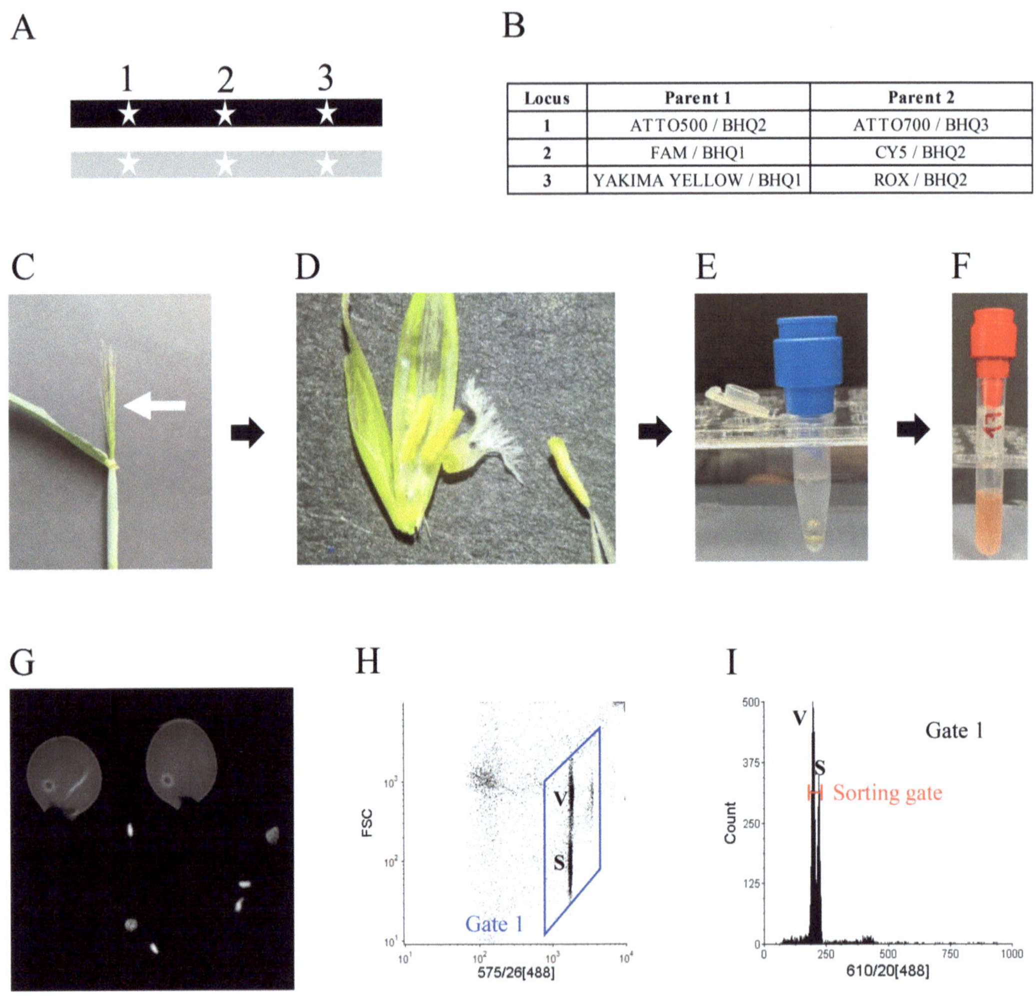

Locus	Parent 1	Parent 2
1	ATTO500 / BHQ2	ATTO700 / BHQ3
2	FAM / BHQ1	CY5 / BHQ2
3	YAKIMA YELLOW / BHQ1	ROX / BHQ2

Fig. 1 Illustration of linked-genetic intervals and key steps during barley pollen nuclei isolation and sorting. (**a**) Illustration of six required markers (white stars) at three loci (depicted as 1, 2, and 3) forming two linked-genetic intervals along parental chromosomes (black and gray blocks) of a hybrid barley plant. (**b**) Dual-labeled fluorescent probes (with corresponding quenchers) used for the genotyping of selected markers. (**c**) Spike selection (spikes with emerged awns; white arrow). (**d**) Collection of mature anthers (yellow and plump) from selected spikes. (**e**) Barley pollen isolation. (**f**) Barley pollen nuclei extraction. (**g**) Bursting of barley pollen to release the nuclei. (**h**) Identification of the pollen nuclei in a dot plot displaying the PI fluorescence vs. FSC signal (log scales). Both types of nuclei are selected with Gate 1 showing the vegetative nuclei (V) above and the sperm nuclei (S) below. (**i**) The sorting gate is defined in a histogram displaying the PI fluorescence signal (lin scale) of the events selected in Gate 1

3. Verify whether selected primers can amplify the desired region using PCR-agarose electrophoresis (*see* **Note 2**).
4. Design dual-labeled probes distinguishing the SNP/InDel markers (*see* Fig. 1b) (*see* **Note 3**) and verify the specificity of the selected primers/probes using qPCR genotyping (*see* **Note 4**).

3.2 Plant Growth and Cultivation

1. Stratify the grains on damp Whatmann filter paper in darkness at 4 °C for 2 days.
2. Germinate the stratified grains at room temperature for up to 4 days, then pot the germinated seedlings in plant germination pots (16 cm Ø) filled with soil (homemade compost combined with substrate 2 and sand in a 2:2:1 ratio) and incubate in a greenhouse with 18 °C day/13 °C temperature regime, 16 h of light at 300 μmolm^{-2} s^{-1} and 80% relative humidity.
3. Verify the zygosity of the hybrid plant by genotyping for the selected markers using qPCR (*see* **Note 4**).
4. Add Plantacote depot 4 m granulate once to the soil at the three-leaf stage.
5. Apply fertilizer (1% Hakaphos blau) once per week after potting seedlings in the soil. During reproductive development, apply fertilizer (1% Hakaphos rot) once per week.

3.3 Pollen Nuclei Isolation Based on Bead-Beating

All steps are performed on ice if not otherwise specified.

1. Collect 30–40 mature anthers (yellow and plump) from one spike with emerged awns (*see* Fig. 1c, d) (*see* **Note 5**) in an empty 2 mL microcentrifuge tube and continue with **step 3** or for longer storage, flash-freeze the anthers using liquid nitrogen and store at −70 °C (*see* **Note 6**).
2. Thaw frozen anther samples on ice for 5 min.
3. Add 500 μL of ice-cold nuclei isolation buffer (*see* **Note 7**).
4. Cut the anthers with sterile scissors into at least three pieces and vortex for 2 min vigorously.
5. Place a 100 μm mesh filter on an empty 1.5 mL microcentrifuge tube and pipette the pollen suspension onto the filter, which allows only the pollen to flow through. Use a micropestle and softly grind the remaining anthers on the surface of the filter for 15 s to release as much pollen as possible. Flush the filter with 250 μL nuclei isolation buffer (*see* Fig. 1e). Dispose the filter (*see* **Note 8**).
6. Add two metal beads (3.5 mm Ø) into the 1.5 ml microcentrifuge tube with the pollen suspension and beat the suspension using a tissue grinder at 30 Hz for 1 min (*see* **Note 9**).

7. Place a 20 μm mesh filter on an empty 5 mL polystyrene sample tube and pipette the suspension onto the filter. Use a micro-pestle to burst the remaining intact pollen by softly grinding on the surface of the filter for 15 s. Flush the filter with 250 μL of nuclei isolation buffer. Repeat the grinding step and flush with an additional 250 μL of nuclei isolation buffer (*see* Fig. 1f). Dispose the filter (*see* **Note 10**).
8. Stain the nuclei suspension with 50 μg/mL of PI, mix gently using a sterile pipette and incubate on ice for 30 min (*see* **Note 11**).

3.4 Flow Sorting of Pollen Nuclei

1. Start and setup the cell sorter according to the manufacturer's instructions. Run the system with 0.5× PBS as sheath fluid.
2. Create in the acquisition software appropriate dot-plot (PI fluorescence log scale vs. FSC log scale/SSC log scale) and histogram windows (PI fluorescence lin scale) (*see* **Note 12**).
3. Place the 5 mL polystyrene sample tube with the nuclei suspension in the sample holder of the cell sorter and run the sample at a rate of 200 to 1000 events per second. Adjust the threshold (PI fluorescence, e.g., 575/40 log scale), the gain of scatter light (FSC/SSC), and PI fluorescence (610/40) detectors that the nuclear populations are displayed in the dynamic range of the instrument.
4. Create a region surrounding the nuclei populations within the dot-plot PI fluorescence vs. FSC/SSC and display the gated region in a histogram of the PI fluorescence (610/40).
5. For sorting, create a sort gate in the PI fluorescence histogram (*see* **Note 12**).
6. Select an appropriate sort device and sort mode (e.g., "1.0 Drop Pure"), place a 1.5 mL microcentrifuge with 23.5 μL of the Crystal dPCR mix containing 1× Buffer A—naica® multiplex PCR MIX, 3% Buffer B—naica® multiplex PCR MIX (*see* **Note 13**) and oligonucleotides at their optimal concentrations (*see* **Note 14**) into a preselected position of the sort device. Define the required number of nuclei to be sorted (e.g., 4000) and start the sorting process. When the sorting is finished, shortly centrifuge the collection tube with a mini centrifuge and store it on ice.

3.5 Loading of Samples into Naica™ Sapphire Chips

1. Switch on the Naica™ Geode before sorting and ensure the pressure generated by the geode is 1200 ± 50 mbar.
2. Place inside a petri dish three layers of tissue wipes and wet the tissue by spraying three times anti-static spray (*see* **Note 15**).

3. Place a Naica™ sapphire chip (*see* **Note 16**) on the wet tissue wipes and remove gently the white flat cap from the inlets of the chips.
4. Pipette at least 27 μL of master mix containing sorted nuclei (*see* **Note 17**) and load 25 μL of the master mix (*see* **Note 18**) into each inlet over the oil phase, without inserting the tip in the oil phase (do not load completely to avoid the injection of air bubbles). In case of air bubble formation, use a sterile 10 μL pipette tip to break the bubble carefully.
5. Gently seal the chips with tall white caps and place them carefully on the thermal plate of the Naica™ Geode.
6. Close and lock the lid of the Naica™ Geode.
7. Run the following program: Sample partitioning step at 40 °C, followed by an initial denaturation of 95 °C for 3 min and then, followed by 45 cycles of 95 °C for 15 s and 58 °C for 30 s with a final pressure release step (*see* **Note 19**).

3.6 Scanning of Naica™ Sapphire Chips

1. Switch on the Naica™ Prism6 instrument. Then, launch the Crystal Reader application.
2. Create a new experiment by clicking on "New experiment." Load the Naica™ Sapphire template suitable for Naica® multiplex PCR Mix (provided by Stilla™ Technologies). Then, define all parameters required for the scanning of the chips, which includes entering the name of the experiment, Chip Unique identifier (it can be entered by scanning the barcode on the chips using the barcode reader provided), chamber details, fluorophore names, and target names.
3. Set the exposure time for fluorescence detection channels (Blue: 100 ms, Teal: 350 ms, Green: 100 ms, Yellow: 125 ms, Red: 500 ms, and Infrared: 500 ms) (*see* **Note 20**).
4. Gently remove the Naica™ Sapphire chips once the dPCR is finished and clean the bottom of the chips gently with the wet tissue wipes (sprayed with the static spray). Load the chips into the Naica™ Prism6 instrument and start the scanning procedure by clicking the "Start" icon (*see* **Note 21**).
5. The results are generated in the Naica Crystal Experimental file (.ncx).

3.7 Recombination Rate Analysis

1. Launch the Crystal Miner application and load the corresponding Crystal Miner Experimental file.
2. Load the Naica Crystal Compensation Matrix (.ncm) generated during the standard configuration setup of the Naica™ Prism6 (provided by Stilla™ Technologies).

3. If there is an air bubble in the chamber, draw a polygon around it in the "Quality Control" panel and exclude the marked area by applying the "Exclude" button.
4. Plot the colors in the 2D plot (in the spillover compensation panel) that two markers correspond to the given locus in X- and Y-axes and verify whether the positive clusters of the given marker pair align perpendicular to each other. Accordingly, adjust the relative position of the positive cluster that deviates from the vertical alignment by changing manually the fluorescence spillover compensation coefficient of the other marker of the given locus in the corresponding row of the spillover compensation matrix of the fluorescence channel that is not vertical (*see* Fig. 2). The modified compensation matrix can be exported and used for further experiments.
5. Click the "Plots and Populations" panel and go to the Positivity combination Zone. Add the population for each haplotype (nonrecombinants and recombinants) as denoted in Fig. 3. The generated analysis Configuration setup can be exported as a Naica Crystal Analysis Configuration file (.nca) and used for further experiments (*see* **Note 22**).
6. Set the fluorescence threshold in the 1D plot to 2000 fluorescence units above the cluster representing negative droplets for any given fluorescence channel (*see* Fig. 4). In some cases, a cluster of droplets appears slightly above the negative cluster of a given fluorescence channel but is separated from the positive clusters as a result of improper spillover compensation. These droplets correspond to positive droplets from other fluorescence channels. Note that these droplets should be considered as a negative cluster (*see* Fig. 4b).
7. Apply the threshold for all fluorescence channels and determine the number of positive droplets (genotyping call) for each channel (can be found in the "View Results" panel).
8. To achieve equal genotyping calls between marker pairs (Blue vs. Red, Teal vs. Yellow, and Green vs. Infrared channels) further increase the fluorescence threshold of the channel with the higher genotyping calls in the 1D plot until you have a balanced call rate. This will ensure the normalization of the genotyping calls among nonrecombinants and recombinants.
9. Obtain the number of positive calls for nonrecombinant and recombinant populations in the "View Results" panel. Use the depicted formula to estimate the recombination rate (*see* Fig. 4c) and the CO interference strength (*see* Fig. 4d) corresponding to a given chamber (*see* **Note 22**).

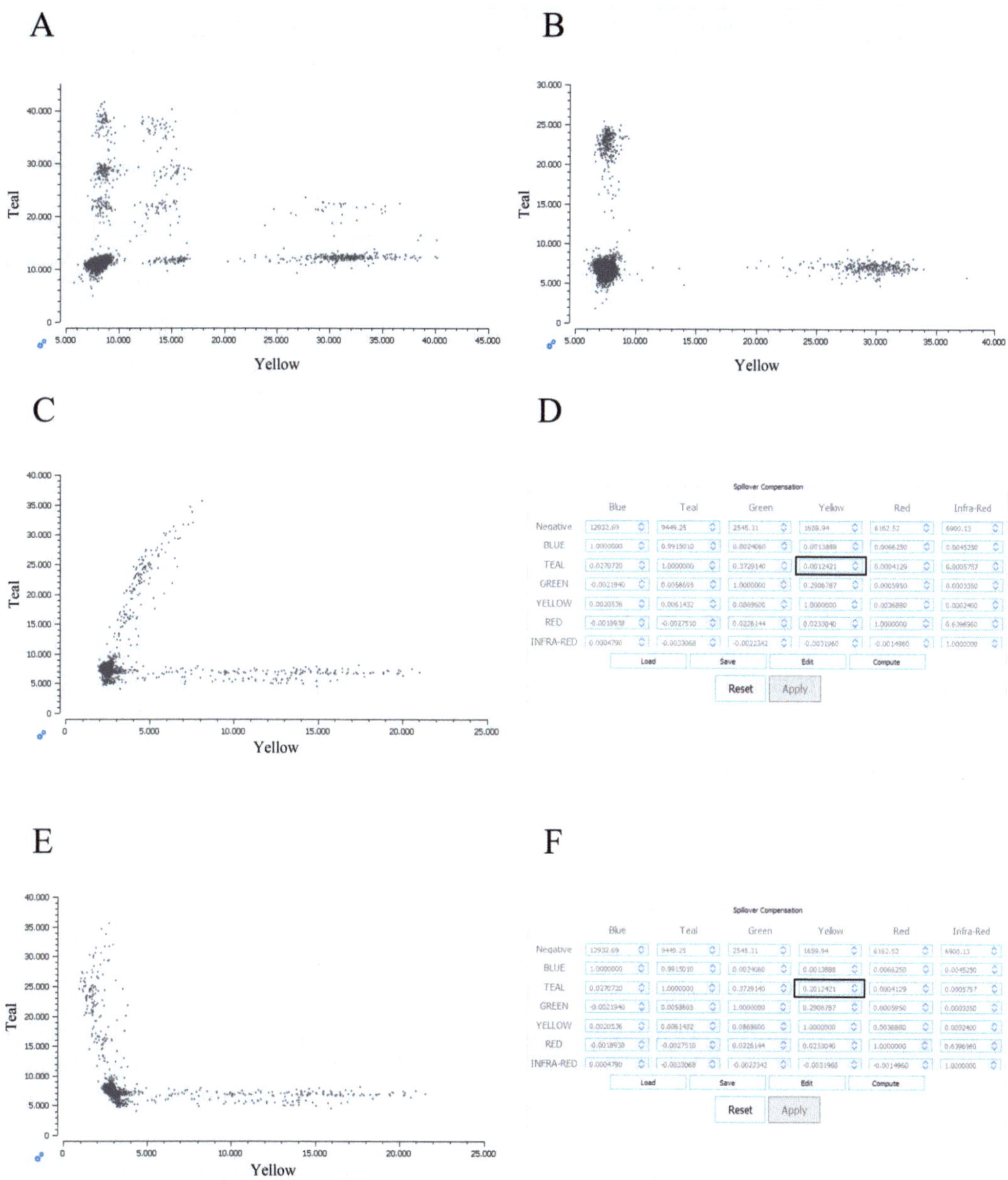

Fig. 2 Key aspects of spillover compensation during recombination rate analysis using Naica™ Crystal Miner application. A marker pair (Teal vs. Yellow) plotted in 2D using the Crystal Miner application before (**a**) and after (**b**) the application of the compensation matrix. The marker pair is properly compensated for fluorescence spillover. (**c**) In this case, even after applying the spillover compensation matrix, the positive cluster corresponding to the Teal channel is not vertical, unlike its fluorescence channel counterpart, the Yellow positive clusters. Manual adjustment of the fluorescence spillover compensation coefficient in the matrix (**d** and **f**; marked with black box) corresponding to the Teal-Yellow channels results in vertical alignment of the teal positive clusters (**e**)

	Interval	Genotype at Locus 1-Locus 2-Locus 3 #		Fluorescence channel					
				Blue	Teal	Green	Yellow	Red	Infra-red
Three-color approach	1 & 2	Non-recombinants	P1-P1-P1	Positive	Positive	Positive	Negative	Negative	Negative
			P2-P2-P2	Negative	Negative	Negative	Positive	Positive	Positive
		Recombinants	P1-P2-P2	Negative	Negative	Positive	Positive	Positive	Negative
			P2-P1-P1	Positive	Positive	Negative	Negative	Negative	Positive
			P1-P1-P2	Positive	Negative	Positive	Positive	Negative	Negative
			P2-P2-P1	Negative	Positive	Negative	Negative	Positive	Positive
			P1-P2-P1	Negative	Positive	Positive	Negative	Positive	Negative
			P2-P1-P2	Positive	Negative	Negative	Positive	Negative	Positive
Two-color approach	1	Non-recombinants	P1-P1-*	Positive	Undefined	Positive	Undefined	Negative	Negative
			P2-P2-*	Negative	Undefined	Negative	Undefined	Positive	Positive
		Recombinants	P1-P2-*	Negative	Undefined	Positive	Undefined	Positive	Negative
			P2-P1-*	Positive	Undefined	Negative	Undefined	Negative	Positive
	2	Non-recombinants	*-P1-P1	Positive	Positive	Undefined	Negative	Negative	Undefined
			*-P2-P2	Negative	Negative	Undefined	Positive	Positive	Undefined
		Recombinants	*-P1-P2	Positive	Negative	Undefined	Positive	Negative	Undefined
			*-P2-P1	Negative	Positive	Undefined	Negative	Positive	Undefined

* any possible genotype at the specified locus

P1 represents Parent 1 genotype whereas P2 represents Parent 2 genotype

Fig. 3 Analysis configuration file setup for estimating different genotypes. Tabular representation of the required positivity combinations for use in the population editor tool of Naica™ Crystal Miner application to estimate the number of droplets genotyped positively for different genotypes for CO analysis using two- or three-color methods

4 Notes

1. Both SNP/InDel markers can be used with our Crystal dPCR setup. However, InDel markers with a polymorphism size of 3–10 bp are preferred due to their higher specificity, especially in combination with TaqMan™ probes.
2. High-resolution agarose is suitable for visualization of small PCR fragments (in our case, 100–200 bp amplicons differing by 3–10 bp in size according to InDel size polymorphisms). The amplicon sequence can be confirmed by Sanger sequencing.
3. Naica™ Crystal dPCR chemistry is compatible only with TaqMan™ dual-labeled fluorescent probes.
4. Fluorescent dyes and quenchers were selected according to Stilla Technologies' recommendation. We recommend using 0.3 μmol of dual-labeled probes, 1 μmol of primers, and 100 ng of genomic DNA template for qPCR assays with 1× PerfeCTa Multiplex qPCR supermix™. Note that the final concentrations of probes and primers should be adjusted for specificity.
5. The time window mature anthers available in the first three emerging tillers varies depending on the barley genotype; for example, 7–9 weeks for B, 9–11 weeks for GP, and 8–10 weeks

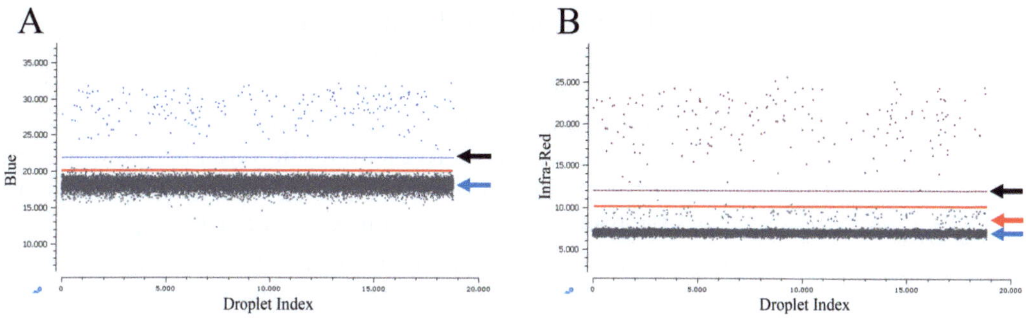

C

$$\text{Meiotic recombination rate (cM)} = \left(\frac{\text{Total recombinants}}{(\text{Total non-recombinants} + \text{Total recombinants})} \right) * 100$$

D

$$\text{CO interference strength} = 1 - \left(\frac{\text{Actual double recombinant frequency}}{\text{Expected double recombinant frequency}} \right)$$

Fig. 4 Fluorescence threshold setup and formulae for CO and CO interference rate analysis using Naica™ Crystal Miner application. (**a**) A properly compensated fluorescence channel (blue) where the negative cutoff (red line) set right above the negative cluster (blue arrow). (**b**) An improperly compensated fluorescence channel (infrared) where a droplet cluster (red arrow) appears slightly above the negative cluster (blue arrow). In this case, the negative cutoff (red line) is set above the additional uncompensated droplets (red arrow). Cutoff fluorescence (black arrow) is 2000 fluorescence units above the negative cutoff. Formulae required to estimate the meiotic recombination rate (**c**) and the CO interference strength (**d**)

for the F1 B × GP hybrids under our standard growth conditions. Plant/spike selection is based on the presence of healthy and mature anthers (yellow and plump) in a random spikelet tested, except for the apical and basal spikelet, which typically contain aborted anthers. In general, spikes with awns that emerge outside the spike sheath contain mature anthers that can be used to isolate pollen nuclei.

6. Barley (F1 B × GP hybrid) anthers stored for up to 12 months were suitable for nuclei isolation, flow sorting, and dPCR genotyping in our case. However, this may be dependent on the initial quality of the anthers during isolation. For other species, we recommend testing our dry-flash frozen method; otherwise, alternative storage methods such as freezing with desiccant (calcium chloride or silica gel) could be considered.

7. Once an aliquot is thawed do not refreeze it afterward, instead, store it in a refrigerator for up to 3 months. Other nuclei isolation buffers might be used as well but their suitability for a given species has to be checked empirically.

8. The bursting of the pollen in the selected nuclei isolation buffer can be controlled microscopically. Pipette 30 μL of the pollen suspension onto a microscopic slide, stain with 4′,6-diamidino-2-phenylindole (DAPI, 1 μg/mL) and add a coverslip. Put a sheet of filter paper onto the slide and carefully apply pressure with the thumb to burst the pollen. The released nuclei are visible using a fluorescence microscope (*see* Fig. 1g).
9. Mechanical disruption at 30 Hz for 1 min using two metal beads of 3.5 mm Ø is suitable for barley with a pollen size of 30 to 40 μm Ø and a nucleus size of 5–7 μm Ø. However, it is recommended to optimize the bead beating parameter and/or to utilize beads of different sizes/types for different plant species.
10. A 20 μm mesh filter is effective for isolating barley pollen nuclei (5–7 μm Ø) from barley pollen (30–40 μm Ø). In the case of other species with different pollen and pollen nuclei sizes, different mesh size filters can be tested.
11. Stained pollen nuclei can be sorted either immediately or stored on ice overnight.
12. Depending on the configuration of the cell sorter also DAPI could be used as a DNA stain. In this case, the SSC signal needs to be collected from the UV laser with a suitable filter/detector configuration or, additionally, a blue laser (488 nm) needs to be used for the FSC and/or SSC signal detection. Depending on the species analyzed and on the quality of the nuclei suspension vegetative and sperm nuclei can be differentiated in a dot plot PI fluorescence vs. FSC. Due to differences in size, condensation, and RNA amount, both nucleus types differ in the scatter and fluorescence light intensities (*see* Fig. 1h). Occasionally, even in a histogram displaying the PI fluorescence intensity alone a double peak can be observed (*see* Fig. 1i). This might be of interest for applications where both nucleus types need to be sorted separately. To observe the stability of the run over time it is recommended to create also a dot plot showing the time vs. PI fluorescence.
13. Naica® multiplex PCR Mix is available in two different concentrations, 5× and 10×. 5× Naica® multiplex PCR Mix is the recommended concentration for Naica™ Sapphire chips (a final volume of 25 μL reaction mix is required for each chamber). Thermostable restriction enzyme(s) not cleaving the amplicons, but cleaving the genome at multiple sites can be used to increase the efficiency of the dPCR [11]. For an appropriate dPCR cycling program using restriction enzyme (s), (*see* **Note 19**).
14. A series of dPCR experiments should be carried out with various primer/probe concentration combinations at varying Tm (PCR melting temperature) for a given number of nuclei

input to ensure an efficient genotyping of each marker so that positive clusters can be separated from negative clusters for each marker. This facilitates the proper application of the fluorescence threshold for each marker during recombination rate analysis.

15. Naica™ Sapphire chips are sensitive to static shock. Placing the Naica™ Sapphire chips on anti-static sprayed tissue wipes during loading ensures the prevention of static shock being applied on the chip.
16. Each Naica™ Sapphire chip can process up to four sample chambers (20,000–25,000 droplets per chamber) while a maximum of three Naica™ Sapphire chips can be loaded into Naica™ Geode Instrument simultaneously. Hence, processing of 12 samples in a single dPCR run is feasible. To further increase the throughput and reduce the costs, Naica™ Ruby chips (13,000–16,000 droplets per chamber) can be employed, which allow the parallel processing of 16 samples per chip, totaling 48 samples in a single dPCR run.
17. Despite having a higher genotyping call per chamber, increasing the nuclei input per chamber results in a higher encapsulation rate of double/more nuclei within a droplet, which leads to increased false-positive calls. Hence, we optimized our dPCR setup using 4000 nuclei as input to ensure a minimal double/more nuclei encapsulation within a droplet.
18. To avoid improper droplet formation during partitioning caused by air bubble introduction during sample loading, pipette a larger volume of master mix (27 μL or more) and load only 25 μL, which is the optimal input volume.
19. The sample partitioning (at 40 °C) and final pressure release step (step down to 25 °C) should not be modified. However, the duration of initial denaturation (at 95 °C) can be varied depending on the DNA template. Further, Tm and annealing time must be optimized for the given primer/probe/master mix combinations to achieve a higher PCR efficiency. The duration of a dPCR run requires about 2 h and 30 min with our current cycling program. In case a restriction enzyme is used, an additional incubation step has to be included in the dPCR cycling program after the partitioning step at optimal conditions for the efficient activity of a given enzyme.
20. Scan the Naica™ Sapphire chip with exposure times for each fluorescence channel according to the Naica™ template used during scanning. Then, adjust the exposure time for a given fluorescence channel in a series of scanning runs using the same chip to enable the separation between positive and negative droplet clusters for each marker without oversaturating the positive droplets.

21. Scanning three Naica™ Sapphire chips (maximum per scanning run) for six fluorescence detection channels requires approximately 41 min.
22. The recombination rate can be estimated using two approaches: In the three-color approach, only droplets containing nuclei that are genotyped for all three loci are included while in the two-color approach, droplets with nuclei genotyped for any two markers are analyzed. The number of genotyping calls at a given locus depends on the efficiency of selected markers. The two-color approach yields higher genotyping calls than the three-color approach. However, CO interference strength can only be estimated using a three-color approach.

Acknowledgments

We would like to thank all lab members and Stilla Technologies®, in particular, Martin Becker for fruitful discussions, Jana Lorenz and Franziska Hartmann for excellent technical assistance, the IPK gardeners for their excellent horticultural support, as well as the IPK Gatersleben (flagship project) and the German Federal Ministry of Education and Research (BMBF) for financial support in the frame of the grant HERBY (FKZ-031B0188) for financial support.

References

1. Wang Y, Copenhaver GP (2018) Meiotic recombination: mixing it up in plants. Annu Rev of Plant Biol 69:577–609
2. Mercier R et al (2015) The molecular biology of meiosis in plants. Annu Rev of Plant Biol 66: 297–327
3. Lambing C, Franklin FCH, Wang CJR (2017) Understanding and manipulating meiotic recombination in plants. Plant Physiol 173(3): 1530–1542
4. Lambing C, Heckmann S (2018) Tackling plant meiosis: from model research to crop improvement. Front Plant Sci 9:829
5. Colas I et al (2016) A spontaneous mutation in MutL-Homolog 3 (HvMLH3) affects synapsis and crossover resolution in the barley desynaptic mutant des10. New Phytol 212(3): 693–707
6. Saintenac C et al (2009) Detailed recombination studies along chromosome 3B provide new insights on crossover distribution in wheat (*Triticum aestivum* L.). Genetics 181(2):393–403
7. Si W et al (2015) Widely distributed hot and cold spots in meiotic recombination as shown by the sequencing of rice F2 plants. New Phytol 206(4):1491–1502
8. Dreissig S et al (2015) Measuring meiotic crossovers via multi-locus genotyping of single pollen grains in barley. PLoS One 10(9): e0137677
9. Dreissig S et al (2017) Sequencing of single pollen nuclei reveals meiotic recombination events at Megabase resolution and circumvents segregation distortion caused by Postmeiotic processes. Front Plant Sci 8:1620
10. Choi K, Yelina NE, Serra H, Henderson IR (2017) Quantification and sequencing of crossover recombinant molecules from Arabidopsis pollen DNA. Haplotyping Methods Protoc 1551:23–57
11. Ahn YJ et al (2021) High-throughput measuring of meiotic recombination rates in barley pollen nuclei using Crystal Digital PCR™. Plant J 107(2):649–661
12. Galbraith DW et al (1983) Rapid flow cytometric analysis of the cell cycle in intact plant tissues. Science 220(4601):1049–1051

Chapter 7

Robust dPCR Hyper-Plexing Approach by Color Combination for Detection of 17 *ESR1* Mutations in Breast Cancer Patient Samples

Sylvain Ursuegui, Julie Donzier, Cécile Jovelet, Etienne Fradet, and Rémi Dangla

Abstract

Digital PCR has seen increasing adoption in recent years since it overcomes most of the main technical limitations of real-time PCR. However, the level of multiplexing remains limited when facing the need to develop approaches that can detect multiple targets in a single reaction with a high sensitivity to obtain a more complete genetic characterization of samples. This need concerns most molecular biology applications, but it is especially valuable in fields such as oncology, for which it is often necessary to detect numerous specific mutations, particularly those involving hotspots, in order to adapt a treatment, monitor the efficacy of a drug, and detect resistance mutations. This is the case of the *ESR1* gene in breast cancer, where numerous mutations can be responsible for mechanisms of resistance to the backbone therapy. In this sense, the implementation of color combination using Crystal Digital PCR® allows high levels of multiplexing to be achieved, while maintaining simplicity and robustness of analysis. This approach has been successfully applied to the detection of 17 *ESR1* mutations in a single assay.

Key words Multiplexing, *ESR1* mutations, Breast cancer, Crystal Digital PCR, Color combination

1 Introduction

1.1 Multiplexing in Digital PCR

Digital polymerase chain reaction (dPCR) has been rapidly adopted as it overcomes most of the main limitations of real-time PCR (qPCR) in terms of detection sensitivity, quantification accuracy, or inhibitor resistance. dPCR enables single-molecule detection and accurate target quantification, even in complex samples such as liquid biopsies [1, 2]. However, as with qPCR, the number of targets detected per assay remains limited. The conventional approach to multiplexing in dPCR consists in using target-specific fluorescent probes to detect a single target per detection channel (Fig. 1a). The multiplexing capacity is thus limited by the number of fluorescence channels available on the instruments, which cannot

Guillaume Pavlovic (ed.), *Digital PCR: Methods and Protocols*, Methods in Molecular Biology, vol. 2969,
https://doi.org/10.1007/978-1-0716-4767-7_7,

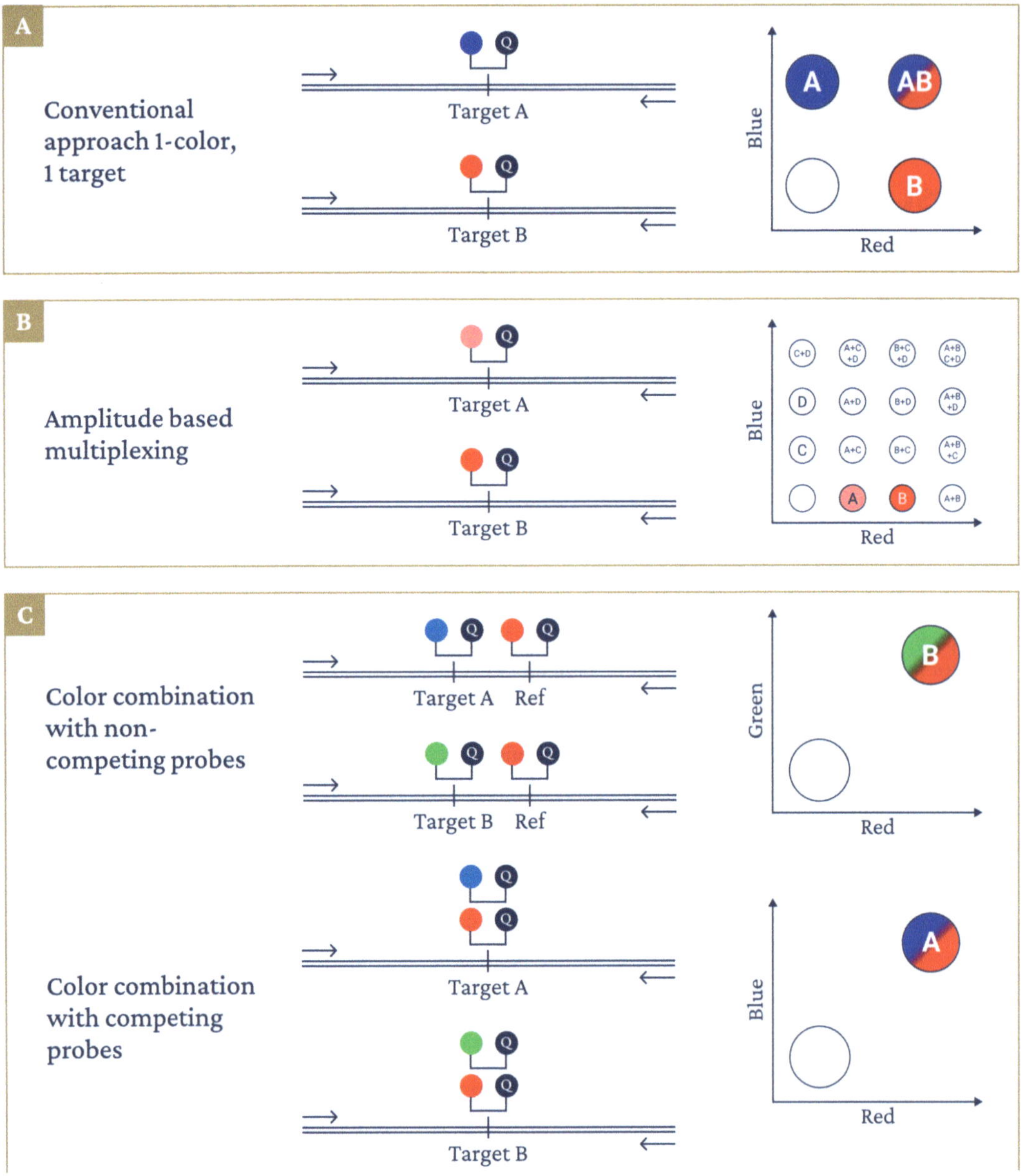

Fig. 1 Multiplexing approaches in Digital PCR. (**a**) Illustration of the 1 color-1 target multiplexing approach. For blue/red duplex assay, there will be four clusters in the 2D dot-plot: all negative partitions, blue for Target A only positive partitions, red for Target B only positive partitions, and blue/red for Target A and Target B positive partitions. (**b**) Illustration of the amplitude-based multiplexing approach. Two distinct targets are detected using the red channel: one with a low intensity cluster (Target A) and one with a high intensity cluster (Target B). In this representation, if two additional targets (Target C and Target D) are detected in another color, results include up to 16 different clusters in the 2D dot plot. (**c**) Illustration of the color combination approach, with two different implementations, noncompeting and competing probes, that lead to equivalent fluorescence data. The cluster of droplets that contain Target A only is blue/red while the cluster of droplets that contain Target B is green/red

be increased infinitely due to the limited bandwidth of optical receptors and the spectrum overlap of commonly available fluorescent reporters.

More sophisticated approaches, known as higher-order multiplexing methods, have been implemented in dPCR to increase the level of multiplexing by detecting more than one single target by detection channel [3, 4]. Among these, the amplitude-based and ratio-based multiplexing approaches consist on encoding different targets with different fluorescence intensities using different concentrations of specific fluorescent probes [5–9]. Nevertheless, these approaches make data analysis much more complex than the conventional approach based on the detection of a single target per fluorescence channel. Whereas the conventional approach requires only one simple thresholding to distinguish negative events from positive ones, these approaches will require several levels of thresholding on a 2D dot plot to separate events according to their intensities (Fig. 1b). Therefore, they make the analysis time-consuming and not very robust, especially when considering that fluorescence intensities vary from sample to sample.

1.2 Color Combination Enables to Expand dPCR Multiplexing

Alternative approaches have been implemented in qPCR around the concepts of multicolor combinatorial probe coding [10–12] or color-coding [13]. In these approaches, each target is encoded using a unique combination of 2 or more distinct colors. However, the implementation of color combination approach in qPCR is limited to applications for which only one single target could be present in the sample. When more than one target is present, the resulting fluorescent signal will comprise more than two colors and will be ambiguous, making reliable target identification impossible. Thanks to partitioning, dPCR can overcome this limitation by precisely identifying the colors in each partition.

In dPCR, co-encapsulation events leading to an ambiguous fluorescence signal can be eliminated from the analysis. Different targets can thus be accurately identified simultaneously in the same sample/reaction, allowing the full potential of the color combination approach to be exploited. However, the ratio between positive and negative partitions forms the basis of quantification in dPCR, and the statistical framework for converting this ratio into a concentration estimate is based on the Poisson approximation. As co-encapsulation events are removed from the analysis, the number of positive droplets for a given target is underestimated if other targets are present in the sample. To counterbalance this bias in the number of positive partitions, a correction must be also applied to the number of negative partitions. Then, to be statistically consistent and obtain accurate quantification, partitions that are negative for the target but positive for another target must also be eliminated from the total number of partitions.

This approach has been recently successfully applied in dPCR using both the naica® and Nio® + system thanks to their software simplicity for data analysis [14]. By using two probes per target, either competing or noncompeting, each population can be revealed in 2 colors expanding the number of targets per assay detectable (Fig. 1c). The maximum level of multiplexing is 15 targets when using the 6-color naica® system and 21 targets when using the Nio® + and its 7-color channels (Fig. 2). One additional target, detected in single color, can be added expanding the multiplexing level to 16 targets with 6 colors and 22 targets with 7 colors. To analyze a color combination assay, it is only required to adjust a

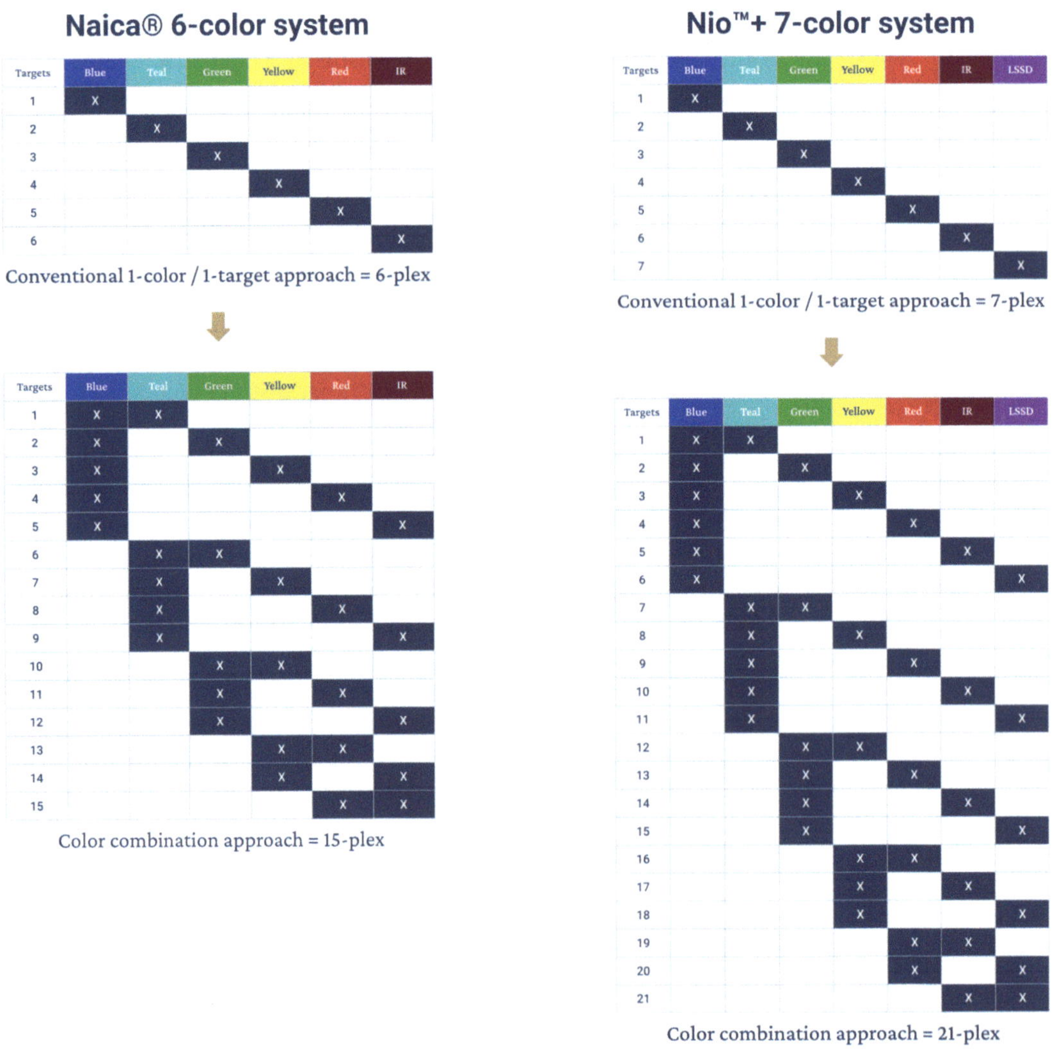

Fig. 2 Multiplexing capacity using conventional approach and color combination. In the color combination strategy, each target is encoded by a unique combination of 2 or more distinct fluorophores, or colors. When 2 colors are used to detect each target, the maximum level of multiplexing is a 15-plex and 21-plex capability, when using a 6-color and 7-color detection instrument, respectively

single threshold for each channel to separate positive droplets from the negative droplets. Based on this 1D thresholding, the analysis software (Crystal Miner and Nio® Analyzer for the naica® and Nio® + systems, respectively) can co-localize fluorescence within the same droplet and thus assign and quantify the presence of a population previously defined with the corresponding color code in the population editing tool. The analysis remains as simple and robust to fluorescence variations as with the 1 color-1 target usual approach.

1.3 Nio® + System and Crystal Digital PCR® Workflow

The Nio® + is an all-instruments-in-one platform for Crystal Digital PCR® (cdPCR) capable of simultaneously and precisely quantifying high numbers of biomarkers in a single sample. The quantification of multiple targets in a single reaction simplifies the detection process, minimizes the detection variability, and significantly reduces the hands-on-time and time to results. This instrument operates with the Ruby Chip, a cdPCR consumable that enables 16 samples to be processed at a time, while maintaining ease of use as just a single pipetting step is required before being transferred into the Nio® + system for partitioning, PCR amplification, and sample readout. In addition, the Ruby Chip can be integrated to automated liquid handling equipment. Using the Ruby Chip, the Nio®+ configuration of the Nio® instrument allows to add up to 8 Chip Plates, each containing 3 chips, whether the instrument is running an experiment or not. This supports a high-throughput capacity of over 768 samples processed per 8-h workday. The Nio® + has 7 separate detection channels, enabling multiplexing of up to 7 targets with the classic 1-color-1-target approach, and up to 22 targets with the color combination approach.

Nio® + is supported by two software:

- The Nio® Reader software is used to set dPCR experiments by defining PCR thermocycling conditions and scanning parameters.
- The Nio® Analyzer enables the analysis of cdPCR data generated with the Nio® + .

The unique features of the Nio® + Analyzer allow to easily analyze color combination data. It contains a population editor function that allows for the definition of the negative and the positive populations detected in a color combination assay (Fig. 3). Using an example of an oncology assay developed via the color combination approach, the *ESR1* E380Q mutation in this example is defined as follows:

- The positive population is first defined in the positivity combination table (Fig. 4b). Teal and Yellow channels are set as

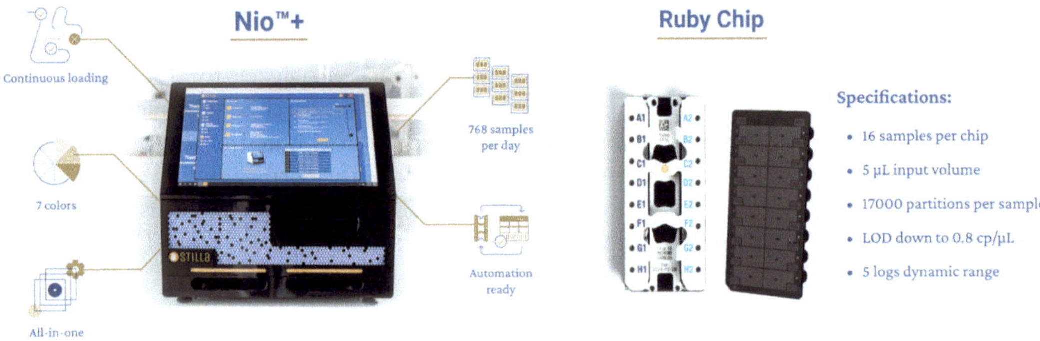

Fig. 3 Nio® + system and Ruby Chip. The Nio® + is an all-in-one digital PCR instrument with continuous loading capabilities and 7-color capabilities, used with the Ruby Chip, that harnesses simplicity and flexibility

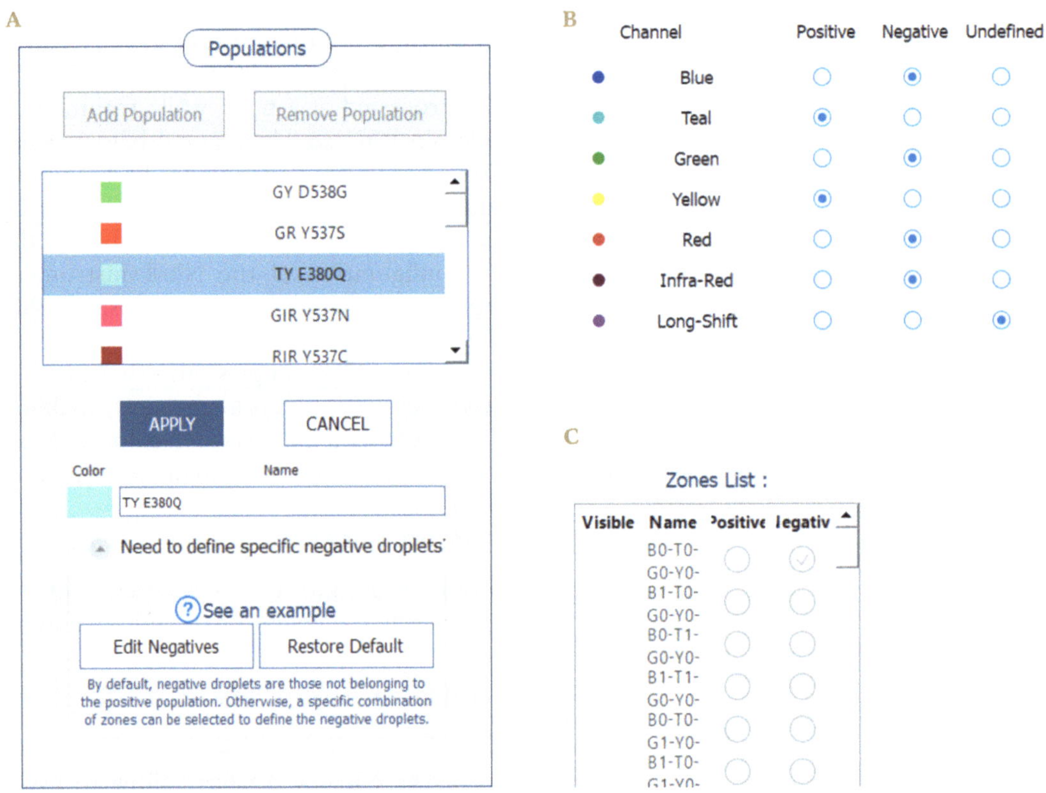

Fig. 4 Population edition in the Nio®+ Analyzer. In a color combination assay, such as the ESR1 12-plex assay, the negative and the positive populations are defined for each target using the population editor function in the Nio® Analyzer. Example for the ESR1 mutation Y537N: the positive channels are defined in the positivity combination table (Green and Infra-Red channels) while the other channels are defined as negative or undefined (Blue channel reserved for the reference gene). Negative populations for this target are defined in the zones list and include the droplets negative for all the channels (B0, T0, G0, Y0, R0, IR0) and the droplets only positive in the Blue channel (B1, T0, G0, Y0, R0, IR0)

positive channels while all other channels are defined as negative (Blue, Green, Red, and IR channels) or as undefined (LSSD channel reserved for the reference gene).

- Negative populations for this target are then defined in the zones list (Fig. 4c) and include the droplets negative for all the channels (B0, T0, G0, Y0, R0, IR0, LS0) that contain no target and the droplets only positive in the LSSD channel (B0, T0, G0, Y0, R0, IR0, LS1) that contains the wild-type DNA of exon 8.

To demonstrate the powerfulness of the color combination approach with the 7 colors of the Nio® + system to detect a high number of targets within a single reaction, we have developed an 18-plex assay for the detection of *ESR1* mutations in breast cancer.

1.4 ESR1 in Breast Cancer

Breast cancer is the most diagnosed cancer worldwide representing one in every six cancer-related deaths in women [15]. The majority of breast cancers are characterized by hormone receptor overexpression (HR+ breast cancer) with the strategy of hormone deprivation through endocrine therapy being the current gold standard treatment. Intrinsic and acquired resistance to endocrine therapy is an inflection point in disease development because it results in relapse and metastasis [16].

A key resistance mechanism to endocrine therapy is characterized by the constitutive activity of the estrogen receptor alpha due to mutations in the estrogen receptor 1 (*ESR1*) gene encoding for this receptor. *ESR1* mutations drastically reduce response to endocrine therapy and particularly to aromatase inhibition [17]. Hence, *ESR1* mutations are emerging clinical biomarkers in metastatic hormone positive breast cancer. Indeed, recent clinical studies have shown the importance of tracking *ESR1* mutations in liquid biopsy samples during first-line treatment as it enables clinicians to switch patient treatments from aromatase inhibition to an *ESR1* receptor antagonist [18].

Due to its high sensitivity and high multiplexing capabilities, Crystal Digital PCR® (cdPCR) using the Nio® + and color combination approach allows for the early detection of *ESR1* mutations in liquid biopsy samples and is therefore an ideal technology for HR + breast cancer patient monitoring.

1.5 ESR1 18-Plex Crystal Digital PCR® Assay

The *ESR1* 18-plex assay is based on the color combination approach and allows for the simultaneous detection and quantification of 18 targets in the same reaction. Fifteen somatic *ESR1* mutations are detected, in color combination, across three exons: exon 5 (E380Q), exon 7 (S463P), and exon 8 (D538G, D538N, L536H, L536P, L536Q, L536R, Y537C, Y537D, Y537H, Y537N, Y537NdelinsTA, Y537S). Two additional mutations of exon 8 coding for the same protein change are co-detected in single color (Y537SdelinsAG and Y537SdelinsCA detected in the red

Target	Mutation	Exon	Base changes	COSMIC ID	Detection channel 1	Detection channel 2
1	ESR1 exon 8 reference	8	N/A	N/A	LSSD	N/A
2	E380Q	5	c.1138G>C	COSM3829320	Teal	Yellow
3	V422del	6	c.1265_1267del	COSM1074628	Blue	IR
4	S463P	7	c.1387T>C	COSM4771561	Teal	Red
5	D538G	8	c.1613A>G	COSM94250	Green	Yellow
6	D538N	8	c.1612G>A	N/A	Blue	Red
7	L536H	8	c.1607T>A	COSM6843697	Yellow	IR
8	L536P	8	c.1607T>C	COSM6906109	Yellow	Red
9	L536Q_delinsAG	8	c.1607_1608delinsAG	COSM4766050	Blue	Teal
10	L536R	8	c.1607T>G	COSM4774826	Teal	Green
11	Y537C	8	c.1610A>G	COSM1074637	Red	IR
12	Y537D	8	c.1609T>G	COSM6918757	Teal	IR
13	Y537H	8	c.1609T>C	COSM7449624	Blue	Yellow
14	Y537N	8	c.1609T>A	COSM1074635	Green	IR
15	Y537N_delinsTA	8	c.1608_1609delinsTA	N/A	Blue	Green
16	Y537S	8	c.1610A>C	COSM1074639	Green	Red
17	Y537S_delinsAG	8	c.1609_1610delinsAG	COSM6948665	Red	N/A
18	Y537S_delinsCA	8	c.1610_1611delinsCA	COSM6971270	Red	N/A

Fig. 5 Gene and mutations targeted by the ESR1 18-plex assay and assay configuration

channel). Finally, a reference probe is detected on an unmutated portion of exon 8 (codons 543–547) allowing the total DNA quantification and the estimation of the Mutant Allele Frequence (MAF). This reference is isolated in a seventh color channel (LSSD channel) to not decrease the sensitivity of the assay (see note "Color combination assay configuration in liquid biopsy analysis"). The targeted mutations and the assay configuration are depicted in Fig. 5.

The ability of the *ESR1* 18-plex assay to detect 18 targets simultaneously in the same sample was first assessed with synthetic DNA using the Nio® + and Ruby Chip (Fig. 6).

The *ESR1* 18-plex assay was characterized on serial dilutions of seventeen DNAs, each bearing one of the targeted *ESR1* mutation in a background of WT DNA (500cp/µL, 1.5 ng/µL). Mutant allele fractions ranging from 10% to 0.3% were detected with an

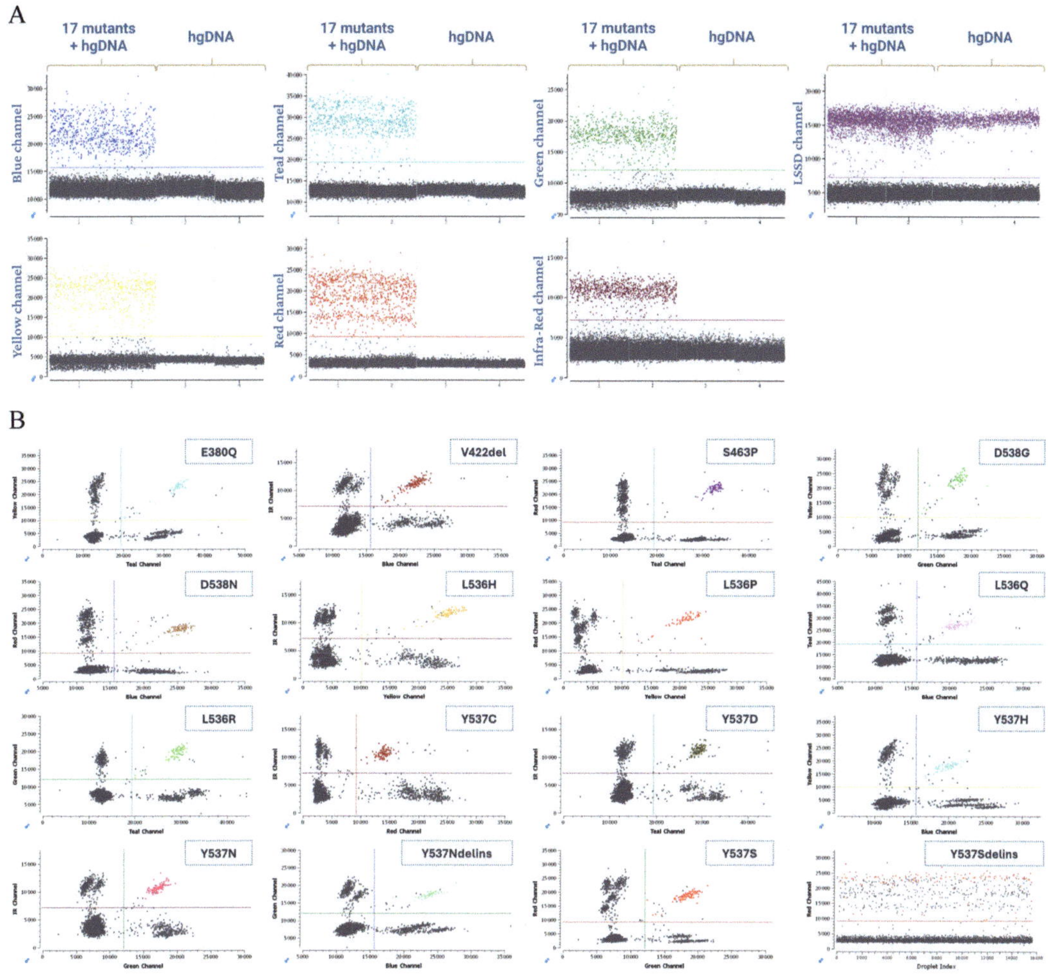

Fig. 6 Validation of the ESR1 18-plex assay with synthetic. (**a**) DNA 1D plots in the 7 detection channels. Chambers 1–2: Positive control = hgDNA (≈150 cp/μL) + the 17 mutants (≈25 cp/μL each) targeted in this assay. (**b**) Visualization of the populations in 1D and 2D plots using Nio® Analyzer

excellent linearity (Fig. 7). The coefficient of determination of linear regressions performed between input and measured copies per reaction ranged from $R^2 = 0.9936$ to $R^2 = 0.9996$, depending on the mutation targeted.

2 Materials

2.1 Equipment and Accessories

1. Barrier/filter tips (sterile and aerosol resistant).
2. Micropipettes.
3. Powder-free gloves.
4. Microcentrifuge.
5. Vortex.

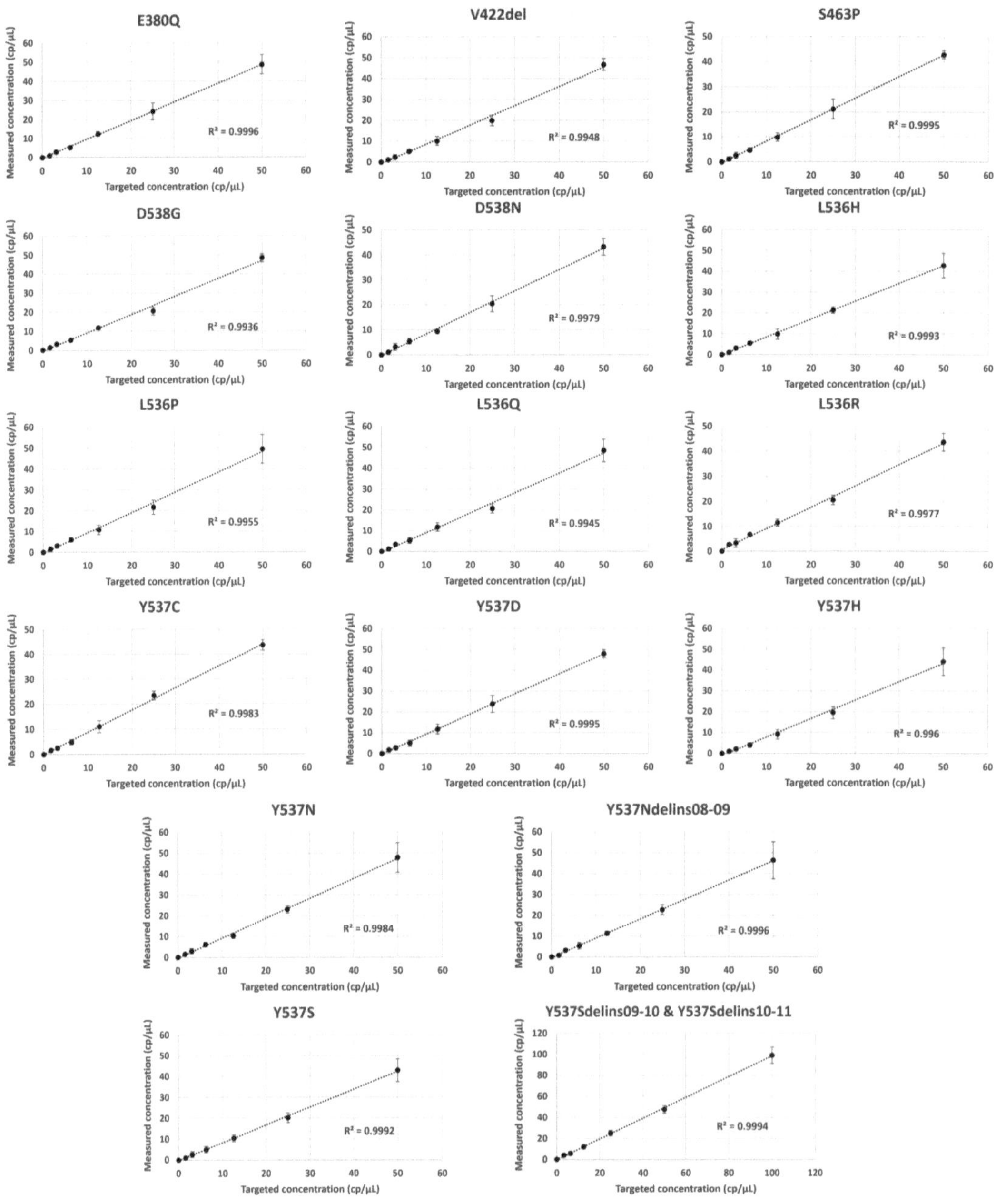

Fig. 7 Evaluation of the performance of the ESR1 18-plex Crystal Digital® PCR assays. The mutations were detected with a 95% confidence level in serial dilutions ranging from 10% to 0.3% of mutant DNA in a background of 500 copies of wild-type DNA per 5 µL reaction. Five replicates were performed for each dilution point. The error bars indicate the standard deviation obtained for the five replicates

6. PCR reaction microtubes (sterile, 1.5 mL).
7. PCR and DNA free flowhood for PCR mix assembly.
8. PCR and DNA free flowhood for DNA template addition.
9. Ruby Chip (Stilla Technologies, C16011).

10. Antistatic wipe (ACL Staticide®, Reference: SW12).
11. Nio® + (Stilla Technologies, N30027).
12. Nio® Reader and Nio® Analyzer software.
13. NanoDrop ND-3300 Spectrophotometer or Qubit fluorometer (Thermo Scientific) or real-time thermocycler.

2.2 Reagents

1. naica® PCR MIX 10X (Stilla Technologies, R10106) containing DNA Hot Start polymerase with 5′-3′ exonuclease activity, DNA polymerase buffer, and optimized concentrations of deoxynucleoside triphosphate (dNTPs) and magnesium chloride (MgCl2).
2. ESR1 18-plex Crystal Digital PCR® Assay (Stilla Technologies, R51011) containing a pool of primers and probes and a pool of positive control including human genomic DNA (hgDNA) and synthetic DNA (gBlocks™) for each *ESR1* mutation.
3. Universal Reporter 7 (Stilla Technologies, R41401 or R41402), including Tube A and Tube B containing 3 and 4 fluorescent reporters, respectively, to activate the 7 channels of the Nio® + .
4. hgDNA template for the preparation of the negative control.
5. Molecular biology grade H_2O.

3 Methods

3.1 General Recommendations to Prevent DNA Contamination

To reduce the risk of DNA carryover contamination that can lead to false positives, the following are crucial

1. All reagents and plastic consumables must be sterile, DNA- and DNAse-free, and of molecular biology grade.
2. Wear gloves while handling reagents, materials, and equipment.
3. Use commercially available DNA decontamination solutions to clean all surfaces dedicated to protocol handling.
4. Areas dedicated to sample extraction, Digital PCR mixture assembly, and Digital PCR amplification must be separated. As PCR products are the leading cause of contamination, reagents and consumables brought in the post-amplification area must not be reintroduced in the pre-amplification areas.

3.2 DNA Template Preparation

1. The *ESR1* 18-plex assay is compatible with DNA resuspended in molecular grade water or 10 mM Tris-EDTA (TE). The assay is also compatible with DNA extracted from frozen tumor tissue, formaldehyde fixed paraffine embedded (FFPE) tissue, and plasma.

2. High quality DNA templates suitable for PCR amplification must be obtained using a dedicated extraction kit (*see* **Note 3**).

3.3 DNA Quantification and Quality Monitoring

1. DNA templates should be quantified using a NanoDrop ND-3300 Spectrophotometer or a Qubit fluorometer or using real-time PCR quantification.
2. Use DNA quantities up to 15 ng per Crystal Digital PCR® reaction equivalent to 5000 copies of DNA per reaction.
3. A DNA template is deemed as high quality when a 260 nm/280 nm absorbance ratio of ~1.8–2 and a 260 nm/230 nm absorbance ratio in the range of 2.0–2.2 are obtained.

3.4 General Guidelines for ESR1 18-Plex cdPCR Assay Controls

1. The positive control is provided in the *ESR1* 18-plex cdPCR assay and includes each *ESR1* mutation at a final concentration of 0.75 ng/μL and hgDNA at a final concentration of 15 ng/μL. The positive control should be considered at a 10× concentration during PCR mix preparation to target a final concentration of 25 cp/μL and 500 cp/μL for mutant and wild-type hgDNA, respectively.
2. Prepare a negative control DNA mix to assess the presence of mutant DNA contaminants using a WT DNA at a final concentration of 15 ng per μL. The resulting solution should be considered at a 10× concentration during PCR mix preparation to target a final wild-type hgDNA concentration of 500 cp/μL.

3.5 PCR Mix Preparation

1. Thaw reagents, mix thoroughly before use, and return unused reagents to the indicated storage temperature.
2. In a clean 1.5 mL microcentrifuge tube, prepare the Digital PCR reaction mix by assembling the reagents (naica® PCR MIX Buffer A and Buffer B, *ESR1* 18-plex pool of primers and probes, Universal Reporter Tube A and Tube B) as outlined in Table 1. Prepare enough master mix for all samples including one positive control and one negative control per run (i.e., for every 16 reactions).
3. Dispense the Digital PCR reaction mix in separate microcentrifuge tubes and close the lids.
4. Open only one tube at a time to avoid cross-contamination, add the DNA template individually to each reaction tube (0.6 μL for positive and negative controls and from 1 to 4.26 μL for the samples) and complete each tube with water to a final reaction volume of 6 μL.
5. Close the lids and vortex the tubes thoroughly.
6. Briefly centrifuge the microcentrifuge tubes to collect the entire volume at the bottom of each tube.

Table 1
ESR1 18-plex cdPCR assay setup for one reaction

Reagents	Initial concentration	Final concentration	Volume (μL)
Naica® PCR MIX buffer A	10×	1x	0.6
Naica® PCR MIX buffer B	25×	1x	0.24
Primers/probes mix	10×	1x	0.6
Universal reporter tube A	40×	1x	0.15
Universal reporter tube B	40×	1x	0.15
DNA template	N/A	N/A	1–4.26
To 6 μL with H_2O			

3.6 Ruby Chip Preparation

1. Antistatic treatment: Open a pouch of the pre-wetted antistatic wipe (ACL Staticide®, Reference: SW12). Wrap the antistatic wipe around a finger and thoroughly wipe the underside and the edges of the Ruby Chip to fully coat it with the antistatic treatment. Without turning the Ruby Chip upside down, lift the Ruby Chip and visually check that the entire bottom surface of the Ruby Chip has been fully covered with antistatic solution. Wrap a piece of precision Wipe (Kimtech™ Science, Reference: 7552) around a finger, and wipe off excess antistatic product with. Place the Ruby Chip back in the Chip Plate and apply the antistatic treatment to the next chip.
2. Ruby Chip piercing: Hold the Piercing Tool in one hand. With the other hand, hold the chip down in the provided Chip plate. Position the Piercing Tool above a column of wells and in contact with the aluminum foil covering the inlet ports. Press the Piercing Tool down to pierce the foil.

3.7 Ruby Chip Loading

1. Load 5 μL of sample mix in the tip of your pipette.
2. Hold the pipette in one hand and with the other hand, hold the chip down in the provided Chip Plate.
3. Position the tip of the pipette above the inlet port, holding the pipette vertically above the chip (90° angle with the chip).
4. Insert the pipette tip inside the inlet well and touch the bottom of the well with your tip.
5. Still keeping your pipette at a 90° angle with the chip, gently lift the pipette tip slightly off the bottom of the inlet well. The displacement of the pipette tip should be small—approximately 1 mm above the bottom of the inlet well—such that the pipette tip remains inside of the oil.

6. Still keeping your pipette steady at a 90° angle with the chip, eject the sample mix by pressing the plunger of the pipette down to the first stop in a continuous and rapid motion. Do not lift your pipette out of the inlet well as you eject. Do not press the plunger of the pipette down to the second stop, to prevent the creation of bubbles.
7. Remove pipette tip out from the inlet well.

3.8 Run cdPCR Experiment on the Nio®+

1. Open Nio® Reader software.
2. Select "Plan & Start" in the menu "RUNS," browse and load the .nioexperiment file for the *ESR1* 18-plex Crystal Digital PCR® Assay (available in the assay technical resources of the Stilla® website).
3. Indicate the names of the samples in each chamber and save the .nioexperiment file, renaming it for the specific experiment. Note that the .nioexperiment file can be prepared on an independent computer and then loaded directly onto the Nio® + .
4. Select the "Start Run" button to start Run 1.
5. Select "Status" in the menu "RUNS" to verify that the first Chip Plate enters the "Thermocycling" step process.
6. At the end of the run, copy the results file .niodata of the two runs on a USB flash drive.
7. To recover the chips, go to "Status" in the menu "RUNS" and select the "Nio Run" to unload. Click on "Unload Chip plate" on the right side of the user interface.

3.9 Data Analysis with Nio® Analyzer

1. Open the first .niodata file generated by the Nio® Reader software in the Nio® Analyzer software.
2. Open Nio® Reader software.
3. Under "ANALYZE DATA," click on "Plots & Populations" and then "1D dot plot." With the selection of positive control chambers, adjust the thresholds for all channels if necessary. The threshold must be centered between the mean fluorescence of the positive population and the mean fluorescence of the negative population for Blue/Teal/Green/Yellow/Infra-Red channels. For LSSD channel, place the threshold just above the negative cluster, meaning the distance between threshold and negative cluster is shorter than the distance between threshold and positive cluster. For the Red channel, do the opposite, place the threshold just below the positive cluster, meaning that the distance between the threshold and positive cluster is shorter than the distance between threshold and negative cluster.
4. Go to "FILE > Save" to save the experiment in .nioresult format (under the same name as the experiment name).

5. Go to "VIEW RESULTS > Result Table" to access quantification results for each target population.
6. Optional: Go to the "EXPORT" menu and click on the "Browse" button to select the output folder, then click on the "Export" button to export the data in the ".xlsx" format (under the same name as the experiment name).
7. Interpretation of the results: a sample is considered positive for a mutation if the quantification is above the Limit of Detection (LoD). The result is considered as undetermined if the concentration is between the Limit of Blank (LoB) and the LoD while the sample is considered as negative if the concentration is below or equal the LoB.
8. MAF (Mutant Allele Frequency) Calculation. The MAF can be directly calculated by dividing the concentration obtained for a mutation by the concentration obtained for the *ESR1* exon 8 reference, which is used to assess the amount of DNA input.

4 Notes

1. Two different strategies can be used to design a color combination assay, either using two competing probes or two noncompeting probes. Both strategies have their inherent advantages and disadvantages. For the competitive approach, it is necessary to design only one probe for each target. However, in this case, the separability between positive and negative droplets is typically reduced due to the competition between the two probes. For the noncompetitive approach, it is necessary to design two probes per target, which may lead to difficulties in oligo design of the entire assay. The benefit is that the separability between positive and negative events is better than with the competitive approach. Depending on the application and the type of assay, one strategy may be preferred over the other. For example, in the case of a target with a hotspot like *ESR1*, the noncompetitive strategy will be limited to the detection of 6 mutations per hot spot using 2 colors. Indeed, the addition of a second probe in a locus adjacent to the hotspot will be detected for all mutations located on that hotspot, and 1 color can be combined with a maximum of 6 others using a 7-color reader. On the other hand, to detect SNPs, the noncompetitive approach can be used, as a reference probe can be shared for the different variants, reducing the number of probes in the assay.
2. As described in the introduction, to quantify accurately a target in color combination, it is necessary to eliminate co-encapsulations in the positive population and positive

partitions for other targets in the negative population. This will reduce the volume analyzed when quantifying a target, and therefore affect the sensitivity of the analysis. The impact on sensitivity will be more significant in cases where one of the detected targets is strongly abundant. In applications such as liquid biopsy analysis for which having the best sensitivity is preferable, the loss of sensitivity due to the background concentration may be problematic. In the case of liquid biopsy, the highest contributor to the background concentration is most likely to be the target used to quantify the wild-type (WT) circulating tumor DNA (ctDNA). A solution to restore sensitivity consists in detecting the WT target using 1 single color, and using color combination with the other available colors to detect the targeted mutant targets. By isolating the wild-type target in a fluorescence channel, the maximum multiplexing capacity with a 7-color system is 17 targets per assay. In this configuration, 1 color is allocated to the detection of the wild-type target, the other 6 colors are combined to detect 15 mutant targets with 2 colors per target, and 1 additional mutant target can be detected with 1 color used in color combination.

3. DNA can be extracted using different types of extraction kits (column or bead-based protocols). The following kits are currently compatible with Crystal Digital PCR® platforms (non-exhaustive list): Maxwell® RSC DNA FFPE Kit (Promega, reference AS1450), Maxwell® RSC ccfDNA Plasma Kit (Promega, reference AS1840), QIAamp DNA FFPE Tissue Kit (Qiagen, reference 56,404), and the QIAamp circulating nucleic acid kit (Qiagen, reference 55,114). Other extraction methods may be suitable. However, the protocol used and sample purity might have an influence on sample compatibility. Individual sample-type and extraction method compatibilities with Crystal Digital PCR® platforms may require a dedicated assay validation by the end user.

   Some extraction techniques with certain types of samples occasionally result in a decrease in fluorescence, likely due to the residual presence of the solvent used during extraction process. Some improvements can be applied to the protocol to restore fluorescence (Fig. 8):

   - For column-based protocol, increase time of the centrifugation following column cleaning with ethanol.
   - For other protocols, an additional drying time at the end of the process minimize the phenomenon and restore the fluorescence.

4. Co-encapsulation phenomena can affect the separability between negative and positive populations required for reliable analysis and quantification. In particular, when

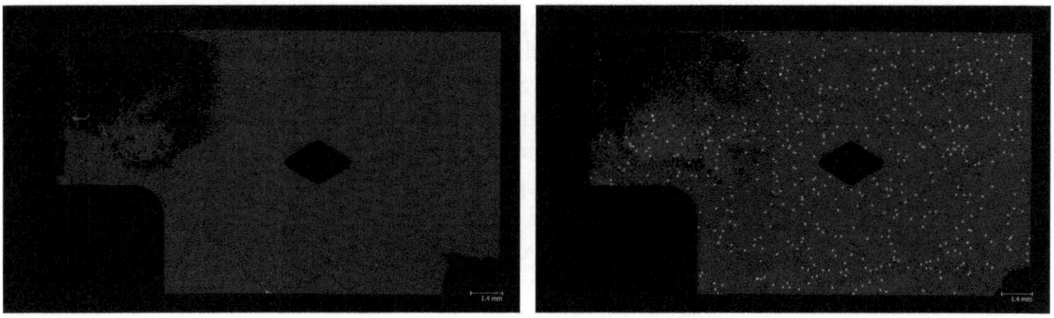

Fig. 8 Example of fluorescence inhibition. The same sample was loaded in two chambers. Left chamber: no additional drying time. Right chamber: additional drying time (1 h, 56 °C), fluorescence restored

co-encapsulations involve variants present on the same amplicon, positive partitions with lower fluorescence intensity may be observed, generating rain below the main positive cluster. This is probably due to competition between the probes and the different amplicons generated during amplification. For this type of assay, it is therefore recommended to test the maximum amount of DNA input to maintain good performance. In the case of the *ESR1* 18-plex assay, this limit has been set at 3 ng/μL corresponding to 18 ng in the 6 μL of PCR mix prepared and 15 ng in the 5 μL of PCR mix loaded in the Ruby Chip.

5. The Ruby chip is also compatible with multichannel pipettes and robotic pipettors, further reducing user time for experiment preparation. For further information on the protocol to be applied with these injection methods, please refer to the chip's IFU.

6. For reliable interpretation of the results, a positive control must be included for each run in order to correctly set thresholds on each fluorescence channel. Similarly, a negative control (extraction blank or negative sample extracted using the same process) must be included to check for contamination during sample preparation.

7. The LoB and LoD values are crucial for interpreting the results. LoB/LoD values obtained using genomic DNA and synthetic controls for each mutation have been determined following the method described in Stilla Technologies' technical note [19], adapted from Clinical and Laboratory Standards Institute (CLSI) EP17-A2 standard (Protocols for Determination Of Limits Of Detection And Limits Of Quantitation; Approved Guideline). Briefly, to determine LoB, this method requires the analysis of at least 30 negative samples to achieve a 95% confidence level. LoB is defined as the 95th percentile of background concentrations measured in blank samples. LoD is

determined using the obtained LoB values and testing replicates of low-level samples, which should be representative positive samples or a representative sample matrix with spiked-in target concentrations within a range of one to five times the LoB concentration previously calculated. LoD is defined as LoD = LoB + Cp * SDL. SDL is the standard deviation of all replicates pooled across unique low-level samples, and Cp is the coefficient to give the 95th percentile of a normal distribution.

LoB/LoD values measured for the *ESR1* 18-plex are depicted in Table 2.

However, it is recommended to determine the LoB under real conditions of use, with the same type of sample and the same preparation process (in particular the extraction step). It is common to observe a variation in LoB values depending on the type of sample and the extraction procedure. For example, fixed samples (FFPE) are known to give a higher LoB than plasma samples. To determine the LoB, it is recommended to follow the method described in the dedicated technical note [19]. If a significant difference in LoB is measured on the real samples, this difference can be directly added to the LoD value, which are directly correlated to the value of LoB, for a more rigorous interpretation of the results.

8. For samples with undetermined results (concentration between LoB and LoD values), it is advisable to retest the sample in a new series and, if possible, increase the sample volume tested directly in the PCR mix or use a pooling strategy of several Ruby chambers (see below).

9. For applications where concentrations may lay close to the LoD values, such as liquid biopsy analysis, Ruby chambers can be pooled to increase the sample volume input and thus assure that a positive outcome of a given mutation is not attributable to randomness. The Ruby Chip reaction volume is 5 μL per chamber, of which 3.4 μL is accounted for the sample volume. By pooling three Ruby chambers, a total sample volume of 10.2 μL can be added to the 15 μL reaction.

   The minimum achievable MAF depends on LoD and sample concentration. Table 3 shows the achievable MAF limits as a function of sample concentration for *ESR1* mutations detected in color combinations. For a concentrated sample (2–4 ng/μ L), a single Ruby chamber achieves very low MAFs of the order of 0.1–0.3%. In the case of a low-concentration sample, pooling several Ruby chambers is an interesting option for achieving low MAFs.

10. To analyze multiplexed cdPCR assays, it is necessary to establish a compensation matrix to correct the spillover of a fluorophore onto neighboring detection channels. With the usual 1-color-1-target approach, single-color controls are performed

Table 2
LoB and LoD of the ESR1 18-plex assay determined following the CLSI EP17-A2 standard (Ruby Chip chamber concentrations in cp/μL)

Number of Ruby chambers pooled	LoB/ LoD	Number of samples	E380Q	V422del	S463P	D538G	D538N	L536H	L536P	L536Q
1	LoB_{95}	91	0.00	0.00	0.00	0.32	0.30	0.00	0.00	0.00
	LoD_{95}	90	1.34	1.23	1.00	1.39	1.45	1.26	1.23	0.99
2	LoB_{95}	45	0.00	0.00	0.00	0.17	0.16	0.00	0.04	0.00
	LoD_{95}	45	0.88	0.78	0.75	0.90	0.95	0.88	0.92	0.69
3	LoB_{95}	30	0.00	0.00	0.00	0.11	0.11	0.00	0.10	0.00
	LoD_{95}	30	0.68	0.54	0.55	0.74	0.70	0.67	0.79	0.54
Number of Ruby chambers pooled	**LoB/ LoD**	**Number of samples**	**L536R**	**Y537C**	**Y537D**	**Y537H**	**Y537N**	**Y537N delinsTA**	**Y537S**	**Y537S delinsAG & delinsAC**
1	LoB_{95}	91	0.00	0.00	0.00	0.00	0.00	0.00	0.00	0.30
	LoD_{95}	90	1.07	1.17	1.00	1.11	1.14	1.07	1.03	2.06
2	LoB_{95}	45	0.00	0.00	0.00	0.00	0.00	0.00	0.00	0.16
	LoD_{95}	45	0.70	0.91	0.66	0.65	0.80	0.68	0.74	1.51
3	LoB_{95}	30	0.00	0.00	0.00	0.00	0.00	0.00	0.00	0.20
	LoD_{95}	30	0.65	0.69	0.55	0.53	0.53	0.46	0.62	1.32

Table 3
Theoretical MAF LoD that can be detected with the ESR1 18-plex assay in a Ruby Chip chamber for typical samples concentrations

Sample concentration (ng/μL)	Theoretical MAF LoD 1 chamber	Theoretical MAF LoD 2 chambers	Theoretical MAF LoD 3 chambers
3.5–4	0.12–0.17%	0.08–0.11%	0.05–0.09%
3–3.5	0.14–0.20%	0.09–0.13%	0.06–0.11%
2.5–3	0.17–0.24%	0.11–0.16%	0.08–0.13%
2–2.5	0.21–0.30%	0.14–0.20%	0.10–0.16%
1.5–2	0.28–0.40%	0.18–0.26%	0.13–0.22%
1–1.5	0.41–0.60%	0.27–0.40%	0.19–0.33%
0.5–1	0.83–1.20%	0.54–0.79%	0.38–0.66%
0.25–0.5	1.65–2.40%	1.09–1.59%	0.77–1.32%

for all the channels by introducing a single target into the PCR mix, thus activating a single probe [20]. The software then calculates the corrections needed to compensate for spillover from one probe into the other detection channels. Given that 2 colors are activated per target in color combination, it is required to adapt this method to establish the compensation matrix. One solution is to prepare specific primer and probe pool for each single-color control. For the given single-color control, only one of the two probes specific to this target is introduced in the primer/probe pool, enabling a single color to be activated. The composition of the rest of the pool remains unchanged. In most cases, this has no impact on the establishment of the compensation matrix. However, if the background level is severely impacted by the removal of a probe, this can have a more significant impact on the correction matrix and is therefore preferable to use another target detected in the same fluorescence channel.

References

1. Salipante SJJ, K. R. (2020) Digital PCR — an emerging technology with broad applications in microbiology. Clin Chem 66:117–123
2. Basu AS (2017) Digital assays part I: partitioning statistics and digital PCR. SLAS Technol 22(4):369–386. https://doi.org/10.1177/2472630317705680
3. Whale AS, Huggett JF, Tzonev S (2016) Fundamentals of multiplexing with digital PCR. Biomol Detect Quantif 10:15–23. https://doi.org/10.1016/j.bdq.2016.05.002
4. Zhong Q, Bhattacharya S, Kotsopoulos S, Olson J, Taly V, Griffiths AD, Link DR, Larson JW (2011) Multiplex digital PCR: breaking the one target per color barrier of quantitative PCR. Lab Chip 11(13):2167–2174. https://doi.org/10.1039/c1lc20126c
5. Appay R, Fina F, Barets D, Gallardo C, Nanni-Metellus I, Scavarda D, Henaff D, Vincent J, Grewis L, Pourquier P, Colin C, Figarella-Branger D (2020) Multiplexed droplet digital PCR assays for the simultaneous screening of

major genetic alterations in tumors of the central nervous system. Front Oncol 10:579762. https://doi.org/10.3389/fonc.2020.579762

6. Du Y, Yan Z, Song K, Jin J, Xiao L, Sun Z, Tan Y, Zhang P, Du Z, Yang R, Zhao Y, Song Y (2022) Development and evaluation of a multiplex droplet digital polymerase chain reaction method for simultaneous detection of five biothreat pathogens. Front Microbiol 13:970973. https://doi.org/10.3389/fmicb.2022.970973
7. Oscorbin IP, Smertina MA, Pronyaeva KA, Voskoboev ME, Boyarskikh UA, Kechin AA, Demidova IA, Filipenko ML (2022) Multiplex droplet digital PCR assay for detection of MET and HER2 genes amplification in non-small cell lung cancer. Cancers (Basel) 14(6). https://doi.org/10.3390/cancers14061458
8. Rowlands V, Rutkowski AJ, Meuser E, Carr TH, Harrington EA, Barrett JC (2019) Optimisation of robust singleplex and multiplex droplet digital PCR assays for high confidence mutation detection in circulating tumour DNA. Sci Rep 9(1):12620. https://doi.org/10.1038/s41598-019-49043-x
9. Zhu X, Liu P, Lu L, Zhong H, Xu M, Jia R, Su L, Cao L, Sun Y, Guo M, Sun J, Xu J (2022) Development of a multiplex droplet digital PCR assay for detection of enterovirus, parechovirus, herpes simplex virus 1 and 2 simultaneously for diagnosis of viral CNS infections. Virol J 19(1):70. https://doi.org/10.1186/s12985-022-01798-y
10. Huang Q, Hu Q, Li Q (2007) Identification of 8 foodborne pathogens by multicolor combinational probe coding technology in a single real-time PCR. Clin Chem 53(10): 1741–1748. https://doi.org/10.1373/clinchem.2007.087502
11. Huang Q, Zheng L, Zhu Y, Zhang J, Wen H, Huang J, Niu J, Zhao X, Li Q (2011) Multicolor combinatorial probe coding for real-time PCR. PLoS One 6(1):e16033. https://doi.org/10.1371/journal.pone.0016033
12. Liao Y, Wang X, Sha C, Xia Z, Huang Q, Li Q (2013) Combination of fluorescence color and melting temperature as a two-dimensional label for homogeneous multiplex PCR detection. Nucleic Acids Res 41(7):e76. https://doi.org/10.1093/nar/gkt004
13. Marras SAE, Tyagi S, Antson DO, Kramer FR (2019) Color-coded molecular beacons for multiplex PCR screening assays. PLoS One 14(3):e0213906. https://doi.org/10.1371/journal.pone.0213906
14. Santos-Barriopedro Irene US, Fradet Etienne, Dangla Remi (2023) Robust higher-order multiplexing in digital PCR by color-combination. bioRxiv. https://doi.org/10.1101/2023.05.10.540190
15. Arnold M, Morgan E, Rumgay H, Mafra A, Singh D, Laversanne M, Vignat J, Gralow JR, Cardoso F, Siesling S, Soerjomataram I (2022) Current and future burden of breast cancer: global statistics for 2020 and 2040. Breast 66: 15–23. https://doi.org/10.1016/j.breast.2022.08.010
16. Lei JT, Anurag M, Haricharan S, Gou X, Ellis MJ (2019) Endocrine therapy resistance: new insights. Breast 48 Suppl 1(Suppl 1):S26–S30. https://doi.org/10.1016/S0960-9776(19)31118-X
17. Brett JO, Spring LM, Bardia A, Wander SA (2021) ESR1 mutation as an emerging clinical biomarker in metastatic hormone receptor-positive breast cancer. Breast Cancer Res 23(1):85. https://doi.org/10.1186/s13058-021-01462-3
18. Bidard FC, Hardy-Bessard AC, Dalenc F, Bachelot T, Pierga JY, de la Motte Rouge T, Sabatier R, Dubot C, Frenel JS, Ferrero JM, Ladoire S, Levy C, Mouret-Reynier MA, Lortholary A, Grenier J, Chakiba C, Stefani L, Plaza JE, Clatot F, Teixeira L, D'Hondt V, Vegas H, Derbel O, Garnier-Tixidre C, Canon JL, Pistilli B, Andre F, Arnould L, Pradines A, Bieche I, Callens C, Lemonnier J, Berger F, Delaloge S, Investigators P (2022) Switch to fulvestrant and palbociclib versus no switch in advanced breast cancer with rising ESR1 mutation during aromatase inhibitor and palbociclib therapy (PADA-1): a randomised, open-label, multicentre, phase 3 trial. Lancet Oncol 23(11):1367–1377. https://doi.org/10.1016/S1470-2045(22)00555-1
19. Stilla Technologies' technical note. How to characterize the Limit of Blank and the Limit of Detection in Crystal Digital PCR™. https://www.stillatechnologies.com/digital-pcr/naica-system-analysis/how-to-characterize-the-limit-of-blank-and-the-limit-of-detection-in-crystal-digital-pcr
20. Stilla Technologies' technical note. How to generate a fluorescence compensation matrix for multiplex assays on the naica® system. https://www.stillatechnologies.com/digital-pcr/naica-system-analysis/1237/

Chapter 8

Digital PCR Analysis of Methylated Circulating Tumor DNA

Rikke Fredslund Andersen, Tina Brandt Christensen, and Torben Frøstrup Hansen

Abstract

Methods for analyzing circulating tumor DNA (ctDNA) in a fast, quantitative, and reproducible way may help clinical implementation. Analyses of ctDNA can be performed without prior analysis of tumor tissue from patients by using tumor-specific methylations as targets. Digital PCR measurements are quantitative and enable the possibility to monitor dynamics of ctDNA during treatment, which has been shown to correlate with patient outcome. The protocol describes methylation analysis of plasma samples where results are available in a timely manner providing the opportunity to adjust treatment of cancer patients.

Key words Methylation, Cancer, Circulating tumor DNA, Plasma, Liquid biopsy, Droplet digital PCR

1 Introduction

Analyses of circulating tumor DNA (ctDNA) are approaching clinical use. Many applications of ctDNA are being tested for clinical utility—including early detection of cancer, monitoring treatment response, and detection of minimal residual disease [1]. Digital Polymerase Chain Reaction (dPCR) is a powerful technique for analyzing ctDNA, providing high sensitivity and precision [2]. The turnaround time of analyses is crucial in a clinical setting because of demand for early interventions and treatment adjustments. In relation to this, a tumor-agnostic approach is desirable. A tumor-informed approach depends on initial analysis of tissue samples, which in some cases are difficult to obtain and may delay the process.

Tumor-specific methylations are early events in cancer development and are often conserved through tumor evolution [3]. It is possible to identify methylations that are present in a very high proportion of tumors and not present in non-cancer tissue. Hence,

Guillaume Pavlovic (ed.), *Digital PCR: Methods and Protocols*, Methods in Molecular Biology, vol. 2969,
https://doi.org/10.1007/978-1-0716-4767-7_8,

tumor-specific methylations can be ideal targets for monitoring disease development in liquid biopsies in a tumor-agnostic way.

HOXA9 is a member of the homeobox gene family and plays a role in embryonic development and differentiation [4]. Aberrant methylation of *HOXA9* has been shown to contribute to the initiation and progression of various diseases, particularly cancer [4]. We have described the presence of *HOXA9* hyper-methylation in high numbers of tumors from ovarian cancer (93%) [5], breast cancer (80%) [6], lung cancer (100%) [7], and bile duct cancer (100%, data not published). This makes hyper-methylation of *HOXA9* a broadly applicable marker that can be used in a tumor-agnostic way for analyzing plasma samples from patients with these types of cancer.

We have investigated hyper-methylated *HOXA9* in plasma from patients with metastatic lung cancer [8], breast cancer [6], ovarian cancer [5, 9], bile duct cancer [10], and pancreatic cancer [11] and have shown that the presence of hyper-methylated *HOXA9* in plasma holds diagnostic and prognostic significance. Hyper-methylated *HOXA9* is found in plasma samples from a high number of patients with metastatic disease and it has been shown that quantitative dynamics of circulating hyper-methylated *HOXA9* is correlated to outcome [12].

The protocol describes droplet digital PCR analysis of hyper-methylated *HOXA9* in plasma samples but is applicable to analysis of other methylation markers. It is developed to enable analysis of circulating cell-free DNA (cfDNA) from 4 mL of plasma. This is a scalable volume but to obtain a high sensitivity of the analysis it is necessary to analyze a large volume of plasma. In a metastatic setting, a lower amount of plasma may be sufficient. A small amount of cfDNA from each sample is tested for quality control parameters to ensure high quality of the DNA prior to ctDNA analysis. This includes determining the concentration of cfDNA in the sample by quantifying two different fragment lengths (amplicon lengths 65 and 250 base pairs) [13]. Increased content of long fragments indicates possible lysis from blood cells during pre-analytical handling. Specific lysis of B-cells is measured with an assay targeting the rearranged immunoglobulin heavy chain gene [14]. A positive signal is an indication of general lysis of blood cells and hence poor quality of the sample. Finally, the protocol describes addition of an exogenous spike-in control to plasma prior to DNA extraction [14]. The level of spike-in is measured after extraction, and if the level is lower than expected, it indicates that the DNA extraction has been suboptimal.

The described method is a simple approach to measuring ctDNA in a variety of cancers. The method is fast and may be performed within 2 working days, which is a relevant timeframe for clinical use. The same ctDNA marker is used for all patients, which makes the laboratory setup simple. Results from dPCR analyses are quantitative and may be used to monitor dynamics of ctDNA as a surrogate for disease evolvement. The results do not provide genetic information from patients and cannot be used to select specific targeted treatments. The dPCR analysis can be supplemented by next-generation sequencing analyses to obtain broader genetic information. Analyzing tumor-specific methylations by dPCR is a way to measure the general level of ctDNA as an additional tool for making treatment decisions in clinical oncology.

2 Materials

2.1 Cell-Free DNA Extraction

1. EDTA lavender-capped vacutainer tubes (Becton and Dickinson #367525_10ml) (*see* **Note 1**).
2. QIAsymphony SP/AS purification system (Qiagen #9001301).
3. QIAsymphony DSP Circulating DNA kit (Qiagen #937556).
4. 15 mL Polypropylene conical-bottom centrifuge tube (VWR #525–1085).
5. Benchtop Centrifuge 5804R (Eppendorf #5805000010).
6. 14 mL Polystyrene round-bottom tube (Falcon #352051).
7. 1.5 mL Polypropylene Screw cap micro tube (Sarstedt #72.692.005).
8. Sample Prep Cartridges, 8-well cartridges (Qiagen #997002).
9. 8-Rod Covers (Qiagen #997004).
10. Filter-Tips, 200 and 1500 μL (Qiagen #990332, #997024).
11. Water nuclease-free molecular biology reagent.
12. Benchtop Vortex mixer.
13. Water bath.
14. *CPP1* exogenous internal control DNA (made in house, concentration 1800 copies/μl in TE buffer [14]).

2.2 Bisulfite Conversion

1. Amicon Ultra-0.5 Centrifugal Filter Unit (Millipore #UFC5030BK).
2. EZ DNA Methylation lightning kit (Zymo Research #D5031).
3. MicroAmp 8-tube strip 0.2 mL (Applied Biosystems #N8010580).
4. MicroAmp 8-cap strip (Applied Biosystems #N8010535).
5. SimpliAmp Thermal Cycler (Thermo Fisher Scientific #A24811).
6. Micro Star 17 centrifuge (VWR #521–1646).

7. 1.5 mL safe-lock tubes (Eppendorf #0030 120.086).
8. Human Methylated and Non-methylated DNA Set (Zymo Research #D5014).

2.3 Droplet Digital PCR

1. Automated Droplet Generator (Bio-Rad #1864101).
2. QX200 Droplet Reader (Bio-Rad #1864003).
3. Safetyspace DNAse and RNAse-free Filter tips (Sartorius #790101F).
4. DG32 Automated Droplet Generator Cartridges (Bio-Rad #1864108).
5. Automated Droplet Generation Oil for Probes (Bio-Rad #1864110).
6. ddPCR Supermix for probes (No dUTP) (Bio-Rad #1863025).
7. Twin.tec semi-skirted PCR plate (Eppendorf, VWR #391–2051).
8. Pierceable Foil Heat Seal (Bio-Rad #1814040).
9. ddPCR Droplet Reader Oil (Bio-Rad #1863004).
10. ddPCR Buffer Control Kit for Probes (Bio-Rad #1863052).
11. Bio-Rad PX1 plate sealer (Bio-Rad #1814000).
12. Veriti Pro Thermal cycler (Thermo Fisher Scientific #A48141).
13. 10× primer/probe mix, *EMC7_250*/*EMC7_65*, PBC/*CPP1* and *HOXA9*/Albumin.

3 Methods

3.1 Plasma Isolation

1. Draw venous blood into a 10 mL EDTA lavender-capped vacutainer tube and immediately invert the tube ten times.
2. Centrifuge the tube at 2000 g for 10 min in a swing bucket centrifuge.
3. Pipette the plasma layer into a sterile 15 mL Polypropylene conical-bottom centrifuge tube.
4. Centrifuge the plasma at 10,000 g for 10 min.
5. The processing should be performed within 4 h of sampling. Continue with cfDNA extraction or freeze the plasma at −20 °C until analysis (*see* **Note 2**).

3.2 cfDNA Extraction

All buffers and magnetic beads are provided in the QIAsymphony DSP Circulating DNA kit (*see* **Note 3**). Other extraction methods may also be used (*see* **Note 4**). Use gloves when handling samples and consumables.

1. Load the QIAsymphony with 200 μL and 1500 μL tips, 8 rod-covers, Sample Prep Cartridges, Reagent box, elution tubes (1.5 mL Sarsted polypropylene Screw cap micro tube) and scan.
2. Proteinase K is transferred to a 14 mL Falcon Polystyrene round-bottom tube and loaded to the instrument.
3. If the plasma samples are frozen, thaw for 30 min in a 30 °C water bath.
4. Transfer 4.1 mL plasma to a 14 mL Falcon Polystyrene round-bottom tube.
5. Add 20 μL *CPP1* exogenous internal control DNA to each tube.
6. Load samples on to the QIAsymphony and start the protocol.
7. Elution volume is 60 μL (*see* **Note 5**).
8. After extraction, add 140 μL nuclease-free water to each sample.
9. Pipet 30 μL DNA into a micro centrifuge tube. This DNA is not bisulfite converted but is used for quality control analysis.

3.3 Concentration of Samples and Bisulfite Conversion

1. Place an Amicon Ultra-0.5 Centrifugal Filter Unit in a micro centrifuge tube.
2. Transfer the remaining 170 μL DNA to an Amicon Ultra-0.5 Centrifugal Filter Unit.
3. Centrifuge at 14,000 g for 30 min at room temperature.
4. Place the filter upside down in a clean microcentrifuge tube.
5. Centrifuge at 1000 g for 2 min.
6. Discard the Amicon Ultra-0.5 Centrifugal Filter Unit.
7. Adjust the volume to 20 μL with nuclease-free water, molecular biology reagent.
8. Transfer 20 μL DNA to a MicroAmp 8-tube strip for bisulfite conversion with the EZ DNA methylation-Lightning kit (*see* **Note 6**).
9. Add 130 μL lightning conversion reagent to each sample and cap the strip.
10. Place in a thermal cycler.
11. Run under following conditions:
    (a) 98 °C for 8 min.
    (b) 54 °C for 60 min.
    (c) Hold at 4 °C for up to 20 h.

12. For clean-up follow the instructions provided in the EZ DNA methylation-Lightning kit handbook. All centrifugation steps are at 14,000 g for 1 min except the second wash step where centrifugation is for 2 min (*see* **Note 7**).
13. Add 15 μL elution buffer and centrifuge at 10,000 g for 1 min.

3.4 Droplet Digital PCR

Prepare ddPCR mastermix in a PCR hood in a separate Pre-PCR room. Wipe down PCR hood with DNA away before and after work in the hood to minimize the risk of contamination.

1. Prepare a mastermix in an Eppendorf 1.5 mL safe-lock tube using primers and probes described in Table 1 and conditions from Table 2. When preparing mastermix for multiple samples, prepare extra mixture to compensate for loss in pipetting.

Table 1
Information on primers and probes

Gene		Sequence	Modification	10× conc. (μM)
*EMC7*_250bp	Forward	AAGTACTACTGAGTATGATGTT		3
	Reverse	CTAGATTTGCCAGATGATTTT		3
	Probe	AGTTGCCTGATGTTTCTGAG TTCAT	5' FAM/ZEN/3' IBFQ	2
*EMC7*_65bp	Forward	CTTTCCCCATGTTGCTTTAT		3
	Reverse	CTGACAACCTCTGATGTTTT		3
	Probe	CAGAGCAAGATATGTGAATTACA TCAA	5' HEX/ZEN/3' IBFQ	2
IGH	Forward	ATCTGCAAATGAACAGYCTGAGA		9
(PBC assay)	Reverse	CTTACCTGAGGAGACGGTGAC		9
	Probe	CYGAGGACACRGCTGTGTATTAC TGTGC	5' FAM/ZEN/3' IBFQ	4
CPP1	Forward	CCATGGATGTATTCGCCAGTTAC		3
(Glycine max)	Reverse	TAAATATTGTGCTTCACCTACTC TAGTG		3
	Probe	TTGGCGTAGTTCTCCCGC TTACCCCG	5' HEX/3' BHQ-1	2
HOXA9	Forward	GAGTATTTCGATTTTAGTTCGTG T		2
	Reverse	CGCGTACACTAAATTCCAC		2
	Probe	TTAGTTTAAGGCGACGGTGTT	5' FAM/3' BHQ-1	2
ALB	Forward	GGGATGGAAAGAATTTTATGTT		4
(Albumin)	Reverse	AAACAAACTAACCCCAAATTCT		4
	Probe	AGGGTTTTTATAATTTA	5' VIC/MGB/3' NFQ	4

Table 2
Mastermix preparation

	Volume per sample
2× ddPCR Supermix for probes (no dUTP)	24 μL
10× primer/probe mix	4.8 μL
Nuclease-free water, molecular biology reagent	7.2 μL
Total	**36 μL**

2. Mix tube with vortexer.
3. Pipet 36 μL mastermix per sample into MicroAmp 8-tube strips.
4. Add 12 μL sample (total volume in strip 48 μL).
5. Cap the strip, vortex 5*2 s and centrifuge briefly.
6. Transfer 2*22 μL into an Eppendorf twin.tec semi-skirted PCR plate using an 8-channel pipette.
7. Seal the plate with a Pierceable Foil Heat Seal for 5 s at 180 °C.
8. Load sample plate, droplet plate, Safetyspace 120 μL Filter tips, and DG32 Automated Droplet Generator Cartridges onto the Automated droplet generator (*see* **Note 8**).
9. Define your setup on the touch screen and press start.
10. After droplets are made, seal the droplet plate with a Pierceable Foil Heat Seal for 5 s at 180 °C.
11. Place the sealed plate in a PCR thermal cycler.
12. Run under following conditions (*see* **Note 9**):
    (a) 95 °C for 10 min.
    (b) 95 °C for 15 s.
    (c) 56 °C for 1 min.
    (d) Repeat b and c for 44 cycles.
    (e) 98 °C for 10 min.
    (f) Hold at 4 °C.
13. Define plate layout in QX Manager Software 2.0.
14. Transfer the sealed plate to QX200 droplet Reader and read the plate (*see* **Note 10**).
15. Data analysis is done with QX Manager Software 2.0. QC parameters for each sample and results from positive and negative controls need to be evaluated and accepted before results from patient samples can be approved. Results from positive and negative samples are shown in Fig. 1 (*see* **Note 11**).

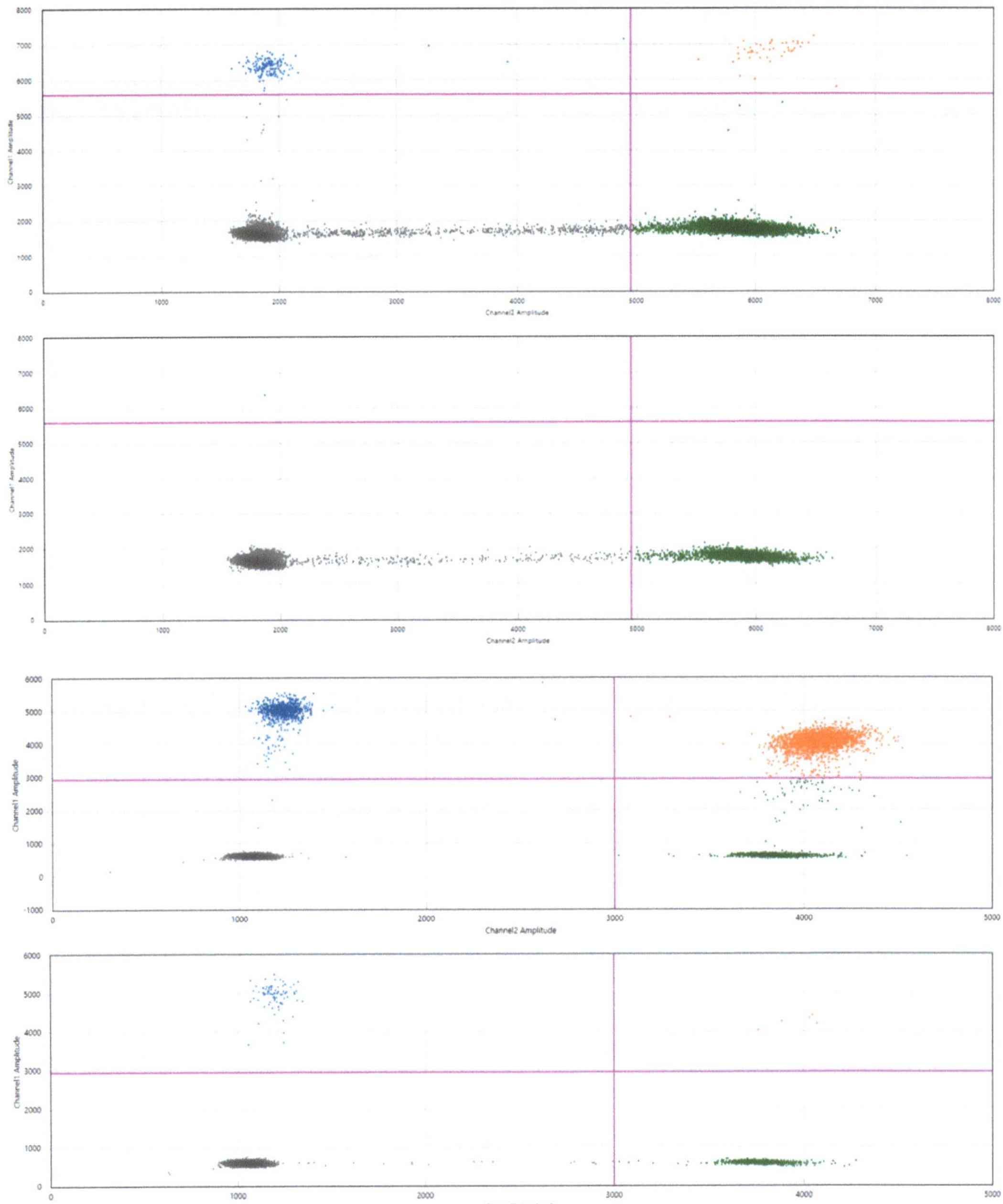

Fig. 1 (**a+b**) ddPCR data from analysis of hyper-methylated *HOXA9* in two plasma samples from ovarian cancer patients. Hyper-methylated *HOXA9* is measured with a FAM-probe (y-axis) and Albumin with a VIC-probe (x-axis) (a) Sample positive for hyper-methylation of *HOXA9* (b) Sample negative for hyper-methylation of *HOXA9*. (**c+d**) ddPCR data from analysis of *EMC7_250*/*EMC7_65*. *EMC7_250* is measured with a FAM-probe (y-axis) and *EMC7_65* with a HEX-probe (x-axis) (c) Genomic DNA sample with a ratio between long and short fragments of DNA equal to 0.95. The majority of extracted genomic DNA fragments are longer than 250 bp and the ratio between long and short fragments is close to 1 (d) cfDNA sample with a ratio of 0.14. The majority of cfDNA fragments are shorter than 250 bp and the ratio between long and short fragments is low. (**e+f**) ddPCR data from analysis of PBC/*CPP1*. PBC is measured with a FAM probe (y-axis) and *CPP1* with a HEX-probe (x-axis) (e) Genomic DNA sample with 1.3% PBC and 800 copies of *CPP1* (f) Plasma sample without PB contamination and with 1208 copies of CPP1 in the analysis. This equals an extraction efficiency of 67%

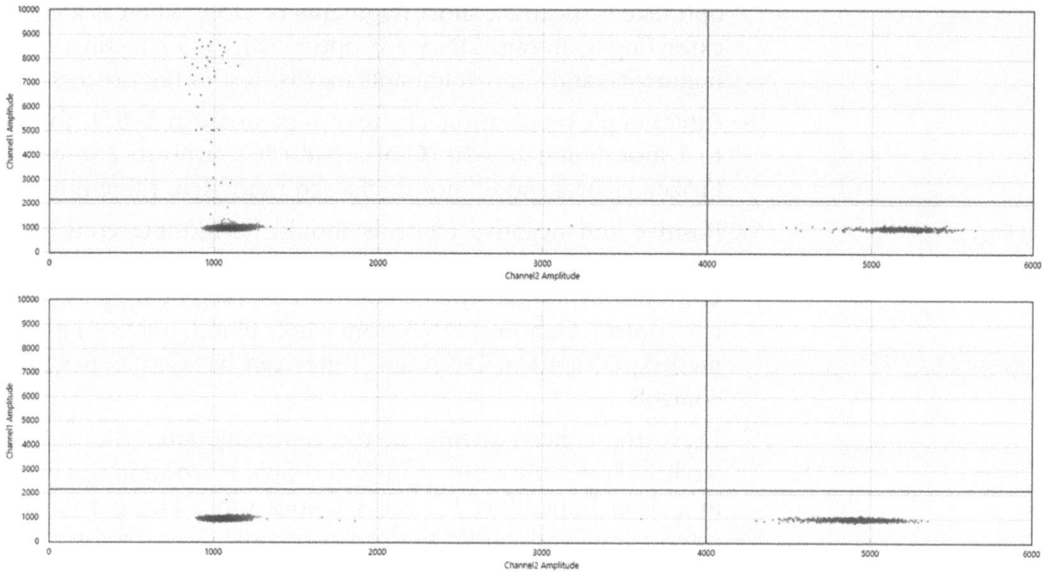

Fig. 1 (continued)

16. Results are reported as [methylated *HOXA9* copies]/[Albumin copies]*100 = % methylated *HOXA9*. Results may also be reported as copies of methylated *HOXA9* per mL plasma.

4 Notes

1. If local processing is not possible, stabilizing tubes such as Streck Cell-Free DNA BCT tubes (Streck #230470) or Roche Cell-Free DNA Collection Tubes (Roche Diagnostics #07785674001) can be used. These preserve samples for at least 7 days by preventing cell lysis.
2. It is important not to disturb the cellular layer when pipetting the plasma. Contamination of the plasma fraction with leukocytes will lead to increased levels of DNA and an incorrect estimate of the concentration of cfDNA and percentage of ctDNA. If plasma is frozen it is important to avoid repeated freeze/thaw cycles. Multiple freeze/thaw cycles can lead to degradation of DNA.
3. Refer to QIAsymphony® DSP Circulating DNA Kit Instructions for Use (Handbook) and QIAsymphony® DSP Circulating DNA Kit Instructions for Use (Protocol Sheet) for additional information and troubleshooting.
4. Automated or manual extraction of DNA can be performed depending on the available equipment. It is important to use an extraction kit intended for cfDNA extraction. These kits are

optimized to extract short fragments of DNA whereas kits for extracting genomic DNA are optimized for extracting long fragments, and short fragments may be lost in the process.

5. After sample preparation, eluates can be stored at 2–8 °C for up to 1 month and at −20 °C or at −80 °C for up to 2 months. Frozen eluates should not be thawed more than three times.
6. Positive and negative controls should be bisulfite converted together with samples to monitor the conversion procedure. Universal hyper-methylated control DNA can be used as positive control. Genomic DNA from whole blood, universal hypomethylated control DNA and water can be used as negative controls.
7. Leave the collection tube in the centrifuge after the second wash and carefully remove the Zymo-Spin IC column and place in a clean Eppendorf 1.5 mL safe-lock tube. This is to avoid carryover of wash buffer to the elution tube. Any carryover will result in inhibition of the PCR amplification.
8. It is important to use tips with no residual micro fragments of plastic. The channels in the DG32 cartridges may clog, which will result in low droplet count.
9. Thermocycling ramp rate should be adjusted to 2 °C/s for all steps.
10. It has been observed that more droplets are accepted if the plate is left for a few hours at 4 °C between PCR amplification and droplet reading. It is often convenient to leave the plate overnight in the thermocycler.
11. Samples from healthy controls or controls with relevant comorbidities can be analyzed in order to determine the cutoff for positivity of patient samples. The cutoff for hyper-methylated *HOXA9* was set at >4 positive droplets per analysis in a study of 100 self-reported healthy controls [9].

References

1. Nikanjam M, Kato S, Kurzrock R (2022) Liquid biopsy: current technology and clinical applications. J Hematol Oncol 15(1):131. https://doi.org/10.1186/s13045-022-01351-y
2. Whale AS, Devonshire AS, Karlin-Neumann G, Regan J, Javier L, Cowen S, Fernandez-Gonzalez A, Jones GM, Redshaw N, Beck J, Berger AW, Combaret V, Dahl Kjersgaard N, Davis L, Fina F, Forshew T, Fredslund Andersen R, Galbiati S, Gonzalez Hernandez A, Haynes CA, Janku F, Lacave R, Lee J, Mistry V, Pender A, Pradines A, Proudhon C, Saal LH, Stieglitz E, Ulrich B, Foy CA, Parkes H, Tzonev S, Huggett JF (2017) International Interlaboratory digital PCR study demonstrating high reproducibility for the measurement of a rare sequence variant. Anal Chem 89(3):1724–1733. https://doi.org/10.1021/acs.analchem.6b03980
3. Feinberg AP, Ohlsson R, Henikoff S (2006) The epigenetic progenitor origin of human cancer. Nat Rev Genet 7(1):21–33. https://doi.org/10.1038/nrg1748
4. Shah N, Sukumar S (2010) The Hox genes and their roles in oncogenesis. Nat Rev Cancer 10(5):361–371. https://doi.org/10.1038/nrc2826

5. Faaborg L, Jakobsen A, Waldstrom M, Petersen CB, Andersen RF, Steffensen KD (2021) HOXA9-methylated DNA as a diagnostic biomarker of ovarian malignancy. Biomark Med 15(15):1309–1317. https://doi.org/10.2217/bmm-2021-0144
6. Hansen T, Bruun SBB, Andersen RF, Madsen JS, Tabor TP, Bechmann T, Kjær IM (2023) Circulating methylated-HOXA9 tumor DNA as a biomarker for mortality in recurrent breast cancer. J Clin Oncol 41(16_suppl):e13106. https://doi.org/10.1200/JCO.2023.41.16_suppl.e13106
7. Wen SWC, Andersen RF, Petersen LMS, Hager H, Hilberg O, Jakobsen A, Hansen TF (2020) Comparison of mutated KRAS and methylated HOXA9 tumor-specific DNA in advanced lung adenocarcinoma. Cancer 12(12). https://doi.org/10.3390/cancers12123728
8. Wen SWC, Andersen RF, Hansen TF, Nyhus CH, Hager H, Hilberg O, Jakobsen A (2021) The prognostic impact of circulating homeobox A9 methylated DNA in advanced non-small cell lung cancer. Transl Lung Cancer Res 10(2):855–865. https://doi.org/10.21037/tlcr-20-826
9. Faaborg L, Andersen RF, Waldstrom M, Henriksen JR, Adimi P, Jakobsen A, Steffensen KD (2022) Prognostic impact of circulating methylated Homeobox A9 DNA in patients undergoing treatment for recurrent ovarian cancer. Cancer 14(7). https://doi.org/10.3390/cancers14071766
10. Andersen LB, Mahler MSK, Andersen RF, Jensen LH, Raunkilde L (2022) The clinical impact of methylated Homeobox A9 ctDNA in patients with non-resectable biliary tract cancer treated with Erlotinib and Bevacizumab. Cancer 14(19). https://doi.org/10.3390/cancers14194598
11. Spindler KLG, Lindgaard SC, Johansen JS, Jensen LH, Hansen TF, Steiniche MM, Andersen RF (2024) 383P ctDNA response evaluation criteria in metastatic pancreatic cancer using aberrant methylation test. Ann Oncol 35:S154. https://doi.org/10.1016/j.annonc.2024.05.524
12. Faaborg L, Andersen RF, Wen SW, Thomsen CB, Raunkilde L, Hansen TF, Jensen LH, Steffensen KD, Jakobsen A (2023) Prognostic impact of early ctDNA dynamics during chemotherapy of metastatic cancer. Future Oncol 19(35):2361–2367. https://doi.org/10.2217/fon-2023-0111
13. Lefevre AC, Pallisgaard N, Kronborg C, Wind KL, Krag SRP, Spindler KG (2021) The clinical value of measuring circulating HPV DNA during chemo-radiotherapy in squamous cell carcinoma of the anus. Cancer 13(10). https://doi.org/10.3390/cancers13102451
14. Pallisgaard N, Spindler KL, Andersen RF, Brandslund I, Jakobsen A (2015) Controls to validate plasma samples for cell free DNA quantification. Clin Chim Acta 446:141–146. https://doi.org/10.1016/j.cca.2015.04.015

Chapter 9

Method for Validation Multiplexed Droplet Digital PCR for the Detection of Copy Number Variants in Cultured Human Pluripotent Stem Cells

Said Assou, Reda Zenagui, Juline Vincent, Laurent Manchon, and John De Vos

Abstract

Human-induced pluripotent stem cells (hiPSCs) have broad applications in basic research and cell-based regenerative medicine. Maintaining a stable genome in culture is crucial for the clinical use of hiPSCs. Indeed, their continuous propagation drives genetic instability, particularly copy number variants (CNV). Among the different methods available to identify gene alterations, digital PCR (dPCR) is a precise, rapid, and cost-effective tool for the sensitive and absolute quantification of CNV. Unlike conventional methods limited to two targets, multiplex dPCR assays allow the simultaneous detection of four targets. In this method paper, we describe strategies for multiplex dPCR assays to detect major CNVs using only small amounts of DNA extracted from hiPSCs. We confirmed this approach applicability for hiPSCs through validation following the International Council for Harmonization (ICH) guidelines that cover different aspects, such as specificity, precision, robustness, and limit of detection. The simultaneous measurement of four targets was possible through the generation of distinct fluorescence endpoints for each target by adjusting the concentrations of primers and/or probes. Compared to the classical dPCR (the "gold standard"), these multiplexed dPCR assays displayed 100% specificity and sensitivity for the simultaneous detection of CNVs. This study highlights the key role of multiplexing for monitoring genetic alterations in hiPSCs and for setting the standards required for the process safety evaluation of hiPSC-based stem cell therapies.

Keywords Multiplexed digital PCR, Human pluripotent stem cells, Genetic abnormalities, G-banding, Sequencing

1 Introduction

The inherent capacity of human-induced pluripotent stem cells (hiPSCs) to renew and to differentiate into various cell types holds immense potential for applications in the field of regenerative medicine, disease modeling, and drug discovery [1, 2]. The successful and safe use of hiPSCs requires the maintenance of a substantial reservoir of undifferentiated, genetically stable cell

Guillaume Pavlovic (ed.), *Digital PCR: Methods and Protocols*, Methods in Molecular Biology, vol. 2969,
https://doi.org/10.1007/978-1-0716-4767-7_9,

populations. However, hiPSCs are susceptible to genetic modifications following cell reprogramming or in vitro culture [3, 4]. During cell culture, exposure to selective pressures might lead to the selection of cell variants with favorable genetic changes that confer a growth advantage [5, 6]. These genetic alterations frequently involve gains of specific segments or entire chromosomes, such as gains of chromosome 1, 12, 17, 20, and X [7, 8] or loss of 18q [9, 10]. Copy number variants (CNVs) are the most frequent genomic abnormalities detected in hiPSCs. These alterations can reflect selective pressures that promote the proliferation and survival of hiPSCs, or hinder their differentiation capacities [4]. Keller et al. investigated the effect on differentiation of 12p and 20q gains, which are among the most frequent CNVs in hiPSCs. They found that these mutations led to a general decrease of the differentiation potential, particularly in cells with a gain of 12p13.31 [11]. Similarly, Markouli et al. demonstrated that gains of 20q11.21 resulted in the nearly complete loss of the capacity to differentiate into neurectoderm due to altered TGF-β and SMAD signaling pathways [12]. Single nucleotide variants (SNVs) also can affect hiPSC differentiation capacities. Cells harboring SNVs can gradually dominate the culture and increase the risk of cancer after transplantation in patients. However, SNVs often remain undetected without suitable assays [13, 14]. The reasons behind the frequent occurrence of genomic abnormalities in hiPSCs are not fully understood, but are typically attributed to suboptimal culture methods, higher susceptibility to culture stress, and prolonged culture compared with somatic cells.

Traditionally, screening for major chromosomal modifications in hiPSC lines is done by karyotyping through chromosome banding of metaphase cells. This method offers valuable insights into numerical and structural changes. However, it cannot be regularly performed, and it is not sensitive enough to detect small genetic alterations [15]. Optical genome mapping (OGM), demonstrated by the Bionano Genomics platform [16], delivers exceptional resolution for detecting structural variants across the genome, far exceeding the capabilities of traditional karyotyping. However, its high cost limits its routine use. Due to the limited resolution of karyotyping (about 5–10 Mb), and the cost challenges associated with OGM, more cost-effective and rapid assays are currently favored. These techniques are often combined to enable more frequent and efficient screening. Specifically, digital PCR (dPCR) provides higher resolution than G-banding, is significantly more cost-effective than OGM, and can identify alterations that cannot be detected by karyotyping, such as sub-karyotypic gene amplifications at the 20q11.21 locus that are the most prevalent genetic alterations in hiPSCs [17]. Through the amplification of targeted genomic fragments in many individual reactions, dPCR facilitates the accurate quantification and detection of nucleic acids

[18, 19]. It is also a rapid testing method, well suited for frequent monitoring. In droplet dPCR (ddPCR), the target DNA is quantified by partitioning the DNA sample into thousands of nanodroplets, each functioning as a miniature PCR reaction [20]. This allows the absolute quantification of target nucleic acids in a sample, without the need of standard curves and calibrators, thus overcoming some of the limitations of real-time quantitative PCR [21]. This approach offers higher sensitivity and precision compared with traditional PCR techniques, two features that are particularly interesting for CNV detection. Additionally, ddPCR can be multiplexed to simultaneously detect different fluorescent signals in different channels. In an advanced form, distinct targets can be assessed by adjusting the concentration of different probes labeled with the same fluorophore. This allows combining different fluorophores and different probe concentrations to detect up to four targets in a single reaction [22, 23].

Here, we describe a highly sensitive multiplex ddPCR assay for the simultaneous detection of frequently occurring genetic alterations (including CNVs) in hiPSCs. Our findings indicate that multiplex ddPCR is efficient for monitoring the emergence of genetic variations. However, this method does not consistently identify variants if they are present in <20% of cells.

Assessing the stability of hiPSCs in culture necessitates employing a variety of analytical methods that complement each other to thoroughly examine the whole genome. Furthermore, to validate these findings, we suggest employing droplet digital PCR (ddPCR) due to its high precision and reliability. Multiplexing enhances the sensitivity for detecting different targets in small-size samples, leading to cost savings and higher sample throughput. Therefore, multiplex ddPCR is a highly efficient approach. Here, we demonstrated the feasibility of multiplexing using just two fluorescent channels, showcasing its simplicity and versatility.

2 Materials

2.1 HiPSC Culture and Characterization

1. HiPSC Lines: UHOMi002-A, UHOMi003-A, Additional, iPS1–5, iPS9, iPS11, iPS14, iPS25, iPS31.
2. Reprogramming Kit: CytoTune®-iPS 2.0 Sendai Reprogramming Kit (Thermo Fisher Scientific).
3. Culture Matrix: Geltrex matrix (Thermo Fisher Scientific).
4. Medium: Essential 8 Medium (Thermo Fisher Scientific).
5. Immunofluorescence Analysis:
   - Fixative: 4% paraformaldehyde.
   - Blocking Solution: 5% donkey serum (Chemicon International) in PBS.

- Primary Antibody: Anti-OCT3/4 (POU5F1) antibody (sc-9081, Santa Cruz Biotechnology, Santa Cruz, CA).
- Secondary Antibody: Alexa 488 secondary IgG (Jackson ImmunoResearch Lab Inc., CA).

2.2 scRNA-Seq

1. Trypsin: Used for cell dissociation.
2. Chromium Controller and Reagent Kit: Chromium Single Cell 3′ Reagent kit V3.1 (10× Genomics).
3. Sequencer: Illumina NextSeq 550 (Transcriptome Core Facility of the Institute in Regenerative Medicine and Biotherapy, Montpellier, France).
4. Software: c-loop, t-distributed Stochastic Neighbor-Embedding (tSNE).

2.3 Nucleic Acid Purification and Quantification

1. DNA Extraction Kit: QIAmp DNA Mini Blood Kit (Qiagen).
2. Quantification Kit: QuBit dsDNA BR Assay Kit (Life Technologies).
3. Fluorometer: Qubit 3.0.

2.4 ddPCR Multiplexing

1. TaqMan Assays: Labeled with FAM or HEX fluorophores.
2. ddPCR Supermix: For Probes (No dUTP) 2X (Bio-Rad).
3. Restriction Enzyme: Added to improve CNV calculation accuracy.
4. ddPCR Platform: QX200 (Bio-Rad).
5. Mosaicism Test Materials: DNA extracted from male and female cell lines, mixed at specific ratios.

2.5 Data Analysis

1. Software: BioRad QX Manager Standard Edition, version 2.1.0.

2.6 Karyotyping

1. Dissociation Reagent: Versene solution.
2. Microscope and Software: For G-banded metaphases analysis (Metasystems).
3. Service Provider: Stem Genomics, Montpellier.
4. The KaryoStat™: Life Technologies Service (Applied Biosystems).

2.7 Next-Generation Sequencing Variant Calling

1. Library Preparation Kits: Twist (Twist Bio-science).
2. Sequencing Instrument: Illumina NovaSeq.
3. Software for Bioinformatic Analysis:
   - Fastp: v0.21.0.
   - FastQC: v0.11.7.

- MultiQC: v1.7.
- Bwa-mem2: v2–2.0.
- Samblaster: v0.1.24.
- Samtools: v1.11.
- GATK: v4.2.0.0.
- CNVkit: v0.9.10.
- PureCN: v3.17.

2.8 Statistical Analysis

1. Software for Statistical Analysis: Tools and software used for Poisson statistics, Student's t-test, and Spearman correlation.

3 Methods

3.1 Characterization of Culture-Expanded hiPSCs

1. Peripheral blood mononuclear cells (PBMCs) were isolated from a healthy male and from patients with chronic obstructive pulmonary disease (COPD) to establish the UHOMi002-A and UHOMi003-A lines, respectively [24, 25]. Isolation was performed using standard Ficoll gradient centrifugation with Ficoll-Paque PLUS (GE Healthcare), followed by thorough washing with PBS containing 10% fetal bovine serum (FBS). The PBMCs were cultured in StemSpan medium supplemented with Erythroid Expansion Medium (SSEM, Stemcell Technologies), following the manufacturer's instructions. Reprogramming was achieved by introducing non-integrating Sendai virus (SeV) vectors expressing the transcription factors OCT4, SOX2, KLF4, and cMYC into the erythroid progenitor cells. Following successful reprogramming and expansion in maintenance E8 medium on a feeder-free Geltrex matrix, undifferentiated hiPSC colonies were identified using a phase-contrast microscope. These colonies were characterized by their distinct morphology, including large size, tightly packed cells, and well-defined borders. Immunofluorescence confirmed the expression of POU5F1 (OCT3/4) (Fig. 1a). Cells were fixed with 4% paraformaldehyde in PBS for 20 min, blocked with 5% normal donkey serum for 30 min, and incubated with a primary antibody against POU5F1 (1:300) for 1 h. After washing, cells were treated with secondary antibody (1:1000) for 1 h at room temperature.
2. To confirm the undifferentiated state and high pluripotency, single-cell RNA sequencing was conducted. iPS cells were dissociated into single cells using Versene, filtered, and centrifuged before resuspension in E8 Medium. Freshly isolated cells were adjusted to the appropriate concentration (1000 cells/μL) in Phosphate Buffered Saline (PBS) with 0.01%

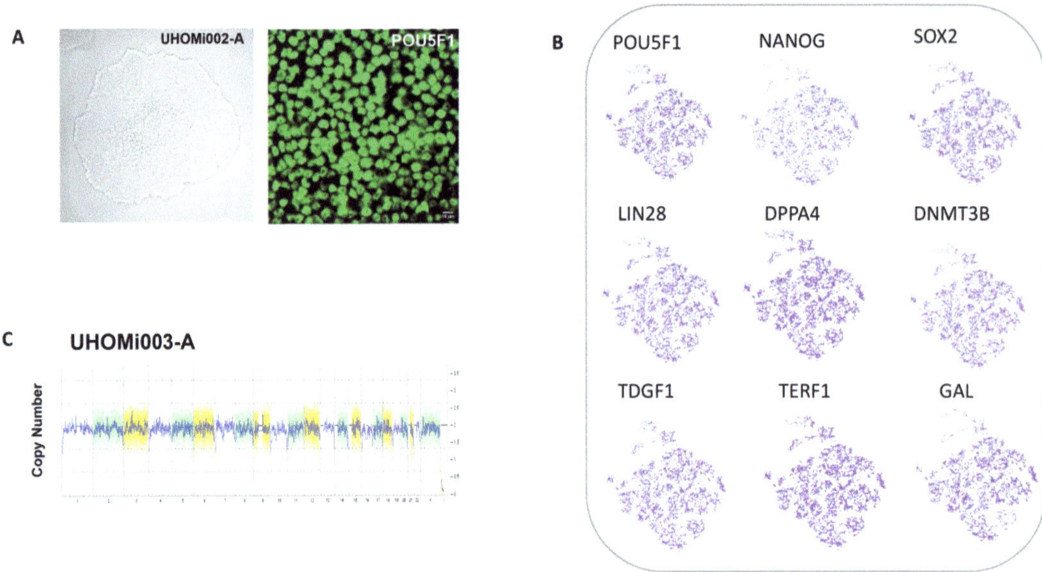

Fig. 1 Characterization of the two hiPSC lines and analysis of undifferentiated state markers. (**a**) Representative images of a hiPSC colony (UHOMi002-A), Magnification: ×4. The hiPSC (UHOMi003-A) undifferentiated state was confirmed by checking the expression of OCT3/4 (POU5F1), Scale bar = 10 μm. (**b**) The pluripotency state was confirmed by single-cell RNA-seq analysis of known pluripotency markers in the two iPSC lines (the panel shows the results for the UHOMi002-A line). The t-SNE plots show the expression of the main pluripotency markers. Each point represents a single cell. The color gradient represents the relative expression level of the gene in a cell across the whole population: (light gray) low; (dark purple) high. (**c**) The KaryoStat™ analysis revealed normal chromosome number and structure in the UHOMi003-A line. The whole genome view displays all somatic and sex chromosomes in one frame with high level copy number. The signal plot, shown on the right *y*-axis, is generated by smoothing the log2 ratios that illustrate the signal intensities of the microarray probes. A value of 2 indicates a standard copy number state (CNV = 2). Conversely, a value of 3 indicates chromosomal amplification (CNV = 3) and a value of 1 a chromosomal loss (CNV = 1). The white, green, and yellow colors indicate the raw signal for each individual chromosome probe, and the blue signal represents the normalized probe signal that is used to identify copy number and alterations (if present). (The KaryoStat image is reprinted from Stem Cell Res. 56:102550, 2021 with permission from Elsevier)

Bovine Serum Albumin (BSA), and 10,000 cells per condition were used for cDNA preparation with the 10X Genomics Chromium platform. Library quality was assessed with an Agilent Bioanalyzer, and sequencing was performed following Illumina protocols. Data were processed with Cell Ranger for demultiplexing, alignment (GRCh38). Expression matrices were analyzed in Seurat, and gene expression were visualized using UMAP. The sequencing revealed strong expression of key stemness markers [26], including POU5F1, NANOG, SOX2, and others (Fig. 1b).

3. Genomic stability was assessed using a DNA microarray method (KaryoStat+). IPS cells were processed using the Genomic DNA Purification Kit (Cat. #K0512) and quantified

with the Qubit™ dsDNA BR Assay Kit (Cat. #Q32850). A total of 250 ng of gDNA was used to prepare the GeneChip® for KaryoStat™ following the manufacturer's instructions. This method did not detect any sub-chromosomal alterations, confirming the stability of the UHOMi003-A line (Fig. 1c). This step was essential before proceeding with the multiplex ddPCR-based analysis.

3.2 Establishment and Optimization of the Multiplex ddPCR Method

This protocol outlines the steps for performing multiplex digital PCR using the iCS-digital™ PSC 24-Probe Test to assess copy number variations (CNVs) and mosaicism. Mosaicism is simulated by mixing male and female DNA in defined ratios to mimic X chromosome gain scenarios.

1. Nucleic acid purification and quantification: DNA was extracted from hiPSC (UHOMi002-A) cell pellets as described below:
   (a) Use the QIAamp DNA Mini Kit following the manufacturer's protocol.
   (b) Quantify the DNA using the Qubit dsDNA BR Assay Kit according to the kit instructions.
   (c) Dilute the DNA to a final concentration of 5 ng/μL using deionized water, and 3 μL per well was used for ddPCR to obtain the amount of 15 ng per well.
2. Multiplex design: Multiplexing was performed using TaqMan Assays with primers and probes labeled with FAM or HEX fluorophores.
   (a) Use four unique copy number assays at the following concentrations:
      - FAM probe: Detect Chr20q11 (FAM high: 1X) and Chr12p (FAM low: 0.5X).
      - HEX probe: Detect ChrXp (HEX high: 1X) and the reference gene RPP30 (HEX low: 0.5X).
   (b) Mix target genes labeled with FAM high, FAM low, and HEX high fluorophores with the reference gene labeled with HEX low.
   (c) Ensure the reference gene is consistent across all 8 wells to minimize variation and calculate CNV values for the 24 targets.
3. PCR Reaction Mix: A mixture with a final volume of 22 μL was prepared for each PCR reaction well.
   (a) 11 μL of ddPCR Supermix for Probes (No dUTP) at 1X concentration.
   (b) Add FAM high, HEX high, FAM low, and HEX low assays at their respective concentrations (1X or 0.5X).

(c) Add 2 units of restriction enzyme per well to enhance CNV calculation accuracy by separating tandem gene copies.
(d) Add 3 μL of DNA sample.
(e) Adjust the final volume to 22 μL with deionized water.

4. ddPCR Setup: perform ddPCR using the QX200 platform (Bio-Rad) following the manufacturer's instructions.

3.3 Mosaicism Test

1. DNA Preparation: DNAs were extracted from a female cell line and a male cell line and mixed as follows:
   (a) 90% male DNA/10% female DNA.
   (b) 80% male DNA/20% female DNA.
   (c) 70% male DNA/30% female DNA.
2. PCR Reaction and Analysis: use the same preparation method described above for the PCR reaction mix.

3.4 Data Analysis of Multiplex ddPCR for CNV Detection

Here we describe the data analysis steps for interpreting multiplex ddPCR results to detect copy number variations (CNVs) using fluorescence amplitude signals. The procedure ensures accurate cluster differentiation and reliable CNV detection through optimized primer/probe concentrations and two-channel fluorescence detection. Results are visualized in dot plots showing distinct droplet clusters, as illustrated in Fig. 2a–c.

1. Cluster Adjustment:
   (a) Analyze fluorescence amplitude signals to spatially adjust droplet clusters in two-dimensional plots.
   (b) Optimize primer and probe concentrations to achieve clear segregation of clusters, ensuring efficient differentiation of positive droplets.
   (c) Results are represented in a dot plot with eight clusters, corresponding to populations of droplets with different positive primer/probe combinations (Fig. 2a).
2. Channel Assignments: Use the following channel assignments for detection:
   (a) Channel 1 (FAM): Detect primers/probes for Chr20q-CNV and Chr12p-CNV.
   (b) Channel 2 (HEX): Detect primers/probes for the reference gene (RPP30-ref) and ChrXp-CNV (Fig. 2b).
3. Validation with Spiked Samples: Spike DNA from iPSC (UHOMi002-A) with synthetic DNAs mimicking the targeted CNVs.
   (a) Perform multiplex ddPCR to detect the targeted CNVs (Chr20q-CNV, Chr12p-CNV, and ChrXp-CNV).

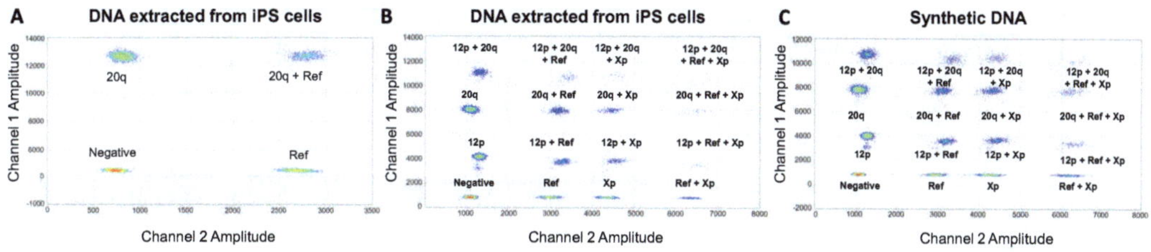

Target	Fluorophore	Concentration (copies/μL)	CNV
Ref	HEX low	252	-
20q	FAM high	255	2.02

Target	Fluorophore	Concentration (copies/μL)	CNV
12p	FAM low	229	1.92
20q	FAM high	234	1.97
Ref	HEX low	238	-
Xp	HEX high	115	0.97

Target	Fluorophore	Concentration (copies/μL)	CNV
12p	FAM low	337	2.72
20q	FAM high	371	3.00
Ref	HEX low	247	-
Xp	HEX high	219	1.77

Fig. 2 2D Amplitude multiplexing plots. (**a**) 2-color graph for the detection of the 20q target (*y*-axis: FAM channel) and the reference gene (*x*-axis: HEX channel) with clusters assigned to each target in a normal hiPSC sample. The negative cluster contains no amplified target; single positive clusters contain the amplified 20q target (20q) or the amplified reference gene (Ref). The double cluster (20q + Ref) contains both targets. The table below represents the CNV value for each target. CNV values were calculated as follows: [Target concentration]/[Reference concentration] ×2. (**b**) 2-color amplitude graph of the multiplex assay (*y*-axis: FAM channel/*x*-axis: HEX channel) with clusters assigned to each target for a normal hiPSC sample. The negative cluster contains no amplified target, single positive clusters contain only one amplified target and the multiple-positive clusters contain 2, 3, or 4 amplified targets. The table below lists the CNV values for each target. (**c**) 2-color graph (*y*-axis: FAM channel/*x*-axis: HEX channel) with clusters assigned to each target for a control human DNA sample spiked with synthetic DNA harboring the target CNVs analyzed by multiplex ddPCR. The negative cluster contains no amplified target, single positive clusters contain only one amplified target, and the multiple-positive clusters contain 2, 3, or 4 amplified targets. The table below lists the abnormal CNV values (CNV $>$ or $<$ 2.00) for each target. Ref = reference

   (b) The 2D plot positions of clusters for these targets are consistent with those obtained using genomic DNA extracted from hiPSCs (Fig. 2c).

4. Reliability Confirmation:
   (a) Confirm that droplet detection is consistent across all duplex assays, as shown in Fig. 2a.
   (b) Validate results by comparing cluster positions from spiked samples with genomic DNA (Fig. 2b, c), confirming the reliability of both multiplex and duplex ddPCR approaches.

3.5 Determining the Performance of the Multiplex ddPCR Assay for CNV Detection

This part of protocol describes the development, optimization, and validation of a multiplex ddPCR assay for detecting recurrent copy number variations (CNVs) in hiPSCs. The assay's analytical performance was evaluated based on repeatability, intermediate precision, reproducibility, limit of detection (LOD), and limit of quantification (LOQ). Visual results, illustrated in Figs. 3a–c and 4a, support the validation process.

A

		REPEATABILITY			INTERMEDIATE PRECISION			REPRODUCIBILITY		
	Target	12p	20q	Xp	12p	20q	Xp	12p	20q	Xp
DNA 5 ng/µL	Mean	1.99	1.92	0.95	1.98	1.95	0.96	1.99	1.99	1.98
	SD	0.07	0.03	0.01	0.07	0.07	0.04	0.05	0.05	0.07
	RSD	3.6%	1.5%	1.5%	3.3%	3.8%	4.5%	2.3%	2.8%	3.3%

B

Sample			12p	20q	Xp	LOD 95%
10% X gain	R1	CNV	1.99	2.45	1.15	66.7%
		p-val	1.00	0.00	0.04	
	R2	CNV	1.94	2.50	1.14	
		p-val	1.00	0.00	0.07	
	R3	CNV	2.05	2.47	1.16	
		p-val	1.00	0.00	0.02	
20% X gain	R1	CNV	2.00	2.52	1.25	100.0%
		p-val	1.00	0.00	0.00	
	R2	CNV	2.00	2.63	1.20	
		p-val	1.00	0.00	0.00	
	R3	CNV	1.95	2.52	1.22	
		p-val	1.00	0.00	0.00	
30% X gain	R1	CNV	2.01	2.70	1.35	100.0%
		p-val	1.00	0.00	0.00	
	R2	CNV	2.15	2.86	1.36	
		p-val	1.00	0.00	0.00	
	R3	CNV	2.04	2.64	1.39	
		p-val	1.00	0.00	0.00	

C

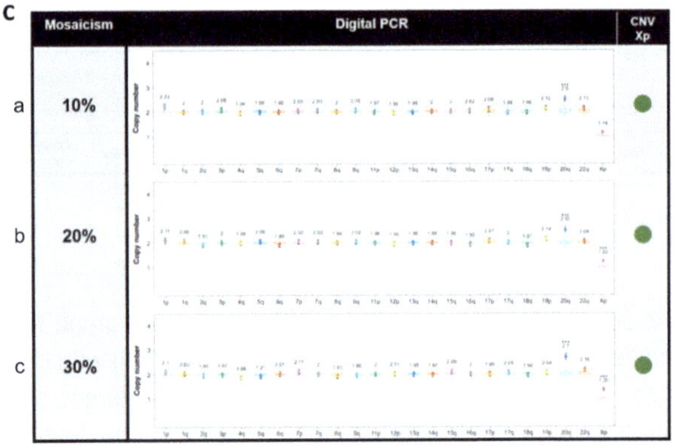

Fig. 3 Precision tests for the 12p, 20q, and Xp targets. (**a**) The mean, standard deviation (SD) and the relative standard deviation (RSD) for the repeatability, intermediate precision and reproducibility tests in a normal hiPSC sample at the concentration of use. In the repeatability test, the same experiment was performed in three replicates. In the intermediate precision test, some settings were intentionally modified and seven replicates were tested. The reproducibility test was carried out using 20 replicates at our 2 facilities in Montpellier (France) and Durham (NC, USA) and represents the condition with the most variable parameters. (**b**) Limit of Detection (LOD) for the 12p, 20q, and Xp targets. The table shows the results obtained with different DNA mixtures consisting of 90% male + 10% female DNA or 80% male + 20% female DNA or 70% male + 30% female DNA that mimic, respectively, 10%, 20%, or 30% of mosaicism for a Xp gain. Three replicates (R1, R2, and R3) were tested for each condition. The LOD is highlighted in blue. (**c**) Varying degrees of mosaicism present in chromosome X as detected by targeted CNV quantification of 24 probes using the same samples described in (**b**) (iCS-digital test, Stem Genomics). (a) CNV quantification with a probe located on chromosome X indicates a level of mosaicism of approximately 10%. (**b**, **c**) Higher levels of mosaicism (~20% and ~30%, respectively)

1. Target Selection and Assay Development: CNV were targeted based on the previously reported repertoire of recurrent alterations in hiPSCs [27]. A multiplex ddPCR assay was subsequently developed and optimized to ensure efficient and accurate detection of these targets. The assay's analytical performance was rigorously validated to confirm its reliability before routine application.
2. Precision Validation:
   (a) Repeatability:
      - Conduct three independent replicates using the same plates, reagents, thermocycler, and reader on the same day.
      - Use a fixed DNA concentration of 5 ng/µL for all replicates.

A

Quantity	Replicate		12p	20q	Xp
1 ng	R1	CNV	1.83	2.35	1.22
		SD	0.77	0.93	0.58
	R2	CNV	1.78	1.83	0.61
		SD	0.85	0.87	0.42
3 ng	R1	CNV	1.80	1.96	1.67
		SD	0.22	0.24	0.21
	R2	CNV	2.26	1.90	1.77
		SD	0.30	0.27	0.25
5 ng	R1	CNV	2.01	1.94	1.76
		SD	0.18	0.18	0.17
	R2	CNV	2.05	2.09	1.77
		SD	0.19	0.19	0.17
10 ng	R1	CNV	2.04	2.02	1.89
		SD	0.13	0.13	0.12
	R2	CNV	1.91	2.03	1.80
		SD	0.12	0.12	0.11
15 ng	R1	CNV	1.92	1.97	1.72
		SD	0.10	0.10	0.09
	R2	CNV	1.89	1.97	1.72
		SD	0.10	0.10	0.09
66 ng	R1	CNV	1.94	1.95	1.78
		SD	0.06	0.06	0.05
	R2	CNV	1.95	1.99	1.83
		SD	0.06	0.06	0.05

B

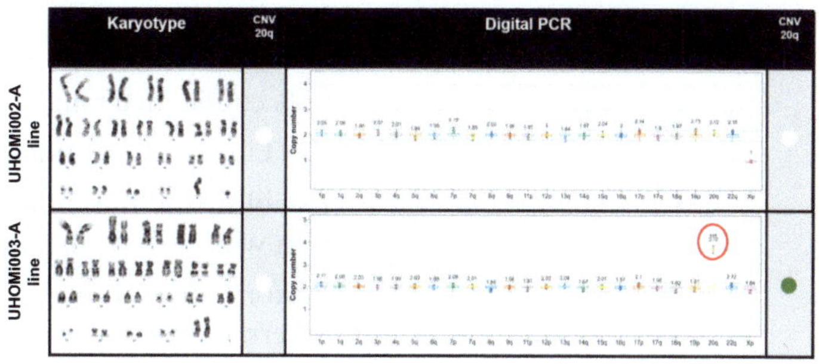

Fig. 4 Limit of quantification and multiplex ddPCR potential. (**a**) The limit of quantification (LOQ) test for the 12p, 20q, and Xp targets. The table shows the results obtained with different DNA amounts, from 1 to 66 ng (the upper limit recommended by the manufacturer of the digital PCR platform). Each amount was tested in duplicate. The LOQ of the test is highlighted in blue. (**b**) Comparison of the G-banding and multiplex ddPCR results to assess the genomic integrity of the UHOMi002-A and UHOMi003-A hiPSC lines. Both techniques yielded similar results and identified a normal karyotype (46, XY) and normal CNVs in the UHOMi002-A line. Conversely, in the UHOMi003-A line, G-banding identified a normal karyotype (46, XX), whereas the multiplex ddPCR found a CNV gain of chromosome 20q11.1 (CNV = 3.72)

- CNV detection demonstrated repeatability with <10% variation (Fig. 3a).

(b) Intermediate Precision:

- Perform seven independent tests using different plates, reagents, and thermocyclers on separate days with two operators.
- Intermediate precision exhibited minimal variance across tests (Fig. 3a).

(c) Reproducibility:

- Ensure replicates are conducted on different days and by different operators, where feasible, to account for variability.
- Perform 20 replicates using identical DNA concentrations.
- Reproducibility showed consistent performances, with <10% variation (Fig. 3a).

3. LOD and LOQ Determination: To establish the reliability of the multiplex ddPCR assay, we determined the LOD and LOQ [28].

   (a) LOD (Limit of Detection):

   - Define the lowest detectable PCR copy number concentration with a 95% confidence range.

- Test mosaicism levels at 20% to determine the detection limit.
- The LOD was determined to be 20% mosaicism with a 95% confidence range (Fig. 3b, c).

(b) LOQ (Limit of Quantification):

- Establish the minimum quantifiable DNA concentration with 95% confidence.
- Perform ddPCR with varying DNA concentrations to determine the threshold.
- The LOQ was identified as 15 ng per well (Fig. 4a).

3.6 The Sensitivity of Multiplex ddPCR Assay for CNVs Detection Were Compared to Karyotyping

In this section, the sensitivity of the multiplex ddPCR assay was evaluated by analysis of two hiPSC lines (UHOMi002-A and UHOMi003-A) using conventional karyotyping (G-banding) and ddPCR, as shown in Fig. 4b. Target CNVs were selected based on previously reported alterations in hiPSCs [27].

1. Multiplex Design: Three targets per well, part of a multiplex set for testing 24 recurrent CNV targets across 8 wells.
2. G-Banding: Performed on 20 metaphases to establish baseline cytogenetic profiles for both hiPSC lines.

The multiplex ddPCR results showed that the UHOMi002-A line exhibited no detectable CNVs, while the UHOMi003-A line demonstrated a gain in the 20q11 region (CNV = 3.72). This result was not detected by G-banding, underscoring the ddPCR assay's ability to identify infra-karyotypic alterations (Fig. 4b).

3.7 Sensitivity of Multiplex ddPCR for CNV Detection Compared to DNA Sequencing

In this section, the sensitivity of the multiplex ddPCR assay for detecting CNVs in hiPSCs and its concordance with next-generation sequencing (NGS), ten hiPSC lines (iPS1–5, iPS9, iPS11, iPS14, iPS25, iPS31) were analyzed. These lines exhibited CNVs on different chromosomes, ranging from gains to losses, with values between 1 and 3. Since the multiplex ddPCR method detects mosaicism only above 20%, NGS was used to explore low-level mosaicism in these samples.

1. DNA investigations: Both multiplex ddPCR and NGS were performed on the same DNA samples to ensure consistency in the results.
2. Analysis workflow: The multiplex ddPCR assay focused on analyzing targeted CNV regions for gains and losses, while NGS provided a broader analysis, detecting low-level mosaicism and single nucleotide variants (SNVs) in addition to CNVs.

Chromosome	Technique	Sample ID									
		iPS-001	iPS-002	iPS-003	iPS-004	iPS-005	iPS-009	iPS-011	iPS-014	iPS-025	iPS-031
1p	Digital PCR	-	-	-	-	-	-	3.21	-	-	-
	NGS	-	-	-	-	-	-	3.05	-	-	-
1q	Digital PCR	-	-	-	-	2.47	-	3.01	-	-	-
	NGS	-	-	-	-	2.51	-	2.95	-	-	-
4q	Digital PCR	-	-	2.7	-	-	-	-	-	-	-
	NGS	-	-	2.8	-	-	-	-	-	-	-
7q	Digital PCR	-	3.17	-	-	-	-	-	-	-	-
	NGS	-	2.7	-	-	-	-	-	-	-	-
12p	Digital PCR	-	-	-	-	-	-	-	-	-	2.85
	NGS	-	-	-	-	-	-	-	-	-	2.96
17q	Digital PCR	-	-	-	-	-	-	-	3.09	-	-
	NGS	-	-	-	-	-	-	-	3.08	-	-
18q	Digital PCR	-	-	2.48	-	-	-	-	1.06	2.25	-
	NGS	-	-	2.44	-	-	-	-	0.99	2.18	-
20q	Digital PCR	3.02	2.21	3.02	-	-	-	-	-	-	-
	NGS	2.97	2.2	2.83	-	-	-	-	-	-	-
22q	Digital PCR	-	-	-	-	-	1.65	-	-	-	-
	NGS	-	-	-	-	-	1.74	-	-	-	-
Xp	Digital PCR	-	-	-	1.25	-	-	-	-	-	-
	NGS	-	-	-	1.22	-	-	-	-	-	-

Fig. 5 Comparison of the multiplex ddPCR and NGS results for CNV detection. All CNVs identified by multiplex ddPCR were subsequently confirmed by the NGS analysis. In the table, copy number gains are highlighted in blue and losses in red

3. A strong correlation was observed between the CNV values obtained from NGS and multiplex ddPCR, with minor variability in the CNV magnitude but no discrepancies in the presence or absence of CNVs (Fig. 5).
4. Spearman correlation analyses showed nearly perfect agreement between NGS and multiplex ddPCR results, with all CNVs (15 out of 15) correctly identified (Fig. 6a).
5. Circos plots provided a visual representation of CNVs and SNVs:
   - *Red regions*: Copy number gains.
   - *Blue regions*: Copy number losses.
   - *Dashed purple lines*: SNVs and impacted genes (Fig. 6b).

This section highlights the exceptional performance of the multiplex ddPCR assay, demonstrating a high degree of sensitivity and concordance with NGS for CNV detection. While NGS provides broader insights, the multiplex ddPCR assay remains a reliable and efficient tool for routine CNV analysis in hiPSCs.

3.8 Summation

The techniques described above are integrated into a comprehensive workflow. Figure 7 presents the organizational chart outlining the key quality control tests for hiPSCs at various stages. A combination of G-banded karyotyping and multiplex ddPCR is recommended to detect both structural rearrangements and subkaryotypic abnormalities. NGS analysis can provide complementary information, particularly for identifying single nucleotide

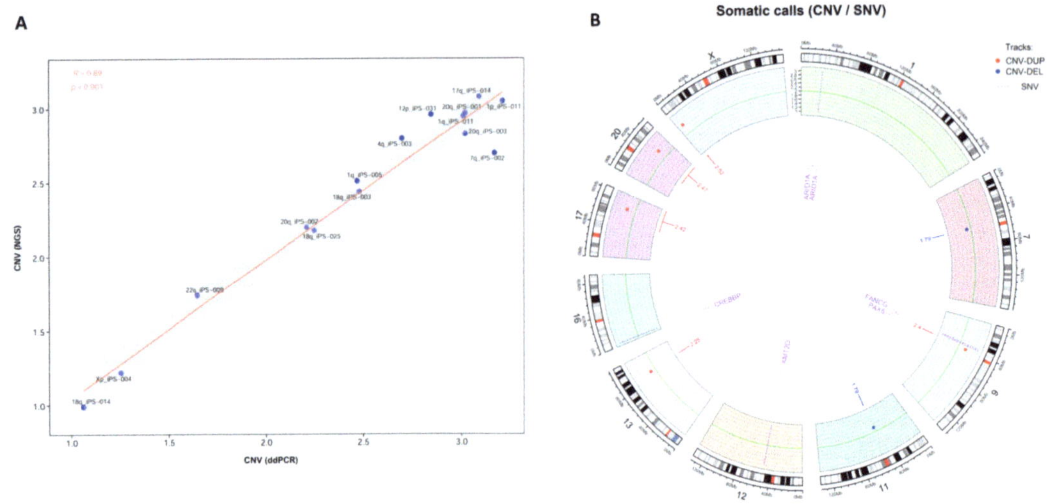

Fig. 6 Multiplex ddPCR assay versus NGS. (**a**) Correlation between the CNV values identified by NGS and multiplex ddPCR. The Spearman correlation coefficient (*R*) and linear regression line are indicated. Results are based on the data from Fig. 6. (**b**) Circos plot of distribution of integration events (CNV & SNV) with their metrics throughout the chromosomes

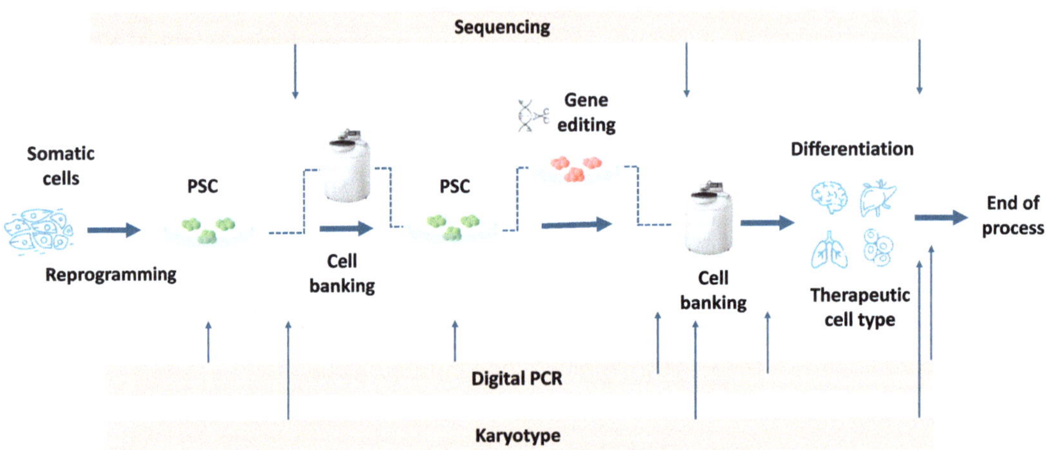

Fig. 7 HiPSC culture workflow and genomic integrity testing techniques. To test the hiPSC integrity at the different culture steps, digital PCR is the only technique that allows in-process sample control, providing real-time insights into genomic stability. G-banding and next-generation sequencing offer a more comprehensive, genome-wide analysis, but are less compatible with in-process control due to their longer turnaround time, typically 4–5 weeks

polymorphisms. Additionally, ddPCR is used for rapid screening of clones before selection and for monitoring genomic stability during clone expansion and routine culture. Due to its speed, cost-efficiency, and high resolution, multiplex ddPCR is employed at multiple stages to verify genomic stability and ensure complete characterization of hiPSC lines prior to banking and differentiation/.

4 Notes

1. Many studies indicate that hiPSCs frequently exhibit genomic instability [7, 16, 27, 29–32]. hiPSC cultures are subjected to selective pressures that lead to the emergence of recurrent genetic alterations, affecting both the undifferentiated state of the cells and their differentiation capacity [33, 34]. For instance, the gain of 20q11.21 impairs commitment to the neuroectodermal lineage [12], and the gain of chr.12 is often found in testicular germ cell tumors [35]. Other examples include the duplication of 1q21.1, which delays the shift from neural progenitors to neurons [36], and chr.17 gains, which are prevalent in neuroblastoma, testicular germ cell tumors, and breast cancer [35]. Loss of chr.18q results in impaired neuroectodermal commitment [37].
2. A recent study listed recurrent aberrations observed in human pluripotent stem cells [27]. This meta-analysis revealed 738 recurrent genetic abnormalities identified in over 100 distinct publications from various laboratories, including chromosome gains and CNVs. This repertoire was obtained by G-banded karyotyping, fluorescence in situ hybridization, microarray, and sequencing [27].
3. There is currently no consensus on the preferred method for monitoring genome integrity in cultured human pluripotent stem cells, although some guidelines are emerging [38]. Beyond CNVs, recurrent SNVs can also affect differentiation capabilities but often remain undetected without appropriate assays. Rapid, highly sensitive, and validated assays are needed for the early detection of abnormal clones.
4. Digital PCR, including standard duplex ddPCR, is increasingly used to assess CNVs in predetermined regions of interest, typically to screen for recurrent acquired aberrations and for precise evaluation of genetic stability in hiPSCs [27, 39]. This study expands the capabilities of ddPCR to detect CNVs in hiPSCs. To the best of our knowledge, this is the first study on multiplex ddPCR for CNV detection in hiPSCs.
5. Multiplexing has been widely used to reduce expenses, increase sample throughput, and optimize detection sensitivity for multiple targets within limited samples. In developing multiplex qPCR assays, individual targets typically require unique labeling with probes to distinguish them. However, ddPCR systems offer a distinct advantage by allowing the detection of more than two targets in a single reaction through higher-order multiplexing.
6. Digital PCR and NGS are complementary techniques with distinct features. NGS can detect unknown genetic

modifications by generating vast amounts of information but is expensive, requires informatic support, and necessitates significant time for deciphering and interpreting results. Conversely, ddPCR is time-saving and easy to perform but requires prior knowledge of the genetic alterations to be identified. These methods can be used in combination, particularly for hiPSC genetic stability analyses [40], where NGS is generally used for the detection of abnormalities [41] and ddPCR in the validation step, for in-process testing, and for screening a large number of clones after gene editing [27].

7. Typical turnaround times for NGS testing are 1–3 weeks, which may limit its usefulness when urgent decisions are needed. Compared to karyotyping, NGS has reduced sensitivity for detecting balanced chromosomal translocations, which require traditional karyotyping or long-read sequencing [42].
8. Considering how quickly recurrent alterations can dominate hiPSC cultures, multiplex ddPCR enables fast and cost-effective detection. This feature is critical for the early detection of genomic abnormalities in hiPSCs used for treatments and research purposes. Regular assessment of genomic stability in hiPSC lines with an efficient testing strategy is crucial.
9. Multiplex ddPCR assays can be particularly useful in various situations:
   (a) Before starting work with hiPSCs, it is imperative to verify the genomic stability of the initial samples. Although most purchased lines include results of genomic stability tests, small-scale alterations may have escaped detection by G-banded karyotyping. Combining G-banded karyotyping and ddPCR at this stage is recommended, as it associates traditional structural rearrangement analysis with the high resolution of ddPCR for detecting sub-karyotyping alterations.
   (b) If the budget permits, NGS testing is advisable when starting a project to include single nucleotide polymorphism detection.
   (c) Gene editing procedures can favor the selection of genomic alterations in hiPSCs. Quick screening of clones using ddPCR is essential before any subsequent use of the selected clone(s).
   (d) Clone amplification and maintenance are stress factors likely to generate genetic alterations in hiPSCs. Given how quickly a recurrent alteration can dominate the culture [43], multiplex ddPCR could become the best routine test with quick turnaround times and cost-effectiveness.

(e) Ensuring genomic stability is crucial for the complete characterization of a hiPSC line before banking. Various methods (karyotyping, ddPCR, and NGS) can be employed, but the choice is often influenced by budget constraints.

(f) Monitoring should continue until full differentiation, with ddPCR being the ideal method for this stage due to its speed.

Acknowledgments

We thank Elena Hauser, Pauline Lacaille, Manon Prevot (Montpellier, France), and Cassandre Frecenon (Durham, USA) for contributing to the collection of hiPSC genomic abnormalities and technical help. We thank Elisabetta Andermarcher for reading the manuscript.

Authors' Contributions S.A. and J.D.V. designed the study and analyzed data. S.A., R.Z., and J.V. performed the experiments. S. A., R.Z., J.V., and L.M collected and analyzed data. S.A., R.Z., J.V., and L.M, analyzed and interpreted the data. S.A., R.Z., J.V., and J.D.V. wrote the paper. S.A. and RZ. revised the manuscript. All authors have read and agreed to the published version of the manuscript.

Conflict of Interests S.A. and J.D.V. filed a patent entitled "Non-invasive methods for assessing genetic integrity of pluripotent stem cells" (priority number: EP20150306389) and are co-founders, shareholders, and scientific advisors of Stem Genomics SAS that acquired the exploitation rights of this patent.

Funding

Supported by grants from the University Hospital of Montpellier and Inserm.

References

1. Thomson JA, Itskovitz-Eldor J, Shapiro SS et al (1998) Embryonic stem cell lines derived from human blastocysts. Science 282:1145–1147. https://doi.org/10.1126/science.282.5391.1145
2. Takahashi K, Tanabe K, Ohnuki M et al (2007) Induction of pluripotent stem cells from adult human fibroblasts by defined factors. Cell 131:861–872. https://doi.org/10.1016/j.cell.2007.11.019
3. Bai Q, Ramirez J-M, Becker F et al (2015) Temporal analysis of genome alterations induced by single-cell passaging in human embryonic stem cells. Stem Cells Dev 24:

653–662. https://doi.org/10.1089/scd.2014.0292

4. Assou S, Bouckenheimer J, De Vos J (2018) Concise review: assessing the genome integrity of human induced pluripotent stem cells: what quality control metrics? Stem Cells 36:814–821. https://doi.org/10.1002/stem.2797
5. Olariu V, Harrison NJ, Coca D et al (2010) Modeling the evolution of culture-adapted human embryonic stem cells. Stem Cell Res 4:50–56. https://doi.org/10.1016/j.scr.2009.09.001
6. Zhang J, Hirst AJ, Duan F et al (2019) Anti-apoptotic mutations desensitize human pluripotent stem cells to mitotic stress and enable Aneuploid cell survival. Stem Cell Rep 12:557–571. https://doi.org/10.1016/j.stemcr.2019.01.013
7. International Stem Cell Initiative, Amps K, Andrews PW et al (2011) Screening ethnically diverse human embryonic stem cells identifies a chromosome 20 minimal amplicon conferring growth advantage. Nat Biotechnol 29:1132–1144. https://doi.org/10.1038/nbt.2051
8. Taapken SM, Nisler BS, Newton MA et al (2011) Karotypic abnormalities in human induced pluripotent stem cells and embryonic stem cells. Nat Biotechnol 29:313–314. https://doi.org/10.1038/nbt.1835
9. Maitra A, Arking DE, Shivapurkar N et al (2005) Genomic alterations in cultured human embryonic stem cells. Nat Genet 37:1099–1103. https://doi.org/10.1038/ng1631
10. Spits C, Mateizel I, Geens M et al (2008) Recurrent chromosomal abnormalities in human embryonic stem cells. Nat Biotechnol 26:1361–1363. https://doi.org/10.1038/nbt.1510
11. Keller A, Spits C (2021) The impact of acquired genetic abnormalities on the clinical translation of human pluripotent stem cells. Cells 10:3246. https://doi.org/10.3390/cells10113246
12. Markouli C, Couvreu De Deckersberg E, Regin M et al (2019) Gain of 20q11.21 in human pluripotent stem cells impairs TGF-β-dependent Neuroectodermal commitment. Stem Cell Rep 13:163–176. https://doi.org/10.1016/j.stemcr.2019.05.005
13. Merkle FT, Ghosh S, Genovese G et al (2022) Whole-genome analysis of human embryonic stem cells enables rational line selection based on genetic variation. Cell Stem Cell 29:472–486.e7. https://doi.org/10.1016/j.stem.2022.01.011
14. Rouhani FJ, Zou X, Danecek P et al (2022) Substantial somatic genomic variation and selection for BCOR mutations in human induced pluripotent stem cells. Nat Genet 54:1406–1416. https://doi.org/10.1038/s41588-022-01147-3
15. Baghbaderani BA, Syama A, Sivapatham R et al (2016) Detailed characterization of human induced pluripotent stem cells manufactured for therapeutic applications. Stem Cell Rev Rep 12:394–420. https://doi.org/10.1007/s12015-016-9662-8
16. Iqbal MA, Broeckel U, Levy B et al (2023) Multisite assessment of optical genome mapping for analysis of structural variants in constitutional postnatal cases. J Mol Diagn 25:175–188. https://doi.org/10.1016/j.jmoldx.2022.12.005
17. Lefort N, Feyeux M, Bas C et al (2008) Human embryonic stem cells reveal recurrent genomic instability at 20q11.21. Nat Biotechnol 26:1364–1366. https://doi.org/10.1038/nbt.1509
18. Vogelstein B, Kinzler KW (1999) Digital PCR. Proc Natl Acad Sci USA 96:9236–9241. https://doi.org/10.1073/pnas.96.16.9236
19. Quan P-L, Sauzade M, Brouzes E (2018) dPCR: a technology review. Sensors (Basel) 18:1271. https://doi.org/10.3390/s18041271
20. Hindson BJ, Ness KD, Masquelier DA et al (2011) High-throughput droplet digital PCR system for absolute quantitation of DNA copy number. Anal Chem 83:8604–8610. https://doi.org/10.1021/ac202028g
21. Sykes PJ, Neoh SH, Brisco MJ et al (1992) Quantitation of targets for PCR by use of limiting dilution. BioTechniques 13:444–449
22. Taly V, Pekin D, Benhaim L et al (2013) Multiplex picodroplet digital PCR to detect KRAS mutations in circulating DNA from the plasma of colorectal cancer patients. Clin Chem 59:1722–1731. https://doi.org/10.1373/clinchem.2013.206359
23. Levy CN, Hughes SM, Roychoudhury P et al (2021) HIV reservoir quantification by five-target multiplex droplet digital PCR. STAR Protoc 2:100885. https://doi.org/10.1016/j.xpro.2021.100885
24. Fieldes M, Ahmed E, Bourguignon C et al (2020) Generation of the induced pluripotent stem cell line UHOMi002-A from peripheral blood mononuclear cells of a healthy male donor. Stem Cell Res 49:102037. https://doi.org/10.1016/j.scr.2020.102037
25. Ahmed E, Fieldes M, Mianné J et al (2021) Generation of four severe early-onset chronic

obstructive pulmonary disease (COPD) patient-derived induced pluripotent stem cell lines from peripheral blood mononuclear cells. Stem Cell Res 56:102550. https://doi.org/10.1016/j.scr.2021.102550

26. Assou S, Le Carrour T, Tondeur S et al (2007) A meta-analysis of human embryonic stem cells transcriptome integrated into a web-based expression atlas. Stem Cells 25(4):961–973. https://doi.org/10.1634/stemcells.2006-0352
27. Assou S, Girault N, Plinet M et al (2020) Recurrent genetic abnormalities in human pluripotent stem cells: definition and routine detection in culture supernatant by targeted droplet digital PCR. Stem Cell Rep 14:1–8. https://doi.org/10.1016/j.stemcr.2019.12.004
28. Armbruster DA, Pry T (2008) Limit of blank, limit of detection and limit of quantitation. Clin Biochem Rev 29(Suppl 1):S49–S52
29. Närvä E, Autio R, Rahkonen N et al (2010) High-resolution DNA analysis of human embryonic stem cell lines reveals culture-induced copy number changes and loss of heterozygosity. Nat Biotechnol 28:371–377. https://doi.org/10.1038/nbt.1615
30. Laurent LC, Ulitsky I, Slavin I et al (2011) Dynamic changes in the copy number of pluripotency and cell proliferation genes in human ESCs and iPSCs during reprogramming and time in culture. Cell Stem Cell 8:106–118. https://doi.org/10.1016/j.stem.2010.12.003
31. Martins-Taylor K, Xu R-H (2012) Concise review: genomic stability of human induced pluripotent stem cells. Stem Cells 30:22–27. https://doi.org/10.1002/stem.705
32. Merkle FT, Ghosh S, Kamitaki N et al (2017) Human pluripotent stem cells recurrently acquire and expand dominant negative P53 mutations. Nature 545:229–233. https://doi.org/10.1038/nature22312
33. Jo H-Y, Lee Y, Ahn H et al (2020) Functional in vivo and in vitro effects of 20q11.21 genetic aberrations on hPSC differentiation. Sci Rep 10:18582. https://doi.org/10.1038/s41598-020-75657-7
34. Vitillo L, Anjum F, Hewitt Z et al (2023) The isochromosome 20q abnormality of pluripotent cells interrupts germ layer differentiation. Stem Cell Rep 18:782–797. https://doi.org/10.1016/j.stemcr.2023.01.007
35. Baker DEC, Harrison NJ, Maltby E et al (2007) Adaptation to culture of human embryonic stem cells and oncogenesis in vivo. Nat Biotechnol 25:207–215. https://doi.org/10.1038/nbt1285
36. Chapman G, Alsaqati M, Lunn S et al (2022) Using induced pluripotent stem cells to investigate human neuronal phenotypes in 1q21.1 deletion and duplication syndrome. Mol Psychiatry 27:819–830. https://doi.org/10.1038/s41380-021-01182-2
37. Lei Y, Al Delbany D, Krivec N et al (2024) SALL3 mediates the loss of neuroectodermal differentiation potential in human embryonic stem cells with chromosome 18q loss. Stem Cell Rep 19:562–578. https://doi.org/10.1016/j.stemcr.2024.03.001
38. Sullivan S, Stacey GN, Akazawa C et al (2018) Quality control guidelines for clinical-grade human induced pluripotent stem cell lines. Regen Med 13:859–866. https://doi.org/10.2217/rme-2018-0095
39. Park JW, Bae SJ, Yun JH et al (2024) Assessment of genetic stability in human induced pluripotent stem cell-derived cardiomyocytes by using droplet digital PCR. Int J Mol Sci 25:1101. https://doi.org/10.3390/ijms25021101
40. Halliwell J, Barbaric I, Andrews PW (2020) Acquired genetic changes in human pluripotent stem cells: origins and consequences. Nat Rev Mol Cell Biol 21:715–728. https://doi.org/10.1038/s41580-020-00292-z
41. Keller A, Tilleman L, Dziedzicka D et al (2019) Uncovering low-level mosaicism in human embryonic stem cells using high throughput single cell shallow sequencing. Sci Rep 9:14844. https://doi.org/10.1038/s41598-019-51314-6
42. Nakano K, Shiroma A, Shimoji M et al (2017) Advantages of genome sequencing by long-read sequencer using SMRT technology in medical area. Hum Cell 30:149–161. https://doi.org/10.1007/s13577-017-0168-8
43. McIntire E, Taapken S, Leonhard K, Larson AL (2020) Genomic stability testing of pluripotent stem cells. Curr Protoc Stem Cell Biol 52:e107. https://doi.org/10.1002/cpsc.107

Chapter 10

Digital PCR Methods for Discovering Molecular Markers for Hanwoo Cattle Breed Genotyping

Songmi Kim, Jinuk Jeong, Seyoung Mun, and Kyudong Han

Abstract

Digital PCR (dPCR) is the third generation of PCR for the absolute quantification of nucleic acids in the reaction without a standard curve by partitioning reactions. In recent years, various dPCR platforms have been developed and used in a variety of research and diagnostic applications. dPCR technology has significant potential as a tool for detecting species identification markers in large samples through accurate quantification of absolute copy numbers. In this chapter, we describe the dPCR detection method for discovering molecular markers used for Hanwoo cattle breed identification using three types of digital PCR platforms (Bio-Rad, Stilla technologies, and OPTOLANE).

Key words Digital PCR, Droplet digital PCR, Molecular marker, Third-generation PCR, Structural variations

1 Introduction

In 1992, Sykes et al. first described the concept of digital PCR (dPCR), which provides an absolute measure of nucleic acid concentration through a combination of limiting dilution, endpoint PCR, and Poisson statistics [1]. In 1999, Vogelstein and Kinzler developed a method for the individual amplification of single template molecules by sample dilution and partitioning, and detecting them using fluorescent probes [2]. Limiting dilutions were performed in an increased reaction chamber using stool samples from colorectal cancer patients, eliminating somatic cell errors, and allowing for the tracking of *KRAS* mutations [2]. This technique is called "dPCR."

Supplementary Information The online version contains supplementary material available at https://doi.org/10.1007/978-1-0716-4767-7_10.

Guillaume Pavlovic (ed.), *Digital PCR: Methods and Protocols*, Methods in Molecular Biology, vol. 2969,
https://doi.org/10.1007/978-1-0716-4767-7_10,

In dPCR, the genetic material, either DNA or RNA, is meticulously partitioned into small compartments with nanometer precision. These compartments are randomly distributed, with each on housing either zero, one, or multiple molecules. After amplification, the absorbance of each compartment is counted and the number of target molecules is calculated from number of PCR positive and negative reactions using Poisson statistics [3, 4]. This process ensures a level of precision and accuracy that is unparalleled in other nucleic acid detection methods [5].

In 2011, droplet digital PCR (ddPCR) based on water–oil emulsion droplet technology was developed, which is characterized by high throughput, automation, and relatively low cost. Various dPCR systems have been developed based on sample partition, including the microfluidic chamber or chip-based digital PCR (cdPCR) and droplet-based digital PCR (ddPCR). Currently, eight dPCR systems, including Quantstudio™ 3D (Thermofisher), BioMark™ HD dPCR system (Fluidigm), Constellation (Formulatrix), Clarity plus (JN Medsys), QX Droplet Digital PCR System (Bio-Rad), Raindrop plus™ Digital PCR System (Raindance), Naica® System (Stilla), Qiacuity digital PCR (Qiagen), and LOAA Analyzer (OPTOLANE) are available [6].

dPCR is a proven method capable of detecting small amounts of DNA, rare mutations, and copy number variations due to the high sensitivity and accuracy of molecular detection [4, 7]. In general, dPCR has been applied to detect rare mutations and pathogens in cancer and disease. dPCR enables precise and accurate detection of even slight differences in copy number, making it useful for analyzing structural variations. Park et al. (2019) reported that structural variation caused by TEs can be used as a robust genetic marker to distinguish between cattle breeds using PCR [8].

A previous study reported the pros and cons of the three dPCR platforms (Stilla Naica® System, Bio-Rad QX200™ droplet Digital PCR System, and OPTOLANE LOAA Analyzer) for Hanwoo-specific molecular marker detection [9]. The Naica® System PCR allows multiplexed detection of three fluorescent channels (FAM, VIC, and Cy5). The Naica system can process 4–12 samples at a time, with data available in 2.5 h. The sample is partitioned by a microchannel network, and this partitioning technique generates 25,000 to 30,000 droplets. The advantage of the Naica® System is that the oil is preloaded into the Sapphire chip simplifying operation and reducing potential contamination [10, 11]. QX200™ ddPCR system uses water–oil emulsion droplet technology with microfluidic systems to generate 20,000 droplets per sample. The QX200™ ddPCR system can run 96 samples in 6 h and detect fluorescence in two color channels (FAM and HEX/VIC). The benefit of the QX200™ ddPCR system is that it can easily, sensitively, and accurately quantify template copy numbers without the need for a standard curve [12]. LOAA analyzer is a unique dPCR platform that combines the characteristics of both real-time PCR and dPCR. The LOAA analyzer is a single-chip format instrument that performs both PCR and data analysis using a microfluidic chip-based chip on a semiconductor chip, providing experimental results in the shortest time (1.5 h) compared to using two platforms (Stilla Naica® System and Bio-Rad QX200™ ddPCR system) [13]. The LOAA analyzer uses Micro Electro Mechanical System (MEMS) to inject one sample per cartridge, partition it into 56–2000 wells, and detect two fluorescence channels (FAM and Cy5) [12]. Interestingly, despite of digital PCR system, the LOAA analyzer provides a digital high-resolution melting analysis that can be used for various applications, including genotyping, mutation scanning, and methylation analysis. A schematic of the workflow for three dPCR systems is shown in Fig. 1.

Hanwoo (*Bos taurus coreanae*) is a unique breed of cattle that has been bred in Korea for 5000 years. Hanwoo is one of the most economically important breeds due to its rapid growth and high meat quality, including marbled fat, tenderness, juiciness, and palatability, which makes it highly consumed [14]. Species identification markers are needed to distinguish Hanwoo from other breeds, and single nucleotide polymorphism (SNP) are common molecular markers for identifying Hanwoo breeds. However, structural variations (SVs) are less abundant and complex than SNPs and may be more powerful biomarkers. A previous study by Park et al. (2019) showed that TE-mediated structural variation could distinguish between the Hanwoo and Holstein genomes. Through comparative genomics between Hanwoo and Holstein breeds, they found an authentic Hanwoo-specific deletion locus [7].

Following this page, we describe three detailed protocols for analyzing Hanwoo-specific SVs with three dPCR platforms to intuitively identify Hanwoo breed [15].

2 Materials

Prepare all solutions using ultrapure water and store reagents at 4 °C until use.

2.1 DNA Preparation

1. 50 ng of genomic DNAs extracted from 18 blood samples of cattle (9 Hanwoo and 9 Holstein) using DNeasy Blood & Tissue kit (Cat. No. 69504) (Qiagen).

2.2 DNA Quantification and Quality Control

1. Horizontal gel electrophoresis unit and power supply; Agarose, TAE buffers, DNA ladder mix.
2. EcoDye™ DNA staining solution (Biofact) and UV light source.

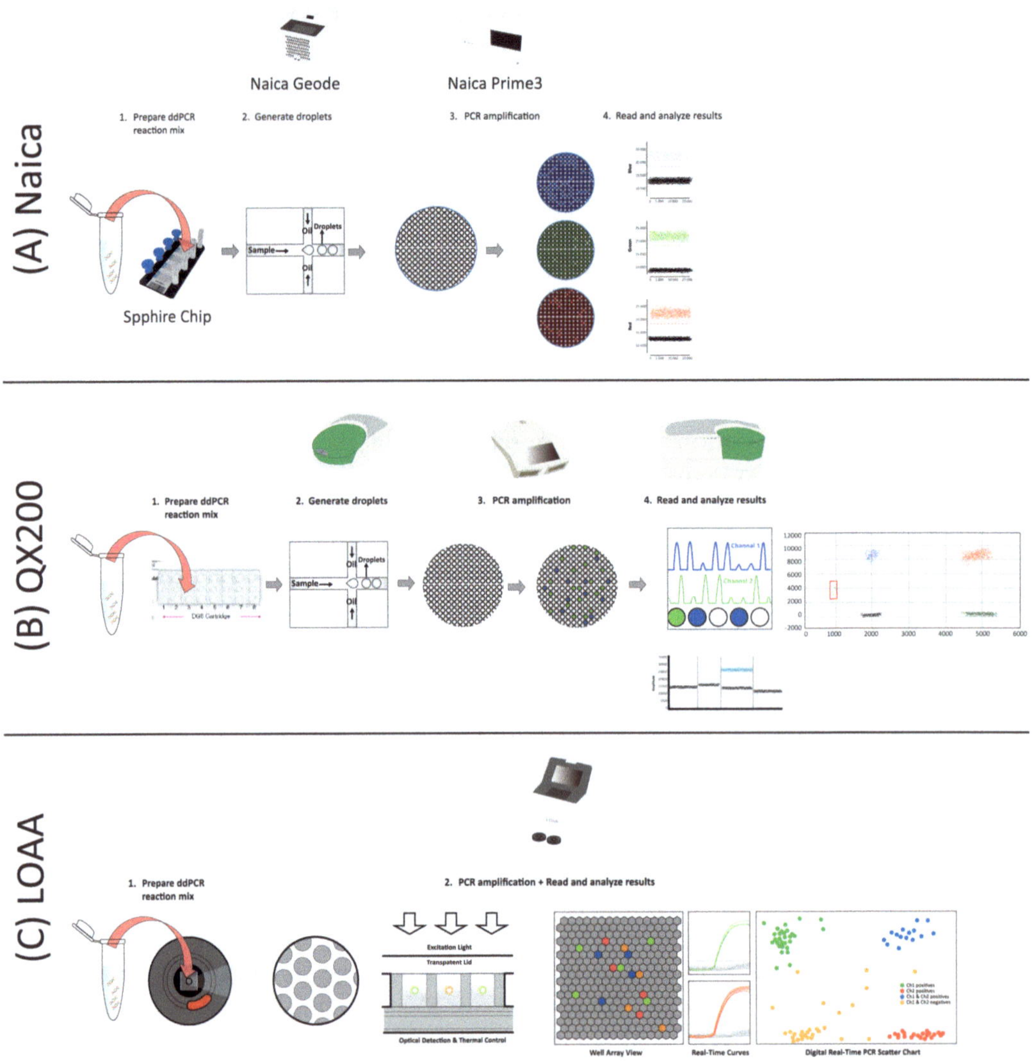

Fig. 1 Schematic illustration of three dPCR systems. This figure illustrates three dPCR workflows. For the assay, the dPCR mixture (probes, primers, and the DNA sample) is loaded in a cartridge. The DNA samples are partitioned into water-in-oil droplets or microchambers. These partitions are individually amplified under specific thermal cycling conditions and analyzed in the fluorescence image. (**a**) Naica® systems (**b**) QX200™ ddPCR system (**c**) LOAA Analyzer

3. Qubit 4.0 Fluorometer (Thermo Fisher Scientific) with 1x dsDNA HS (high sensitivity) assay kit (Thermo Fisher Scientific) for dsDNA measurement.
4. NanoDrop Spectrophotomter (Thermo Fisher Scientific).
5. Nuclease-free water.

2.3 Primer/Probe Design and PCR

1. Identify authentic Hanwoo-specific deletion locus (Position: chr8:32,892,529-32,893,208) that distinguish between Hanwoo cattle breed and Holstein breeds through comparative genomics [7].
2. Downloaded cattle genome and obtain 1Kb of DNA sequences at both sides of the Hanwoo-specific deletion region using UCSC genome browser (bosTau9): http://genome.ucsc.edu/cgibin/hgGateway).
3. Oligonucleotide primer and hydrolysis probe synthesized by IDT oligo analyzer: http://www.idtdna.com/calc/analyzer [16].
4. Design the primer/probes with 18–25 bps in length that can detect 80–150 bp of Hanwoo-specific deletion region using Oligo analysis tool: http://www.operon.com/tools/oligo-analysis-tool.aspx and Oligo Calc: http://biotools.nubic.northwestern.edu/OligoCalc.html [17].
5. Primer/probe sequences we used for the digital PCR are as follows:

   Deletion_Primer-FW: 5′-AGTTCATGTCCACTGCATTGGT-3′.

   Deletion_Primer-RV: 5′-GGAGAAGAGAGCATCAGCAGATG-3′.

   Deletion_Probe: 5′-/FAM/ATGCCATCAGCCATC/MGB/-3′.

   Boundary_Primer-FW: 5′-TTCATATATTTTGTGATTGACTGTCCTATC-3′.

   Boundary_Primer-RV: 5′-TGTGGTGCTGGACAAGATTCC-3′.

   Boundary_Probe: 5′-/VIC/CTTGCTTTTCCCTCCTT/MGB/-3′.
6. Standard PCR reagents.
7. A thermal cycler PCR machine.
8. Single channel pipettes.

2.4 Stilla Naica System for Crystal Digital PCR

1. 5x Naica® multiplex PCR mix Buffer A (Cat. No. R10054) (Stilla Technologies).
2. 100% Naica® multiplex PCR mix Buffer B (Cat. No. R10054) (Stilla Technologies).
3. 25x FAM- and VIC-labeled TaqMan probes (Final 1× concentration of each primer is 450 nM, and the final 1× concentration of 250 nM for each of the probes).
4. 2x PerfeCta qPCR ToughMix UNG (Cat. No. 95138–012) (Quanta Biosciences).
5. 100 nM of Fluorescein (VWR International).
6. Sapphire Chip (Cat. No. R10054) (Stilla Technologies).

7. Naica Geode (automated droplet generator and thermocycler) (Stilla Technologies).
8. Naica Prism 3 equipment (multicolor fluorescence imager) (Stilla Technologies).
9. Crystal Miner software (Stilla Technologies).

2.5 Bio-Rad for QX200™ Droplet Digital PCR System

1. QX200™ 2x ddPCR Supermix for Probes (No dUTP) (Cat. No. 186–3023) (Bio-Rad).
2. 20x FAM- and HEX- labeled TaqMan probes (Final 1× concentration of each primer is 450 nM, and the final 1× concentration of 250 nM for each of the probes).
3. ddPCR™ Droplet Reader Oil (Cat. No. 1863004) (Bio-Rad).
4. Droplet Generation Oil for probe (Cat. No. 1863005) (Bio-Rad).
5. QX200™ droplet generator and reader (Bio-Rad).
6. DG8 Cartridges (Cat. No. 1863006) and Gaskets for QX200™ Droplet Generator (Cat. No. 1863009) (Bio-Rad).
7. ddPCR 96-well semi-skirted plates (Cat. No. 17005224) and PX1 PCR plate sealer (Bio-Rad).
8. C1000 Touch™ Thermal Cycler (Bio-Rad).
9. Quanta Soft analysis software (Bio-Rad).

2.6 Optolane for LOAA Analyzer

1. 3x Dr. PCR Master Mix (OPTOLANE).
2. 20x FAM- and Cy5-labeled TaqMan probes (Final 1× concentration of 500 nM for each primer and the final 1× concentration of 250 nM for each of the probes).
3. Semiconductor-based Dr. PCR cartridge (Cat. No. CTR002-R) (OPTOLANE).
4. LOAA POSTMAN (OPTOLANE).
5. LOAA dPCR system (OPTOLANE).
6. LOAA Dr. PCR software 3.0.0 (OPTOLANE).

3 Methods

3.1 DNA Extraction

Hanwoo and Holstein Bloods are obtained from domestic livestock in Korea. Genomic DNAs of cattle (Hanwoo and Holstein breed) are extracted from blood samples using the DNeasy Blood & Tissue kit, a silica-based membrane method (*see* **Note 1**).

1. Add 20 μL Proteinase K into a 1.5 mL microcentrifuge tube containing 200 μL of Whole blood cells.
2. Add 200 μL Buffer AL. Mix thoroughly by vortexing. Incubate blood samples at 56 °C for 10 min.

3. Add 200 μL ethanol (96–100%). Mix thoroughly by vortexing.
4. Transfer the mixtures into a DNeasy Mini spin column placed in a 2 mL collection tube.
5. Centrifuge at ≥6000 ×*g* (8000 rpm) for 1 min. Discard the flow-through and collection tube.
6. Place the spin column in a new 2 mL collection tube and add 500 μL Buffer AW1.
7. Centrifuge for 1 min at ≥6000 ×*g*. Discard the flow-through and collection tube.
8. Place the spin column in a new 2 mL collection tube and add 500 μL Buffer AW2.
9. Centrifuge for 3 min at 20,000 ×*g* (14,000 rpm). Discard the flow-through and collection tube.
10. Transfer the spin column to a new 1.5 mL or 2 mL microcentrifuge tube.
11. Elute the DNA by adding 200 μL Buffer AE and incubate for 1 min at room temperature.
12. Centrifuge for 1 min at ≥6000 ×*g* (8000 rpm).
13. Stored at −20 °C or −70 °C until use.

3.2 DNA Quantification and Quality Control

DNAs are measured absorbance from 220 nm to 340 nm, to evaluate DNA purity using NanoDrop spectrophotometers.

1. Select the nucleic acid tab and tap dsDNA samples to be measured.
2. Lift the NanoDrop arm and pipet 1 μL of blanking solution on top of the Nanodrop sensor.
3. Lower the arm and tap Blank for measurement.
4. Lift the arm and wipe on top of the sensor.
5. Pipet 1 μL of sample solution on top of the Nanodrop sensor.
6. Lower the arm to automatically start measuring the sample.
7. Confirm DNA concentration and two absorbance ratios (A260/230 and A260/A280).

For quantification, DNAs are measured using the Qubit 4.0 fluorometer with 1x ds DNA HS (high-sensitivity) assay kit.

1. Set up and add 10 μL of Qubit™ standard solution (*see* **Note 2**).
2. Set up and add 1–20 μL of each sample in the sample assay tube.
3. Add the Qubit™ 1x dsDNA working solution to each tube so that the final volume is 200 μL.

4. Vortex vigorously for 5 s and incubate at room temperature for 2 min.
5. Read the standard and samples.

3.3 Primer/Probes Design and PCR

1. Design primers with a GC content of 40–60%, a Tm value of 50–65 °C, and a size of 80–150 bp using Oligo analysis tool and Oligo Calc web tool (*see* **Note 3**).
2. Prepare primer/probes by diluting them with ultrapure water to a working concentration.
3. Prepare a ready-to-use PCR reaction mixture that contains DNA template, primer pairs, standard PCR reagents, and water.
4. PCR amplification is conducted by the following process: pre-denaturation at 95 °C for 5 min, followed by 35 cycles of denaturation at 95 °C for 30 s, annealing step at 60 °C for 30 s, and extension at 72 °C for 1 min, followed by a final extension at 72 °C for 1 min.
5. Through PCR amplification and gel electrophoresis, confirm the polymorphic pattern of the PCR product (680 bp/310 bp) using Hanwoo and Holstein samples (Fig. 2) [8, 9].

3.4 Stilla Naica® System for Crystal Digital PCR

3.4.1 Partition Generation and Thermocycling

1. ddPCR reaction mixture (25 μL) contains 12.5 μL of PerfeCta qPCR ToughMix UNG 2x, 2.5 μL of 100 nM of Fluorescein, 1.25 μL of primer set/FAM- and VIC-probe (final concentration of 900 nM/250 nM, respectively), 50 ng of DNAs, and nuclease-free water up to 25 μL (*see* **Notes 4** and **5**).
2. Open the Sapphire chips and load the reaction mixtures (25 μL) into the inlet port of Sapphire chips (*see* **Notes 6** and **7**).
3. Seal the inlet port with the white PCR caps.
4. Place up to three Sapphire chips in the Naica Geode.
5. Launch the partitioning and thermocycling system.
6. The PCR thermocycling program: 95 °C for 10 min, followed by 45 cycles of 95 °C for 10 s and 60 °C for 15 s.

3.4.2 Analysis of Fluorescence Image

1. Launch the Naica Prism3 software.
2. Transfer the sapphire chips into the Naica Prism3 for fluorescence imaging.
3. After the run has completed, analyze the droplets using Naica Prism3.

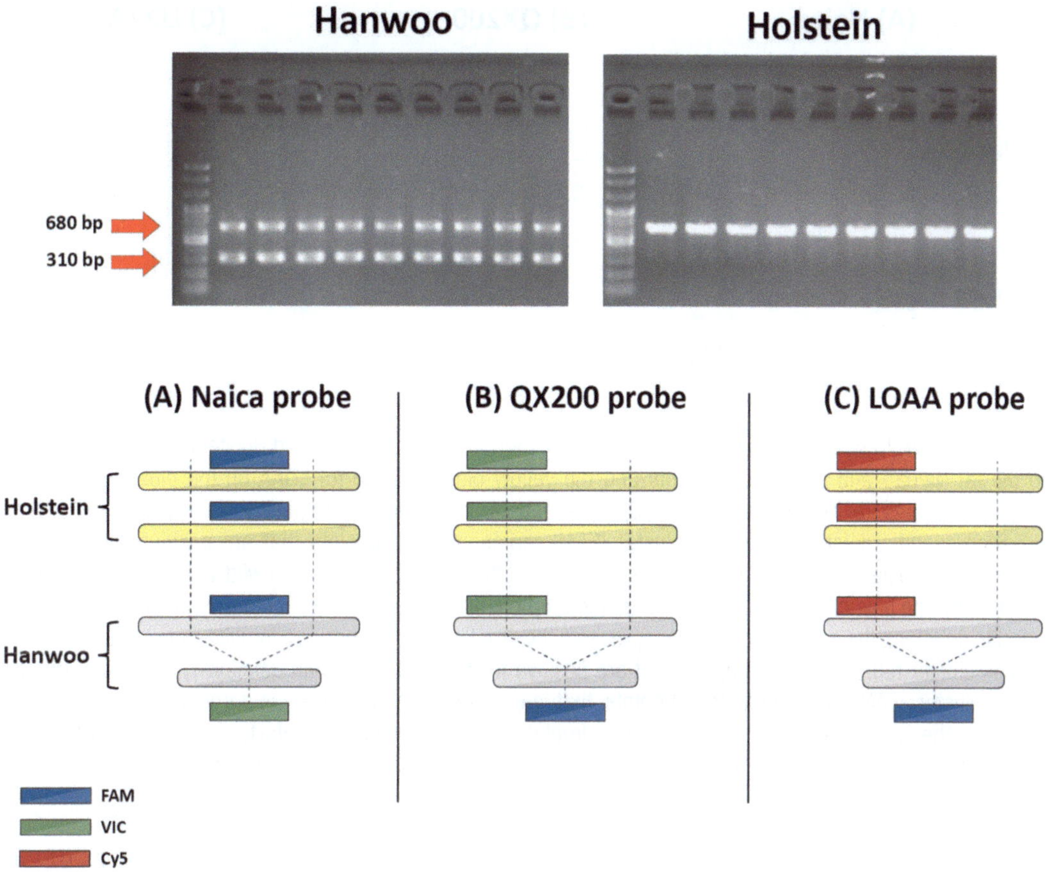

Fig. 2 Structural variation with polymorphic patterns in the Hanwoo and Holstein genomes. Gel photograph of Del96 locus showing heterozygous alleles (680 bp and 310 bp) in Hanwoo samples and homozygous alleles (680 bp) in Holstein samples [9]. Multiple probe sets detected the alleles. (**a**) For the Naica® probe, the FAM dye and the VIC dye are designed to detect all bovine genomes and only the Hanwoo genomes, respectively. (**b**) For the QX200™ probe, the VIC dye and FAM dye are designed to detect all bovine genomes and only the Hanwoo genomes, respectively. (**c**) For the LOAA probe, the Cy5 dye and FAM dye are designed to detect all bovine genomes and only the Hanwoo genomes, respectively

4. High-resolution images for each droplet crystal are analyzed fluorescence values using Crystal Miner software, adjusting the thresholds (Fig. 3).
5. In the 2D plot view of the software, manually set the fluorescence thresholds for the FAM and VIC fluorophores to determine the positive population for each fluorophore in each sample.
6. Determine copy number and concentration of targeted nucleic acids.
7. Export the data from the Crystal Miner software into a Comma Separated Values (CSV) file.

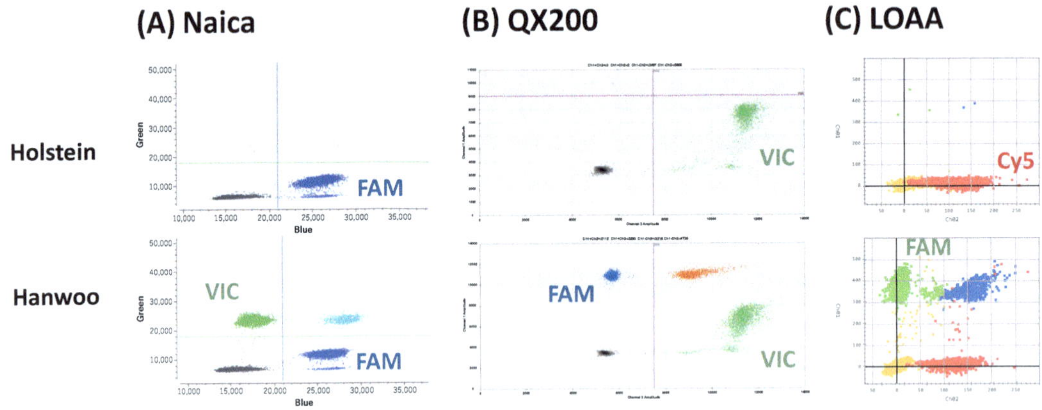

Fig. 3 Example 2D scatter plots of three digital PCR experiments for Hanwoo and Holstein. The black clusters in the (**a**) Naica® and (B) QX200™ plots and the yellow cluster in the (**c**) LOAA on the plot represent the negative droplets. (**a**) The green cluster (VIC) represents the droplets that are positive for mutant-type targets (Hanwoo-specific deletion), the blue cluster (FAM) represents the droplets that are positive for wild-type targets, and the light blue cluster represents the droplets that are positive for both wild-type and deletion-type targets. (**b**) The blue cluster (FAM) represents the droplets that are positive for mutant-type targets (Hanwoo-specific deletion), the green cluster (VIC) represents the droplets that are positive for wild-type targets, and the orange cluster represents the droplets that are positive for both wild-type and deletion-type targets. (**c**) The green cluster (FAM) represents the droplets that are positive for mutant-type targets (Hanwoo-specific deletion), the red cluster (Cy5) represents the droplets that are positive for wild-type targets, and the blue cluster represents the droplets that are positive for both wild-type and deletion-type targets

3.5 Bio-Rad for QX200™ Droplet Digital PCR System

3.5.1 Partition Generation

1. ddPCR reaction mixture (20 μL) contains 10 μL of 1x dPCR supermix for probes, 2 μL of primer set/FAM- and HEX-probes (final concentration of 900 nM/250 nM, respectively), 50 ng of DNAs, and nuclease-free water up to 20 μL (*see* **Notes 4**, **5**, and **8**).
2. Open the QX200™/QX100™ Droplet Generator and insert the DG8™ Cartridges cartridge into the cartridge holder.
3. Transfer 20 μL of PCR mix in the middle wells of the cartridge.
4. Transfer 70 μL of droplet generation oil in the bottom wells of the cartridge (*see* **Notes** 7 and **9**).
5. Close the cartridge with one dedicated gasket.
6. Place the cartridge holder and close the QX200™ Droplet generator.
7. When droplet generation is complete, all three indicator lights are solid green.

3.5.2 Partition Transfer

1. Recover the DG8™ Cartridges holder from the Droplet Generator and remove the gasket.
2. Pipette the 40 μL of generated emulsion (top wells of the cartridge) and smoothly transfer it into a single column of ddPCR 96-Well plate (*see* **Note** 7).

3. Place and seal the Pierceable foil on the ddPCR 96-well plate (*see* **Note 10**).
4. Set up and launch the thermocycling system.
5. The PCR thermocycling program: Enzyme reaction step of 3 min at 50 °C, initial denaturation step of 15 min at 95 °C, followed by 40 cycles of denaturation of 10 s at 95 °C, 20 s at 60 °C for annealing and extension.

3.5.3 Analysis of Fluorescence Image

1. Set up and launch QX200 Droplet Reader and QuantaSoft software.
2. Create a new template and assign a plate template.
3. Open the reader door and check the indicator lights on the front of the droplet reader (*see* **Note 11**).
4. Place the ddPCR 96-well plate into the plate holder.
5. Load the plate holder into the droplet reader.

3.5.4 Data Analysis

1. After the run has completed, analyze the droplets using QuantaSoft (*see* **Note 11**).
2. The scatter plot is automatically analyzed, and the thresholds for each detection channel are adjusted (Fig. 3).
3. Determine the positive population for each fluorophore in each sample by manually setting the fluorescence thresholds for FAM and HEX fluorophores in the 2D plot view of the software.
4. Determine copy number and concentration of targeted nucleic acids.
5. Export the data from the QuantaSoft software into a Comma Separated Values (CSV) file.

3.6 Optolane for LOAA Analyzer

1. ddPCR reaction mixture (30 μL) contains 10 μL of 3x Dr. PCR Master Mix, 10 μL of primer set/FAM- and Cy5- probe mix (final concentration of 900 nM/250 nM, respectively) 50 ng of DNA, and nuclease-free water up to 30 μL (*see* **Notes 4** and **5**).
2. Place in a semiconductor-based cartridge and spread on the chip using a sample loader, LOAA POSTMAN.
3. Open the red rubber at the cartridge inlet and inject the sample into the cartridge (*see* **Note** 7).
4. Press the activating button on the LOAA POSTMAN for about 1 s. The sample is automatically loaded into the cartridge.
5. Place the cartridge in the LOAA analyzer (*see* **Note 12**).
6. Press the "Next" button on the screen, and the loader automatically moves back into the equipment.

7. Launch the thermocycling system (*see* **Note 13**).
8. The PCR thermocycling program: Enzyme reaction step of 3 min at 50 °C, initial denaturation step of 15 min at 95 °C, followed by 40 cycles of denaturation 10 s at 95 °C, and 20 s at 62 °C for annealing and extension.
9. Analyze the fluorescence image automatically with LOAA Dr. PCR software 3.0.0.
10. The result is automatically displayed on the screen after completing the analysis. The scatter chart shows the result of the distribution of fluorescence values (Fig. 3).
11. By manually adjusting the FAM and Cy5 fluorescence thresholds in the software's 2D plot view, determine the positive population for each fluorophore in each sample.
12. Determine copy number and concentration of targeted nucleic acids.
13. Export the data from the LOAA Dr. PCR software into a Comma Separated Values (CSV) file.

Additional details on digital experiments and genetic information are available as Electronic Supplementary Material, in line with the dMIQE (Minimum Information for Publication of Quantitative Digital PCR Experiments) and LAG-R (Laboratory Animal Genetic Reporting) frameworks.

4 Conclusion

Structural variations are genetic differences between individuals or species, which contribute to phenotypic differences. Three different dPCR platforms are applicable to detect species identification marker due to high accuracy and sensitivity. In this protocol, we show that dPCR is an ideal platform for detecting genomic variation and can effectively detect structural variation in the genome. Therefore, our protocol compares the pros and cons of each platform, allowing researchers to choose the right dPCR platform based on their goals and resources.

5 Notes

1. Fresh blood is placed in cold storage at +2 °C or +8 °C within 48 h. Whole blood should be immediately frozen and stored at −20 °C or −80 °C until use.
2. The Qubit™ 1x dsDNA HS Assay requires two standards.
3. Amplicons should ideally be no longer than 130 bp in length.

4. All samples are run in triplicates.
5. Negative controls should always be included in every experiment.
6. The Sapphire chips are already prefilled with oil.
7. Gently pipette to dispense droplets. Pipette only until the first stop, and do not push the pipette down to the final stop.
8. Use only Supermix for ddPCR. SYBR Green is incompatible with the dPCR system.
9. Use different oils for probes and EvaGreen.
10. Do not spin the plate.
11. The first two lights on the left should be solid green. If the second light is flashing green, you can start the run. When the droplet reading is complete, all four indicator lights should be solid green.
12. The cartridge must be locked, and Optolane company logo on the top of the cartridge must be aligned with triangle mark on the top of the loader of the LOAA analyzer.
13. The cartridge is automatically checked and completed within about 30 s. The experiment will start automatically when the cartridge and analysis are done successfully.

Acknowledgments

This research was supported by Basic Science Research Program through the National Research Foundation of Korea (NRF) funded by the Ministry of Education (NRF-RS-2023-00275307) and Basic Science Research Capacity Enhancement Project through Korea Basic Science Institute (National research Facilities and Equipment Center) grant funded by the Ministry of Education (Grant No. 2019R1A6C1010033).

References

1. Sykes PJ, Neoh SH, Brisco MJ, Hughes E, Condon J, Morley AA (1992) Quantitation of targets for PCR by use of limiting dilution. BioTechniques 13(3):444–449
2. Vogelstein B, Kinzler KW (1999) Digital PCR. Proc Natl Acad Sci USA 96(16):9236–9241. https://doi.org/10.1073/pnas.96.16.9236
3. Basu AS (2017) Digital assays. Part I: Partitioning statistics and digital PCR. Slas Technol 22(4):369–386. https://doi.org/10.1177/2472630317705680
4. Mao X, Liu C, Tong H, Chen Y, Liu K (2019) Principles of digital PCR and its applications in current obstetrical and gynecological diseases. Am J Transl Res 11(12):7209–7222
5. Majumdar N, Wessel T, Marks J (2015) Digital PCR modeling for maximal sensitivity, dynamic range and measurement precision. PLoS One 10(3):e0118833. https://doi.org/10.1371/journal.pone.0118833
6. Tan LL, Loganathan N, Agarwalla S, Yang C, Yuan W, Zeng J, Wu R, Wang W, Duraiswamy S (2023) Current commercial dPCR platforms: technology and market review. Crit Rev Biotechnol 43(3):433–464. https://doi.org/10.1080/07388551.2022.2037503

7. Day E, Dear PH, McCaughan F (2013) Digital PCR strategies in the development and analysis of molecular biomarkers for personalized medicine. Methods 59(1):101–107. https://doi.org/10.1016/j.ymeth.2012.08.001
8. Park J, Shin W, Mun S, Oh MH, Lim D, Oh DY, Bhak Y, Bhak J, Park YS, Han K (2019) Investigation of Hanwoo-specific structural variations using whole-genome sequencing data. Genes Genomics 41(2):233–240. https://doi.org/10.1007/s13258-018-0772-3
9. Lee CJ, Shin W, Song M, Shin SS, Park Y, Srikulnath K, Kim DH, Han K (2023) Comparison of digital PCR platforms using the molecular marker. Genomics Inform 21(2): e24. https://doi.org/10.5808/gi.23008
10. Madic J, Zocevic A, Senlis V, Fradet E, Andre B, Muller S, Dangla R, Droniou ME (2016) Three-color crystal digital PCR. Biomol Detect Quantif 10:34–46. https://doi.org/10.1016/j.bdq.2016.10.002
11. Shin W, Kim H, Oh DY, Kim DH, Han K (2020) Quantitative evaluation of the molecular marker using droplet digital PCR. Genomics Inform 18(1):e4. https://doi.org/10.5808/GI.2020.18.1.e4
12. Lee SS, Park JH, Bae YK (2021) Comparison of two digital PCR methods for EGFR DNA and SARS-CoV-2 RNA quantification. Clin Chim Acta 521:9–18. https://doi.org/10.1016/j.cca.2021.06.016
13. Lee H, Lee CJ, Kim DH, Cho CS, Shin W, Han K (2021) High-accuracy quantitative principle of a new compact digital PCR equipment: lab on an array. Genomics Inform 19(3):e34. https://doi.org/10.5808/gi.21035
14. Chung Y, Jang SS, Kang DH, Kim YK, Kim HJ, Chung KY, Choi I, Lee SH (2023) Identification of potential biomarkers associated with meat tenderness in Hanwoo (Korean cattle): an expression quantitative trait loci analysis. Anim Genet 54(6):786–791. https://doi.org/10.1111/age.13360
15. Choi JW, Lee KT, Liao X, Stothard P, An HS, Ahn S, Lee S, Lee SY, Moore SS, Kim TH (2013) Genome-wide copy number variation in Hanwoo, Black Angus, and Holstein cattle. Mamm Genome 24(3–4):151–163. https://doi.org/10.1007/s00335-013-9449-z
16. Owczarzy R, Tataurov AV, Wu Y, Manthey JA, McQuisten KA, Almabrazi HG, Pedersen KF, Lin Y, Garretson J, McEntaggart NO, Sailor CA, Dawson RB, Peek AS (2008) IDT SciTools: a suite for analysis and design of nucleic acid oligomers. Nucleic Acids Res 36(Web Server issue):W163–W169. https://doi.org/10.1093/nar/gkn198
17. Kibbe WA (2007) OligoCalc: an online oligonucleotide properties calculator. Nucleic Acids Res 35(Web Server issue):W43–W46. https://doi.org/10.1093/nar/gkm234

Chapter 11

A Duplex Digital PCR Method for Quantification of Transgene Soybean Line MON40-3-2 in Complex Matrices

Alexandra Bogožalec Košir, Tina Demšar, Dejan Štebih, and Mojca Milavec

Abstract

The widespread use of genetically modified organisms (GMOs) since 1994 has led to increased regulatory scrutiny to monitor their presence in supply chains. For many countries, strict adherence to regulatory thresholds for labeling GMO-containing products is crucial. However, detecting and quantifying GMOs in complex food and feed matrices poses challenges. While real-time quantitative PCR (qPCR) has been the gold standard for GMO detection, it can face limitations, especially in the presence of inhibitors. Recognizing these challenges, digital PCR (dPCR) emerges as a promising alternative, offering absolute quantification without the need for standard curves and mitigating the impact of inhibitors. This chapter outlines a duplex dPCR method targeting a specific genetically modified soybean line MON40-3-2 and soybean reference gene (*Le1*), providing several practical advantages over simplex dPCR. By allowing simultaneous amplification of two target sequences, the duplex method reduces time, reagents, and resources, while ensuring consistency and reliability of results. Although described for the QX200 platform, the method is adaptable to other platforms with caution.

Key words Genetically modified organisms, Digital PCR, GMO quantification, Complex matrices, Soybean

1 Introduction

Since the introduction of the first genetically modified (GM) plant in 1994 [1], the utilization of GM organisms (GMOs) in food and feed products has grown significantly. This expansion has prompted stringent regulatory measures worldwide to monitor and control the presence and quantities of GMOs within supply chains [2–4]. Within the European Union (EU), adherence to regulatory thresholds for labeling food and feed products containing, consisting of, or produced from authorized GMOs is essential (Regulation [EC] 1829/2003; European Commission, 2003) [5]. Given the complexity of food and feed matrices, ensuring accurate detection and quantification of GMOs presents challenges.

Guillaume Pavlovic (ed.), *Digital PCR: Methods and Protocols*, Methods in Molecular Biology, vol. 2969,
https://doi.org/10.1007/978-1-0716-4767-7_11,

In the field of GMO detection and quantification, real-time quantitative PCR (qPCR) methods have long been regarded as the gold standard. In the EU, National Reference Laboratories (NRLs) and control laboratories utilize methods provided and validated by the European Union Reference Laboratory for Genetically Modified Food and Feed (EURL GMFF). However, despite their established efficacy, qPCR methods can encounter challenges when applied to complex food and feed matrices, primarily due to the presence of inhibitors. Recognizing these limitations, digital PCR (dPCR) emerges as a promising alternative. By partitioning the reaction mixture into numerous individual reactions, dPCR not only offers absolute quantification without the need for standard curves, but also mitigates the impact of inhibitors.

The duplex dPCR method outlined in this chapter targets both the soybean reference lectin gene (*Le1*) and the widely distributed soybean line MON40-3-2 ("Roundup Ready" soybean) within a single reaction (assays taken from the GMOMETHODS database [6]). Duplex dPCR offers several practical advantages over simplex PCR reactions. By allowing the simultaneous amplification of two target sequences within a single reaction, it reduces the time, reagents, and resources required for analysis, which reduces costs of analysis. Furthermore, duplex dPCR ensures consistency by subjecting both target sequences to identical reaction conditions, thereby reducing experimental variability and improving the reliability of results. This protocol describes the application of the method on the QX200 Droplet Digital PCR System but was also tested on an older version, the QX100 Droplet Digital PCR System. While the duplex has not been tested on the newer QX ONE or QX600 Droplet Digital PCR Systems, based on previous experience, the methods should be easily transferred. However, for any key deviation from the protocol, we advise to proceed with caution, and to include additional controls such as certified reference materials or samples that have been previously tested and have known GMO content. For specific guidelines for method verification on other platforms see also the Definition of Minimum Performance Requirements for Analytical Methods of GMO Testing part 2 [7].

2 Materials

Make sure that all reagents are thawed and have reached room temperature unless stated otherwise by the manufacturer. Exposure of probes to light should be minimized.

2.1 Samples

1. DNA extracted from food or feed sample (*see* **Note 1**).
2. Positive and negative controls (*see* **Note 2**).

Table 1
Primers and probes used for quantification of MON40-3-2

Target	Method[a]	Type of oligonucleotide	5′-sequence-3′	Final concentration in PCR (nM)
Le1	QT-EVE-GM-002	*Le1* forward primer	CCAGCTTCGCCGCTTCCTTC	600
		Le1 reverse primer	GAAGGCAAGCCCATCTGCAAGCC	600
		Le1 probe	FAM-CTTCACCTTCTATGCCCCTGACAC-BHQ	180
MON40-3-2	QT-EVE-GM-005	MON40-3-2 forward primer	TTCATTCAAAATAAGATCATACATACAGGTT	650
		MON40-3-2 reverse primer	GGCATTTGTAGGAGCCACCTT	650
		MON40-3-2 probe	FAM-CCTTTTCCATTTGGG-MGBNFQ	200

[a]GMOMETHODS database [6]

2.2 Reaction Setup and Thermal Cycling

1. UV chamber (*see* **Note 3**).
2. Pipettes with corresponding tips.
3. Vortex and microcentrifuge.
4. ddPCR™ Supermix for Probes (No dUTP) (Bio-Rad) (*see* **Note 4**).
5. Nuclease-free water.
6. Primers and probes (detailed in Table 1).

2.3 Partitioning and Thermal Cycling

1. UV chamber (*see* **Note 3**).
2. Pipettes (Rainin) with corresponding tips (*see* **Note 5**).
3. Manual QX200 droplet generator (DG8) or Automated droplet generator (AutoDG) (Bio-Rad).
4. Droplet Generation Oil for Probes or Automated droplet Generation Oil for Probes (Bio-Rad).
5. DG8 Cartridge Holders, Cartridges and Gaskets for manual droplet generator or DG32 Automated Droplet Generator Cartridges (Bio-Rad).
6. Pipette Tips for the AutoDG System (Bio-Rad) (*see* **Note 6**).
7. Pipette Tip Waste Bins for the AutoDG™ System (Bio-Rad).
8. Cooling block.
9. 96-well semi-skirted microtiter plate (*see* **Note 7**).
10. Pierceable Foil Heat Seal (Bio-Rad) or equivalent.

Table 2
Thermal cycling conditions for amplification of both simplex and duplex methods

No. of cycles	Time	Temperature (°C)
1	10 min	95
40	30 s 60 s	94 60
1	10 min	98
1	Infinite	4

Lid temperature: 105 °C, reaction volume: 40 μL, Ramp rate: 2.5 °C/s

11. Heat Sealer (Bio-Rad PX1 PCR Plate Sealer or equivalent).
12. C1000 Touch or T100 Thermal Cycler (Bio-Rad) or equivalent (cycling conditions detailed in Table 2).
13. Nuclease-free microcentrifuge tubes.

2.4 Partition Readout and Analysis

1. Droplet Reader Oil (Bio-Rad).
2. QX200 Droplet Reader (Bio-Rad) (*see* **Note 8**).
3. QuantaSoft analysis software (Bio-Rad) (*see* **Note 9**).
4. Excel (Microsoft) or similar software.

3 Methods

3.1 Preparation of Reaction Mixture

1. Each final PCR reaction should consist of 10 μL of ddPCR Supermix for probes (No dUTP), 4 μL of DNA, and primers and probes in appropriate working concentration to achieve final concentration as described in Table 1. Due to the loss of volume in the pipetting steps, 20% surplus should be prepared, when calculating volumes.
2. Vortex the mixture and centrifuge it briefly (*see* **Note 10**).
3. Distribute the mixtures into a 96-well polypropylene plate or an 8-well strip, with a final volume of 17.6 μL per reaction (*see* **Note 11**).
4. Add 4.4 μL of sample or control into the wells containing the reaction mixtures (*see* **Note 11**). Seal the 96-well plates with pierceable foil or close the 8-well strip with corresponding caps.
5. Vortex and centrifuge briefly.

3.2 Partitioning and Thermal Cycling

1. Turn on the droplet generator by connecting it to a power source.
2. Droplet can be generated using manual or automated droplet generator:
   (a) When using the manual droplet generator, insert the DG8 cartridge into the cartridge holder and transfer 20 μL of each reaction sample into the sample wells (*see* **Note 12**) and 70 μL of droplet generation oil into each oil well on the cartridge. Using the holes on the sides, hook the gasket over the cartridge holder. Place the holder with the cartridge into the instrument and check if the cartridge status indicator light (middle) turned green. Close the instrument by pressing the button on the top of the instrument. When the lid is closed, partition generation begins, indicated by a flashing green light on the right. Partition generation is finished when all three indicator lights are solid green. Place the holder with the cartridge on a flat surface and remove and discard the gasket. Transfer 40 μL of suspension with droplets (partitions) from droplets well (top) into a 96-well plate placed on cooling block according to the plan (*see* **Note 13**). Repeat droplet generation and plate transfer steps for all the samples.
   (b) When using the automated droplet generator, configure the sample plate in the instrument software, place the DG32 cartridges, pipette tip boxes without box lids, and the tip waste bin into their dedicated holders in the instrument. Place the sealed 96-well plate containing the prepared reactions into its dedicated holder (named "sample plate") in the instrument. Place the cooling block (*see* **Note 14**) into the holder named "droplet plate" and place an empty 96-well plate into the cooling block for droplet collection. Check if there is enough automated droplet generation oil present and change the oil if needed (*see* **Note 15**). Tap "START Droplet Generation" and "Start Run" to initiate droplet generation. After the run is finished and the instrument door is unlocked, remove the plate with droplets from the instrument and discard the used consumables.
3. After the transfer of all the samples to a 96-well plate, seal the plate with a pierceable foil using PCR plate sealer. Do not vortex or centrifuge the plate.
4. Place the sealed 96-well plate in the PCR cycler. Run the program described in Table 2 (*see* **Note 16**).

3.3 Partition Reading and Data Analysis

1. Transfer the 96-well plate into the QX200 Droplet Reader (*see* **Note 17**).
2. Run the QuantaSoft software on the connected computer.
3. Flush the instrument by clicking "Setup" → "Instrument routines" → "Flush System" in the QuantaSoft software. If the instrument was not used for longer than a week, priming should be done instead of flushing before running a plate by clicking "Setup" → "Instrument routines" → "Prime."
4. To configure the plate layout, select "Setup" → "Template" → "New" and enter the necessary information. Save the created template.
5. Click "Run" to initiate droplet (partition) reading (*see* **Note 18**).
6. After reading is finished, save the experiment data, and remove the plate from the instrument.
7. Open the experiment in the QuantaSoft software and click "Analyze" to start data analysis.
8. Set the fluorescence thresholds for distinguishing between positive and negative partitions. The thresholds should be set simultaneously for all samples. NTCs can be used as a guide for determining the negative population and setting the threshold just above the negative cluster found in NTC(s) (*see* **Note 19**). Both 1D and 2D views should be checked (Fig. 1).
9. Before further analysis, quality control should be performed:
   (a) Wells with an accepted number of partitions <10,000 should not be considered for further analysis.
   (b) The reaction in each well is considered positive if three or more partitions show the fluorescent signal above a threshold.
   (c) NTC(s) should be negative (*see* **Note 20**).
   (d) Positive control reaction(s) should be positive (*see* **Note 20**).
10. The software calculates the concentration of target copies in each sample based on the set thresholds. The results can be viewed under the "Concentration" tab and can be exported in . csv format for further analysis in spreadsheet software such as Excel (Microsoft).

4 Notes

1. Different DNA extraction methods can be used for food and feed samples; for the development of the assays described here, two methods were used. DNA was extracted either from 2 g of

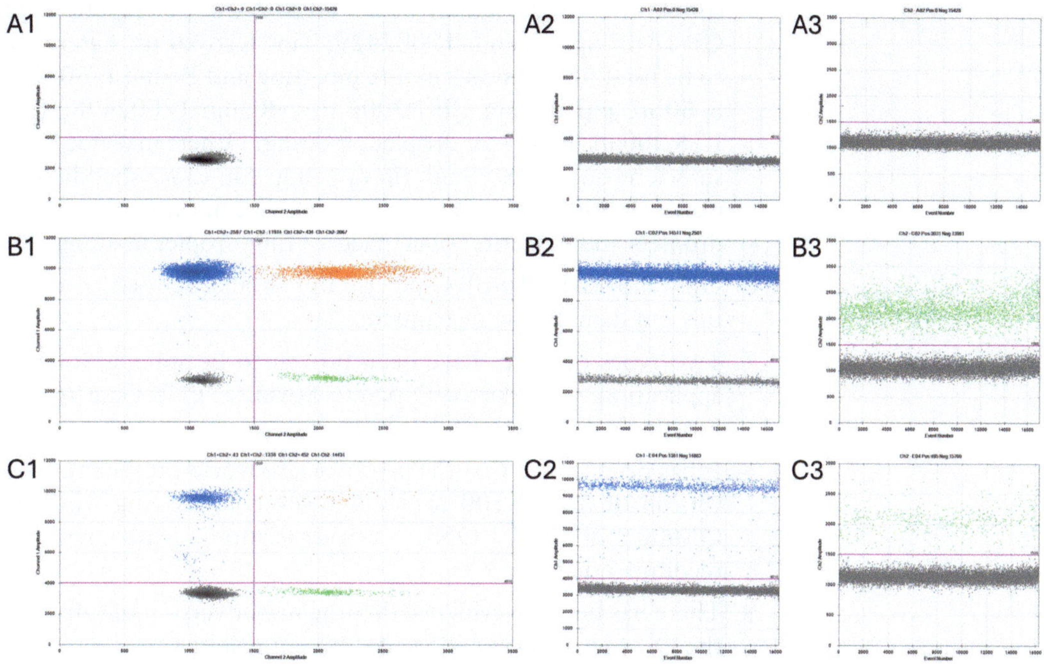

Fig. 1 Amplification plots in 1D and 2D view. Panel A1 presents the negative template control, in a 2D plot, while A2 and A3 present the negative template control in a 1D plot for FAM and 1D plot for HEX channel, respectively. Panels B and C present positive reactions, with B1, B2, and B3 presenting the positive control (DNA from certified reference material), and panels C1, C2, and C3 presenting a real-life sample (feed). The pink line represents the set threshold

sample using a cetyltrimethyl-ammonium bromide (CTAB) protocol (as described in Annex A.3 of ISO21571; International Organization for Standards, 2005) [8] or from 200 mg of sample using the NucleoSpin Food kit (Macherey Nagel GmbH & Co. KG, Düren, Germany).

2. Controls should include at least a non-template control (NTC) and a positive control, containing target DNA extracted preferably from certified reference materials. It is recommended to also include a negative extraction control to test for potential contamination during nucleic acid extraction.
3. To prevent any cross contamination of reagents and the reaction mixture, the pre-reaction mixture (primers, probes, mastermix) and addition of samples should be performed at separate locations. Ideally, both steps should be prepared in separate UV chambers.
4. ddPCR Supermix for probes (No dUTP) is stable at −20 °C until the expiration date. Once thawed, it can be stored at 4 °C for up to 2 weeks.
5. According to recommendations from the dPCR platform manufacturer, Rainin tips in combination with Raining pipettes

should be used. Rainin L-20 XLS 2–20 μL pipette and GP-L10F tips (Rainin 17002429; Tips LTS 20 μL Filter) are to be used for reaction mixture pipetting and Rainin L-50 XLS 5–50 μL pipette and GP-L200F tips (Rainin 17002428; Tips LTS 200 μL Filter) for droplet pipetting. Others may not be of sufficient quality, as rough tip openings can cause shedding of plastic debris in the reaction mixture that can cause shearing of droplets and detection of artifacts during droplet imaging.

6. Only Bio-Rad's AutoDG pipette tips should be used, as other tips can damage the instrument.
7. Two types of plates have been proven to be suitable for this application: semi-skirted Eppendorf twin.tec PCR Plate 96 and Bio-Rad's ddPCR™ 96-Well Plates.
8. It has been shown that the protocol also works on the previous version of the QX100 reader. However, for testing with the QX600 or QX ONE, a verification experiment is recommended.
9. There has been no evidence that the newer version of software, the QX Manager, would yield different results.
10. Reaction mixture should be mixed well to ensure repeatability of results. Due to the viscosity of the mastermix, vortexing at lower speeds is recommended to prevent bubble formation.
11. The volumes given include 10% excess volume to offset pipetting losses.
12. For best droplet generation air bubbles should be avoided (for pipetting techniques see manufacturer's instruction manual (Catalog Number 1864002, Bio-Rad).
13. To avoid evaporation, the 96-well plate should be placed in a cooling block. To avoid shearing or coalescing the partitions, pipette tips should be positioned at a 30–45° angle to the junction where the side wall meets the bottom of the wells. The suspension with the droplets (partitions) should be slowly drawn from the wells into the pipette tip. The pipette tip should be then positioned along the side of the wells of the 96-well plate, near the bottom of the well, and the suspension with the droplets should be slowly dispensed. Drawing and aspirating of droplets should take 5 s each (see also the manufacturer's instruction manual (Catalog Number 1864002, Bio-Rad).
14. Dark purple color indicates that the cooling block is at the correct working temperature; to achieve this, the block should be placed in a freezer for at least 2 h.
15. To save time during experiments, the oil bottle should be prepared ahead of time.

16. Thermal cycling should start within 30 min of sealing the plate. Alternatively, sealed plates can be stored at 4 °C for up to 4 h before thermal cycling.
17. The droplet reader can be switched on in advance to warm up the instrument.
18. Set the dye (FAM/HEX in our example) when the Run Options window appears.
19. In general, thresholds should be set just above the negative cluster in NTCs. Sometimes the negative cluster in a particular reaction may have higher fluorescence than the negative cluster in NTCs. The thresholds should be set separately for such reactions.
20. The analysis is not valid if the NTC(s) and/or negative extraction controls are positive or if the positive control reaction is negative. It is sufficient to test the negative extraction controls only for the target that is expected in the highest concentration (in this case *Le1*).

References

1. U.S. Food and Drug Administration (1994) Agency summary memorandum Re: Consultation with Calgene, Inc., concerning FLAVR SAVR™ tomatoes
2. Gruère GP (2007) A review of international labeling policies of genetically modified food to evaluate India' s proposed rule. AgBioforum 10: 51–64
3. Vigani M, Raimondi V, Olper A (2012) International trade and endogenous standards: the case of GMO regulations. World Trade Rev 11:415–437. https://doi.org/10.1017/S1474745612000262
4. Žel J et al (2012) How to reliably test for GMOs. Springer, Boston
5. European Commission (2003) Regulation (EC) No 1829/2003 of the European parliament and of the council of 22 September 2003 on genetically modified food and feed
6. European Commission's Joint Research Centre EU database of reference methods for GMO analysis. http://gmo-crl.jrc.ec.europa.eu/gmomethods/
7. Gatto F et al (2023) Definition of minimum performance requirements for analytical methods of GMO testing – Part 2. Publication Office of the European Union, Luxendourg. https://doi.org/10.2760/63656. JRC 125975
8. International Organization for Standards (2005) ISO 21570:2005 Foodstuffs – Methods of analysis for the detection of genetically modified organisms and derived products – Quantitative nucleic acid based methods. 1–103

Chapter 12

DNA Methylation Analyses of Tissue Samples and Cell-Free DNA by Droplet Digital PCR

Heidi Pharo, Sara Brandt-Winge, Evy Marie Thorkildsen, Hege Marie Vedeld, and Guro E. Lind

Abstract

Droplet digital PCR (ddPCR) allows highly sensitive detection of minimal amounts of a target of interest. The most common applications of ddPCR have traditionally been detection of genetic variants, copy number variations and gene expression, but the method has great potential also for DNA methylation analyses. We here describe an optimized and standardized protocol for methylation-specific ddPCR analyses, including use of an internal control; the 4Plex, and an automated algorithm for scoring of positive droplets; PoDCall. We also describe a protocol for bisulfite treatment using either of two kits from QIAGEN, utilizing cell-free DNA (cfDNA) or tissue samples as staring material. Implementation of these protocols across laboratories could contribute to more robust and consistent ddPCR methylation results.

Key words 4Plex, Bisulfite treatment, cfDNA, DNA methylation, Droplet digital PCR, Optimization, PoDCall, Protocol, Standardization

1 Introduction

Droplet digital PCR offers sensitive detection of minimal amounts of a target of interest. The sensitivity is in principle only limited by the number of droplets analyzed [1], and the method has been demonstrated to detect down to 0.01% of a given genetic variant [2]. In ddPCR, the sample mixture is divided into thousands of droplets, and individual PCRs are performed inside each droplet, significantly reducing competition from any background DNA. The absolute quantity of the target is calculated from the fraction of fluorescence-positive droplets.

DdPCR may be applied to analyze several molecular and genomic alterations, including DNA methylation, which is a chemical modification to the DNA strand where a methyl group is added to the 5′ position of cytocine residues. DNA methylation controls

Guillaume Pavlovic (ed.), *Digital PCR: Methods and Protocols*, Methods in Molecular Biology, vol. 2969,
https://doi.org/10.1007/978-1-0716-4767-7_12,

gene expression, is essential for normal development and differentiation, and plays a key role in e.g. x-chromosomal inactivation, genomic imprinting, and silencing of retrotransposons [3]. The normal DNA methylation pattern is altered in many diseases, including in cancer, characterized by genome-wide hypomethylation and region-specific hypermethylation.

While ddPCR has been frequently applied in genetic research for more than a decade, its use in epigenetic research has lagged behind, potentially due to lack of a standardized comprehensive analysis protocol. In recent years, our group has developed tools for increased standardization, robustness, and accuracy of methylation-specific ddPCR. First, we have developed the 4Plex internal control, which allows adjusting for varying or unknown DNA input [4]. Although an endogenous control is, in theory, not needed in ddPCR, we still find it essential to include in our methylation analyses. A control ensures that enough DNA is present for amplification, with a negative test result indicating the absence of methylation rather than lack of DNA. Using the 4Plex internal control for normalization reduces the variability in methylation values, corrects for differences in template amount, and diminishes the effect of chromosomal aberrations [4]. Second, we have developed the Positive Droplet Calling (PoDCall) algorithm, with an accompanying shiny app, which ensures automated and standardized calling of positive droplets [5]. These tools have already been successfully applied in published research from both our group [6–8] and from other research groups [9].

As DNA methylation is lost following PCR amplification, detection of DNA methylation by ddPCR relies on some sort of pretreatment, including bisulfite treatment. This harsh treatment, involving incubation of DNA with sodium bisulfite in combination with high temperature and pH, causes degradation and loss of DNA, and contributes to complicate DNA methylation analyses. The challenge is most evident when the amount of starting DNA is limited, as is often the case with cell-free DNA (cfDNA). Our group has evaluated several different commercial kits for bisulfite conversion of cfDNA from blood, and identified the kit giving the highest yield of amplifiable bisulfite converted cfNDA from plasma [8]. The protocol for this kit; the EpiTect Plus DNA Bisulfite Kit (QIAGEN), in addition to the EpiTect Bisulfite Kit for FFPE and fresh-frozen sample (QIAGEN), will be described in the following chapter, together with a detailed protocol of how to perform DNA methylation analyses using ddPCR with the 4Plex internal control and PoDCall. Hopefully, these standardized protocols, including useful tips for optimal performance, will be valuable for others wishing to perform ddPCR for DNA methylation analyses.

2 Materials

Follow the manufacturers' guidelines regarding storing conditions of the reagents. Read all material safety data sheets (MSDS) for correct handling, use, and disposal of chemicals.

2.1 Bisulfite Conversion

2.1.1 Reagents

1. DNA samples.
2. Positive control: Universal Methylated Human DNA standard (IVD). Cat#D5011 (BioSite).
3. Negative control: Human WGA Non-Methylated DNA (Non-Meth). Cat#D5013-1 (BioSite).
4. Bisulfite conversion kit: Either the EpiTect Bisulfite Kit. Cat#59104 (QIAGEN) for FFPE and fresh-frozen samples, or the EpiTect Plus Bisulfite Kit. Cat#59124 (QIAGEN) for cfDNA (*see* **Note 1**).
5. Ethanol (molecular biology grade, 96–100%).

2.1.2 Equipment and Accessories

1. Pipettes and pipette filter tips.
2. PCR tubes; Cat#951010022 (Eppendorf) or 96-well PCR plate; Cat#AB-0600 (Thermo scientific) and 8 Domed Cap Strips; Cat# AB-0602 (Thermo scientific).
3. Thermal cycler with heated lid.
4. Vortexer.
5. Microcentrifuge.
6. QIAcube Connect (QIAGEN).
7. 2 mL sample tubes. Cat#990381 (QIAGEN).
8. Rotor adapters and 1.5 mL microcentrifuge tubes. Cat#990394 (QIAGEN).
9. 200 μL filter tips for QIAcube. Cat#990332 (QIAGEN).
10. 1000 μL filter tips for QIAcube. Cat#990352 (QIAGEN).
11. Recommended: Heating block, thermal mixer, or heated orbital incubator.

2.2 Droplet Digital PCR

2.2.1 Reagents

1. Bisulfite treated samples, including controls.
2. Primers and probes.
3. ddPCR Supermix for Probes (no dUTP; Bio-Rad). Cat#186–3023(2 × 1 mL)/#186–3024(5 × 1 mL)/ #186–3025(5 × 5 mL).
4. Automated Droplet Generation Oil for Probes (Bio-Rad). Cat#186–4110(140 mL).
5. ddPCR Droplet Reader Oil (Bio-Rad). Cat#1863004(2 × 1 L).
6. RNase-free water.

2.2.2 Equipment and Accessories

1. Pipettes and pipette filter tips.
2. ddPCR Plates 96-well, Semi-Skirted (Pkg of 25). Cat#12001925 (Bio-Rad).
3. Vortexer.
4. QX200 Automated Droplet Generator (Bio-Rad).
5. PX1 Heat Sealer (Bio-Rad).
6. Pierceable Foil Heat Seal (Pkg of 100). Cat#181–4040 (Bio-Rad).
7. PCR machine (recommended: PTC Tempo Deep-well Thermal Cycler (Bio-Rad)).
8. QX200 Droplet Reader (Bio-Rad).
9. Plate holder for the QX200 Droplet Reader (Bio-Rad).
10. DG32™ Automated Droplet Generator Cartridges (Pkg of 30). Cat#186–4108 (Bio-Rad).
11. Pipet Tips for the AutoDG™ System (Pkg of 20, boxes of 96 each). Cat#1864120 (Bio-Rad).

3 Methods

3.1 Bisulfite Conversion

The bisulfite conversion is described for the EpiTect Bisulfite Kit for FFPE and fresh-frozen samples and for the EpiTect Plus Bisulfite Kit for cfDNA (QIAGEN) (*see* **Note 1**).

3.1.1 Conversion Reaction

1. Equilibrate DNA/IVD/NonMeth samples and protection buffer to room temperature.
2. Dissolve the bisulfite mix by adding 800 μL RNase-free water (one tube is enough for ten samples) and vortex until fully dissolved (*see* **Note 2**). Alternatively, use a thermal mixer (for 10 min at 27 °C and 1400 rpm).
3. In 200 μL PCR tubes or a 96-well PCR plate add:
   (a) 20–40 μL DNA/IVD/NonMeth samples (1 ng–2 μg) (*see* **Note 3**).
   (b) 85 μL Bisulfite mix.
   (c) 15–35 μL DNA protection buffer (*see* **Note 3**).

   The amount of sample and DNA protection buffer should equal 55 μL.
4. Seal the tubes/plate. To ensure proper sealing, use a PCR plate rack and a PCR strip cap tool when capping each well. Mix thoroughly by vortexing until the solution is uniformly blue (approximately 5 s), and then briefly centrifuge the tubes/plate.

Table 1
Thermal cycling conditions for bisulfite treatment[a]

Cycling step	Temperature (°C)	Time (min)
Denaturation	95	5
Incubation	60	25
Denaturation	95	5
Incubation	60	85
Denaturation	95	5
Incubation	60	175
Hold	20	Indefinite[b]

[a]Use a heated lid set to 100 °C. Set the sample volume to 140 μL or max volume
[b]Samples can safely be left overnight in the thermal cycler (*see* Note 4)

5. Place the PCR tubes/plate in a thermal cycler with heated lid, and run the reaction according to Table 1 (*see* **Note 4**).

3.1.2 Cleanup Procedure Using the QIAcube Connect System

The procedure is explained for cleanup using the QIAcube Connect instrument. For manual cleanup, see the manufacturers' (QIAGEN) guidelines.

1. Equilibrate the spin columns and the buffer BD to room temperature.
2. Add 27 mL ethanol (96–100%) to buffer BD and mix by inversion.
3. Add 30 mL ethanol (96–100%) to buffer BW and mix by inversion.
4. Add 310 μL RNase-free water to carrier RNA and transfer 90 μL of dissolved carrier RNA into n=13 1.5 μL microcentrifuge tubes. One tube will be used now, while the remaining 12 aliquots may be stored in the freezer (−20 °C) for later use. For next cleanup on the QIAcube, thaw one aliquot of dissolved carrier RNA, vortex and briefly centrifuge.
5. If buffer BL contains precipitation, use a thermal mixer (e.g., 30–60 min at 50 °C and 300 rpm) to dissolve the precipitate.
6. Briefly centrifuge the PCR tubes/plate containing the bisulfite reactions, and transfer the solutions into 2 mL sample tubes.
7. On the QIAcube Connect instrument: Choose "Cleanup" and select the EpiTect protocol suited for the samples; either the EpiTect Bisulfite or EpiTect Plus Bisulfite (*see* **Note 1**). Follow the instructions on the screen, including (i) fill up reagent bottles and place them in their respective position in the bottle rack, (ii) fill up 1000 μL and 200 μL filter tips, (iii) place Carrier

RNA in the micro-centrifuge tube slot, (iv) place rotor adaptors with EpiTect spin columns and 1.5 mL labeled elution tubes into the centrifuge, (v) place sample tubes in the shaker rack of the QIAcube—ensure that the lids are folded down into their designated slots, and (vi) press start.

8. After completed cleanup (approximately 70 min for 12 samples) collect the bisulfite treated DNA and store at −20 °C (*see* **Note 5**).

3.2 Droplet Digital PCR

3.2.1 Sample and Reaction Mix Preparations

1. Thaw all reagents for the reaction mix and equilibrate to room temperature: probes, primers, and ddPCR Supermix.
2. Prepare all samples at the desired concentration (*see* **Notes 6** and 7).
3. Vortex the ddPCR Supermix thoroughly (10–15 s) and spin down briefly.
4. Flick and briefly vortex the primers, then spin down briefly.
5. Flick the probe tubes and spin down briefly.
6. Prepare the reaction mix according to the volumes specified in Table 2 (*see* **Note 8**).
7. Vortex the reaction mix thoroughly and spin down briefly.
8. In a ddPCR 96-well plate pipette i) the desired sample volume, and ii) the correct volume of reaction mix, to each well (*see* **Note 9**). The combined volume of DNA sample and reaction mix should equal 22 μL (Table 2).
9. Seal the plate in the Plate Sealer, by doing the following: (i) turn on the instrument and take out the support block before heating, (ii) check that temperature is set to 180 °C and time to 5 s, (iii) let the instrument heat up, (iv) place the plate onto the support block and cover the plate with foil (*see* **Note 10)**, (v) place the plate in the support block and place the block in the sealer, and (vi) press "Seal" on the screen (the tray closes automatically, the sealing is initiated, and the tray opens automatically once the sealing is completed).
10. IMPORTANT: Mix the plate **thoroughly** by vortexing (e.g., 30 s at 1800 rpm), preferably in a thermal cycler, but a regular vortex mixer can also be used (*see* **Note 11**).
11. Centrifuge briefly to collect all liquid at the bottom of the reaction tubes. Make sure that there are no air bubbles on the bottom of the tube. If air bubbles are still present: repeat centrifugation.
12. Allow equilibration of the plate at room temperature for 3 min before initiation of droplet generation.

Table 2
Preparation of the ddPCR mix

	Volume per well (μL)	Total volume reaction mix (μL)
Number of wells[a]		106
Supermix for probes (no dUTP)	11	1166 (=11 × 106)
Primer s TARGET gene (100 μM)	0.18	19.1 (=0.18 × 106)
Primer as TARGET gene (100 μM)	0.18	19.1 (=0.18 × 106)
Probe TARGET gene (10 μM)	0.4	42.4 (=0.4 × 106)
Primer s 4Plex[b] (100 μM)	0.72 (=0.18 × 4)	76.3 (=0.72 × 106)
Primer as 4Plex[b] (100 μM)	0.72 (=0.18 × 4)	76.3 (=0.72 × 106)
Probe 4Plex[b] (10 μM)	1.6 (=0.4 × 4)	169.6 (=1.6 × 106)
H20	Variable (=22–11–0.18-0.18-0.4-0.72-0.72-1.6-DNA template)	Variable
DNA template	Variable[c] (up to 7.2 μL)	–
Total volume	22	–
Reaction mix added to each well	Variable (=22-DNA template)	

[a]Number of sample wells +10% surplus. In this table, a full 96 well plate is used: 96 wells ×1.10 ≈ 106 wells

[b]The 4Plex consists of four single-copy genes/assays (*EPHA3*, *KBTBD4*, *PLEKHF1*, and *SYT10*) that are all VIC-labeled, and that have been combined into three separate stock tubes containing equal amounts of (i) the four sense primers, (ii) the four antisense primers, and (iii) the four probes

[c]DNA amount <13.5 ng. When using the 4Plex internal control, up to 7.2 μL of DNA template may be included for analyses. If higher DNA input volume is desired, adjust/increase the concentration of the primers and/or probes, or alternatively use the ddPCR Multiplex Supermix instead of the Supermix for probes (no dUTP) (*see* **Note 8**)

3.2.2 Droplet Generation Using the AutoDG Instrument

The procedure is explained for droplet generation using the AutoDG instrument. For manual droplet generation, see the manufacturers' (Bio-Rad) guidelines.

1. Press "Configure Sample Plate" and choose the number of columns with samples for your plate. Orange light will light up inside the instrument at positions that need to be filled.
2. Load AutoDG with:
   (a) Oil (bottle is permanently placed in the instrument) (*see* **Note 12**).
   (b) Cartridges.
   (c) Pipette tips.
   (d) Blue cold adapter (*see* **Note 13**) loaded with new (empty) ddPCR 96-well plate (unsealed).
   (e) Sealed 96-well ddPCR plate with sample + reaction mix.

Table 3
Thermal cycling conditions[a]

Cycling step	Temperature (°C)	Time (min)	Ramp rate (°C/s)	Number of cycles
Enzyme activation	95	10	2	1
Denaturation	94	0.5		40[b]
Annealing/extension	60	1		40[b]
Enzyme deactivation	98	10		1
Hold	4	Indefinite[c]		1

[a]Use a heated lid set to 105 °C. Set the sample volume to 40 μL
[b]40 PCR cycles are enough for an optimized ddPCR assay
[c]30 min to 48 h

When all equipment is correctly loaded, the light, both inside the instrument and on the AutoDG screen, turns green.

3. Press "Start Droplet Generation."
4. After droplet generation is completed (approximately 45 min for a full 96-well plate), seal the plate with droplets in the Plate Sealer (explained in Subheading 3.2.1) (*see* **Note 14**).

3.2.3 Thermal Cycling

1. Place the plate in a thermal cycler, and start the PCR immediately (or no later than within 30 min) after droplet generation. The thermal cycling conditions recommended from the manufacturer are shown in Table 3.
2. After thermal cycling is completed, leave the plate at 4 °C (either in the thermal cycler or in a refrigerator) for at least 30 min in order to stabilize the droplets. The plate should preferably be run on the Droplet Reader within 48 h.

3.2.4 File Preparations in the QuantaSoft Software

The protocol refers to analyses on the QX200 Droplet Reader using the QuantaSoft software. Analyses on the QX600 Droplet Reader require the QXManager software, and necessitate adjustments to the following protocol.

1. To generate a setup file/template (.qlt file) double click on a well, and fill in the sample information for this well:
   (a) Name: Sample ID of choice.
   (b) Experiment (drop down list): "ABS"—*for absolute quantification.*
   (c) Supermix (drop down list): "ddPCR Supermix for Probes (no dUTP)" (*see* **Note 15**).
   (d) Target 1 name: Name of the assay in channel 1 (FAM-labeled), e.g., *the methylated target of interest.*
   (e) Target 1 type (drop down list): "Ch1 Unknown" (*see* **Note 16**).

(f) Target 2 name: Name of the assay in channel 2 (VIC-labeled), e.g., *the "4Plex" internal control.*

(g) Target 2 type (drop down list): "Ch2 Unknown" (*see* **Note 16**).

2. Click "Apply" and repeat the procedure explained above for the rest of the wells on the plate (*see* **Notes 17** and **18**).
3. Once the information has been registered for all sample-containing wells, click "OK" and save the template.

3.2.5 Droplet Reading: Using the QX200 Droplet Reader

1. Place the sealed plate in the plate holder for the QX200 Droplet Reader.
2. Place the plate holder with the sealed plate into the Droplet Reader.
3. Open QuantaSoft and load your previously prepared setup file/template (.qlt file; *see* Subheading 3.2.4).
4. Press the "Run" button (*see* **Note 19**).
5. In the pop-up window, choose if you want to acquire the data row-wise or column-wise and your probe dye set (e.g., FAM/-HEX or FAM/VIC).
6. Press "OK" and the run will initiate (*see* **Note 20**).

Reading of one full 96-well plate takes approximately 2.5 h.

3.2.6 Data Analysis Using Positive Droplet Calling (PoDCall)

1. Open the QuantaSoft result file and export amplitude files containing information about the droplet fluorescence levels by doing the following: i) in the "Setup" tab, mark the relevant wells (i.e., the wells you would like to analyze (*see* **Note 21**), and ii) press "Options" at the top and then "Export Amplitude and Cluster Data." Select an empty folder of choice to store the .csv files (*see* **Note 22**).
2. In the QuantaSoft result file export the sample sheet file (*see* **Note 23**) by doing the following: (i) in the "Setup" tab make sure that the same wells as in point 1.i) are marked, (ii) open the "Analyze" tab and press "Export CSV" at the bottom of the sample list, top left, and select a folder of choice to store the .csv files.
3. Open the PoDCall shiny GUI interface (*see* **Note 24**) by opening R studio and type:

```
> library(PoDCall)
> podcallShiny()
```

The PoDCall GUI interface will open in your internet browser.

4. From the dynamic dropdown list that appears, choose if your data has been exported from QXManager or from QuantaSoft.
5. Upload the amplitude files from QuantaSoft by pressing "Browse..." under "Select amplitude (.csv) file(s)."
6. Upload the sample sheet file from QuantaSoft by pressing "Browse..." under "Optional: select samplesheet file (.csv)."
7. Set the position of a reference well from the dynamic dropdown list (used to set a reference distribution of droplets). Ideally, the reference well should contain a methylation positive control such as an IVD, but any well with a representative droplet distribution may be used as a reference well.
8. Choose the number of permutations; $n = 200$ is recommended as default value (*see* **Note 25**).
9. Choose the Q value. The default Q value is 9, but can be changed by the user (*see* **Note 26**).
10. Press "Set thresholds." For a full plate (=96 csv files), the algorithm uses approximately 3–5 min to generate the results (*see* **Note 27**).

3.2.7 Data Interpretation

1. When the algorithm is finished, inspect the results, including i) the number of generated droplets for each well (*see* **Note 28**) in the "Total Droplets" tab, and ii) the thresholds in each channel (by selecting several wells under "Well name") in the "Visualization" tab. Adjust thresholds manually if desired/needed (*see* **Notes 29–33**).
2. When all thresholds are inspected and set correctly, download the result table by going to the "Thresholds" tab and "Download results" (*see* **Note 34**).
3. Open the result .cvs file in excel and split the column with comma separator.
4. Process the results (*see* **Note 35**). Consider reruns if the total droplet counts or the DNA concentrations are too low (*see* **Note 36**).

4 Notes

1. The EpiTect Bisulfite Kit and the EpiTect Plus Bisulfite Kit utilize the same reagents and similar protocol for bisulfite treatment, but feature different spin column sizes. The EpiTect Bisulfite Kit employs larger columns with a recommended elution volume of 20–40 μL, while the Plus kit, with smaller columns, allows for smaller elution volumes down to 10 μL, making it suitable for low concentrated samples such as cfDNA. For cleanup of the bisulfite treated samples, there are minor

differences in the protocols depending on the kit. For automated cleanup on the QIAcube, we use the EpiTect Bisulfite Kit with the FFPE program for DNA from patient-derived tissue and cells, and the EpiTect Plus Bisulfite Kit with the Plus program “up to 100 ng DNA” for low concentrated samples such as cfDNA samples. Customized protocols can be bought from QIAGEN. Both kits support an input range of 1 ng–2 μg.

2. The bisulfite mix is dissolved when there are no visual particles settling at the bottom of the tube upon inversion.
3. The recommended amount of DNA protection buffer in the conversion reaction is 35 μL, but the amount may be reduced down to 15 μL to allow for up to 40 μL DNA solution Alternatively, DNA samples can be up-concentrated using filtrate tubes (Cat#VN01H22, Sartorius).
4. Bisulfite treated samples can be stored at 20 °C for 24 h. This allows the bisulfite converted samples to be safely left overnight in the thermal cycler, with the possibility to perform the cleanup protocol the following day.
5. In our laboratory, we have good experience with storage of bisulfite treated samples [10]. If samples are not to be used immediately, we recommend storing your samples at −20 °C in screw cap tubes to prevent evaporation. Evaporation may lead to an increased DNA concentration, which potentially saturates the ddPCR results.
6. In our experience (mainly based on urine samples), 1.6–13.5 ng bisulfite-treated genomic DNA per ddPCR is optimal (theoretically calculated from Qubit measurement before bisulfite treatment, simplified with 100% recovery). This will result in a theoretical 4Plex (i.e., internal control) concentration of 10–500 cop/μL.
7. We recommend to always include the following controls on every ddPCR plate: bisulfite treated IVD (in duplicate), bisulfite treated NonMeth, and non-template control (NTC; in duplicate). The IVDs should be in the same concentration range as the other samples on the plate in order to function well as the reference well for PoDCall analyses.
8. The ddPCR Multiplex Supermix is 2x more concentrated than the ddPCR Supermix (No dUTP). If using this mix, adjust the Supermix volume to 0.5x and supplement with more sample volume or water accordingly.
9. It is essential that each well contains exactly 22 μL reaction volume. A lower volume in one well is likely to negatively affect the droplet generation of all other wells in that column. Also importantly, droplet generation is performed in an eight-

chamber format, and each column must consequently either be completely empty or filled up with either (i) 22 μL sample/control and reaction mix, or (ii) 22 μL 1x ddPCR Buffer Control for probes Cat#186–3052 (Bio-Rad). The ddPCR Buffer Control is cheaper than the Supermix.

10. Place the foil with the red line along the upper edge, facing upward, to facilitate orientation and identification of the upper and lower rows of the plate after sealing. If several plates are under preparation it is advisable to mark the side of the plate with plate ID.
11. Thorough vortexing of the sample plate before droplet generation is essential in order to achieve a high droplet number. If using a thermal cycler, keep a hand on the plate while vortexing to avoid the plate escaping the cycler.
12. The QX200™ Droplet Digital PCR System also supports DNA-binding dye (EvaGreen®) assays, which requires use of a different supermix and droplet generation oil. See www.bio-rad.com for more information.
13. Blue cold adapter (for empty ddPCR 96-well plates) should be stored upside down in freezer (−20 °C) to keep the cold water and ice in the upper part of the adapter, where the ddPCR plate is placed.
14. Visually inspect the droplets. If wells contain significantly less droplets or are completely clear of droplets, this will explain absence or low droplet count in the QuantaSoft results. Consistent low droplet count/absence of droplets may be caused by plastic debris from the pipette, which can clog the capillaries on the cartridges used for droplet formation. Contact Bio-Rad if experiencing consistent low droplet formation.
15. In QuantaSoft there is no option to choose the ddPCR Multiplex Supermix. If using this mix, choose the ddPCR Supermix (No dUTP) in the dynamic dropdown list.
16. The dynamic dropdown list gives potential to mark wells as unused, unknown, reference, positive, negative, blank, or NTC. This will not affect the reading of the droplets. For convenience we choose unknown for all samples and controls.
17. It is possible to fill in the same information to multiple wells simultaneously by selecting multiple wells at once. By pressing "Apply" or "OK," all the selected wells will be given the same filled-in/ticked-off information. NB! There is no "regret" button in QuantaSoft. Therefore, double-check that the information ticked-off are indeed supposed to be applied to the marked wells before pressing "OK"/"Apply," i.e., make sure that you do not overwrite the sample ID of choice.

18. The QX200 reader will only read wells with sample information, and empty wells should not be given any information in QuantaSoft. Attempts to read empty wells may cause air bubbles in the system. Unlike the droplet generator, the QX200 reader does not require completely full columns or rows and can start and stop wherever, including skipping blank wells and continue reading the next well with sample information.
19. If there is more than 1 week since the last run on the Droplet Reader, the instrument should be "primed" before use to prevent air bubbles in the system. In the "Setup" tab, press "Prime" under "Instrument Routines" in the top right corner. The procedure takes ~1 min.
20. It is possible to stop the instrument in the middle of a run, but you will then loose the reading information for the well that is currently being read. To start again you need to blank out/-delete the well information for the samples that are already read, save the new template, and start the run from the first well that is not yet pierced by the QX200 reader.
21. If different assays are run on the same plate, it might be necessary to upload and analyze them in separate sessions, especially if different Q values are used (*see* **Note 26**).
22. The folder containing the exported amplitude files should not contain any other files. This will make it easier to upload amplitude files to PoDCall.
23. This is not required to perform analyzes in PoDCall, but it will connect your amplitude readings to your sample name (registered in QuantaSoft). Without the sample sheet your results will be linked to the well name (e.g., A01, B01...).
24. The PoDCall algorithm is available in Bioconductor and must be installed on your computer before first time use: https://www.bioconductor.org/packages/release/bioc/html/PoDCall.html
25. Demonstrated in Jeanmougin et al. [5]: 200 permutations ensure the most optimal combination of consistent methylation concentrations and computational time.
26. For assays with small differences in amplitude values for negative and positive droplets, the Q value could preferentially be reduced. We recommend that the user defines one default Q value per assay, and that this value is used across all experiments for consistency.
27. In some cases, PoDCall fails to set a threshold. In such cases, select another reference well; preferentially one containing IVD, or alternatively another well with a representative droplet distribution.

28. Inspect the amplitude plot visually and evaluate if the droplet pattern is valid and the results trustworthy, or if the reaction should be rerun. Look carefully at wells with droplet count below 10,000. We consider droplet counts <6000 as "not valid" in our data interpretation.
29. By selecting several wells simultaneously and comparing the amplitude plots it is easier to detect deviations in droplet patterns and thresholds.
30. Adjust thresholds manually by marking the well and fill in the desired amplitude in the "Threshold" box. Press "Set New Threshold" when satisfied. If "Undo changes" in the "Thresholds" tab is pressed at a later time, all manually adjusted thresholds are set back to PoDCall thresholds.
31. If ddPCR reader amplitudes are saturated (i.e., no clear separation between the positive and negative droplet cluster) and PoDCall struggles with setting thresholds, the DNA concentration in the initial reaction may have been too high. Repeat the experiment with lower amount of DNA input.
32. If thresholds are not set correctly in a significant number of wells, run PoDCall again. Go back to the "Import files" tab and press "Set thresholds." If the same problem appears, use another well as reference; ether an IVD or a sample. Make sure that the width of the negative droplet band in the reference well is similar to the remaining samples.
33. A sudden and unexpected poor separation between negative and positive droplet for a well-known assay could be due to a bad primer and/or probe batch.
34. It is possible to upload a previously calculated result table to the PoDCall shiny app. Upload the amplitude files and the sample sheet files in the "Import files" tab as described under Subheading 3.2.6. Instead of pressing "Set thresholds" go to the "Thresholds" tab and upload your previously PoDCall results file (.csv) in the "Upload results file" browser. For easier access to inspect the amplitude plots and total droplet plots in a later time, we recommend downloading the plots to a folder of choice.
35. PoDCall does not differentiate between a real signal and "noise" when calculating the normalized DNA methylation levels; i.e., you have to manually set the normalized concentration to 0 in cases where the number of positive droplets is below the limit of blank (LOB; often <3 positive droplets) for the target gene.
36. The 4Plex values correlates well with DNA concentrations as measured by Qubit. Inspect your samples and decide a lower limit threshold for the 4Plex for excluding samples. A low

4Plex concentration indicates low amounts of DNA, and may cause false negatives. Use the same threshold for all samples in your experiment and be aware that the optimal threshold may vary depending on, e.g., the material being analyzed, DNA amount, and limit of detection (LOD) of your assay.

References

1. Hindson BJ, Ness KD, Masquelier DA, Belgrader P, Heredia NJ, Makarewicz AJ, Bright IJ, Lucero MY, Hiddessen AL, Legler TC, Kitano TK, Hodel MR, Petersen JF, Wyatt PW, Steenblock ER, Shah PH, Bousse LJ, Troup CB, Mellen JC, Wittmann DK, Erndt NG, Cauley TH, Koehler RT, So AP, Dube S, Rose KA, Montesclaros L, Wang S, Stumbo DP, Hodges SP, Romine S, Milanovich FP, White HE, Regan JF, Karlin-Neumann GA, Hindson CM, Saxonov S, Colston BW (2011) High-throughput droplet digital PCR system for absolute quantitation of DNA copy number. Anal Chem 83(22):8604–8610. https://doi.org/10.1021/ac202028g
2. Nikanjam M, Kato S, Kurzrock R (2022) Liquid biopsy: current technology and clinical applications. J Hematol Oncol 15(1):131. https://doi.org/10.1186/s13045-022-01351-y
3. Jin B, Li Y, Robertson KD (2011) DNA methylation: superior or subordinate in the epigenetic hierarchy? Genes Cancer 2(6):607–617. https://doi.org/10.1177/1947601910393957
4. Pharo HD, Andresen K, Berg KCG, Lothe RA, Jeanmougin M, Lind GE (2018) A robust internal control for high-precision DNA methylation analyses by droplet digital PCR. Clin Epigenetics 10:24. https://doi.org/10.1186/s13148-018-0456-5
5. Jeanmougin M, Brodal HP, Dietrichson Pharo H, Vedeld HM, Lind GE (2023) PoDCall: positive droplet calling and normalization of droplet digital PCR DNA methylation data. Bioinformatics 39(1). https://doi.org/10.1093/bioinformatics/btac766
6. Pharo HD, Jeanmougin M, Ager-Wick E, Vedeld HM, Sørbø AK, Dahl C, Larsen LK, Honne H, Brandt-Winge S, Five MB, Monteiro-Reis S, Henrique R, Jeronimo C, Steven K, Wahlqvist R, Guldberg P, Lind GE (2022) BladMetrix: a novel urine DNA methylation test with high accuracy for detection of bladder cancer in hematuria patients. Clin Epigenetics 14(1):115. https://doi.org/10.1186/s13148-022-01335-2
7. Vedeld HM, Grimsrud MM, Andresen K, Pharo HD, von Seth E, Karlsen TH, Honne H, Paulsen V, Färkkilä MA, Bergquist A, Jeanmougin M, Aabakken L, Boberg KM, Folseraas T, Lind GE (2022) Early and accurate detection of cholangiocarcinoma in patients with primary sclerosing cholangitis by methylation markers in bile. Hepatology 75(1):59–73. https://doi.org/10.1002/hep.32125
8. Kresse SH, Brandt-Winge S, Pharo H, Flatin BTB, Jeanmougin M, Vedeld HM, Lind GE (2023) Evaluation of commercial kits for isolation and bisulfite conversion of circulating cell-free tumor DNA from blood. Clin Epigenetics 15(1):151. https://doi.org/10.1186/s13148-023-01563-0
9. Loi E, Zavattari C, Tommasi A, Moi L, Canale M, Po A, Sabato C, Vega-Benedetti AF, Ziranu P, Puzzoni M, Lai E, Faloppi L, Rullán M, Carrascosa J, Amat I, Urman JM, Arechederra M, Berasain C, Ferretti E, Casadei-Gardini A, Avila MA, Alonso S, Scartozzi M, Zavattari P (2022) HOXD8 hypermethylation as a fully sensitive and specific biomarker for biliary tract cancer detectable in tissue and bile samples. Br J Cancer 126(12):1783–1794. https://doi.org/10.1038/s41416-022-01738-1
10. Pharo HD, Honne H, Vedeld HM, Dahl C, Andresen K, Liestøl K, Jeanmougin M, Guldberg P, Lind GE (2016) Experimental factors affecting the robustness of DNA methylation analysis. Sci Rep 6:33936. https://doi.org/10.1038/srep33936

Chapter 13

Quantification of Tissue and Circulating MicroRNAs by Droplet Digital PCR

Elisabetta Broseghini, Beatrice Fontana, Giorgio Durante, Roberta Roncarati, and Manuela Ferracin

Abstract

MicroRNAs (miRNAs) are gene expression regulators that play a fundamental role in developmental and biological processes. Dysregulated expression of miRNAs has been associated with most human conditions, and their accurate quantification is an essential activity in molecular biology laboratories. Indeed, the levels of specific tissue or circulating miRNAs have been tested as disease biomarkers in many clinical and experimental settings. Droplet digital PCR (ddPCR) technology is a sensitive and accurate method to obtain the absolute or relative quantification of specific miRNAs, bypassing several issues related to low abundance targets and PCR efficiency. This chapter addresses the workflow and methods for miRNA quantification in formalin-fixed paraffin-embedded (FFPE) samples, cells, and plasma/serum/extracellular vesicles using EvaGreen-based ddPCR or probe-based ddPCR, as well as how to analyze and interpret results. In addition, we provide a ddPCR method to quantify miRNA isoforms (isomiRs).

Key words MicroRNA, Droplet digital PCR, FFPE, Cell, Serum, Plasma, Cancer, Biomarkers, IsomiR

1 Introduction

MiRNAs are small noncoding RNAs of about 18–24 nucleotides in length that regulate target messenger RNAs at the post-transcriptional level. Their activity is fundamental in all developmental and biological processes. Most human diseases, including tumors, have been associated with a dysregulated expression of miRNAs [1, 2]. MiRNA dysregulation can be due to chromosomal abnormalities, transcriptional control changes, epigenetic changes, and defects in the miRNA biogenesis machinery [3, 4]. MiRNAs modulate several and specific target genes, in a regulatory network that is both tissue- and disease-specific, and therefore they are implicated in the regulation of all cellular processes, including proliferation, cell death, and metabolism. Although miRBase

Guillaume Pavlovic (ed.), *Digital PCR: Methods and Protocols*, Methods in Molecular Biology, vol. 2969,
https://doi.org/10.1007/978-1-0716-4767-7_13,

database [5] reports a unique mature sequence for each miRNA, the so-called canonical form, there is evidence from deep-sequencing experiments that miRNAs could have modifications in length and sequence in human tissues. These miRNA isoforms are called isomiRs [6].

Normal and pathological cells and tissues can be faithfully classified using both canonical miRNA [7, 8] and isomiR global expression [9, 10]. Considering their chemical structure, miRNA molecules are the object of post-translational modifications, which make these small RNAs more stable than longer RNAs, resistant to extreme temperatures, pH, and formalin-fixed paraffin-embedding (FFPE) process, thus allowing their detection also in FFPE tissues [11–13]. In addition, miRNA molecules can be actively and passively released outside cells and these miRNAs are known as cell-free or circulating miRNAs. Circulating miRNAs are detectable in all human body fluids [14] and inside extracellular vesicles (EV) [2, 15, 16]. Both tissue and circulating miRNAs can be used as diagnostic and prognostic biomarkers in patients.

Mature miRNAs were initially quantified using Northern Blot technique, microarray technology [7], and small RNA sequencing [17]. Later, reverse transcription (RT) quantitative PCR (qPCR) assays were designed for miRNA-specific quantification [18]. A sensitive and accurate method for miRNA quantification is the use of miRNA assays and droplet digital PCR (ddPCR) technology. With a ddPCR system it is possible to obtain the relative and absolute quantification of specific miRNAs, bypassing several issues related to low abundance of the target, miRNA normalization, PCR inhibition, and PCR efficiency [19].

This chapter focuses on how to perform miRNA quantification in FFPE, cell, or plasma/serum samples using probe-based or intercalating dye-based ddPCR chemistry with QX200 system (Bio-Rad) (Fig. 1).

We provide two protocols for RNA extraction for each sample type, one that combines phenol/guanidine-based lysis of samples and silica-membrane-based purification of total RNA (Qiagen kits, namely, miRNeasy FFPE Kit, miRNeasy Mini Kit, and miRNeasy Serum/Plasma Kit) and an automated system that uses paramagnetic particles (Promega kits, namely, Maxwell RSC miRNA Tissue Kit, Maxwell RSC RNA FFPE Kit, and Maxwell RSC miRNA Plasma and Serum Kit). Reverse transcription, also known as cDNA generation, is carried out according to the selected ddPCR chemistry.

For tissue/circulating canonical miRNA quantification, we provide a protocol based on LNA primers, the miRCURY LNA microRNA Assay (Qiagen), and EvaGreen detection, named EvaGreen-based ddPCR. For miRNA quantification based on TaqMan probes, we provide different protocols, including one for isomiR detection based on Dumbbell-PCR, which is a

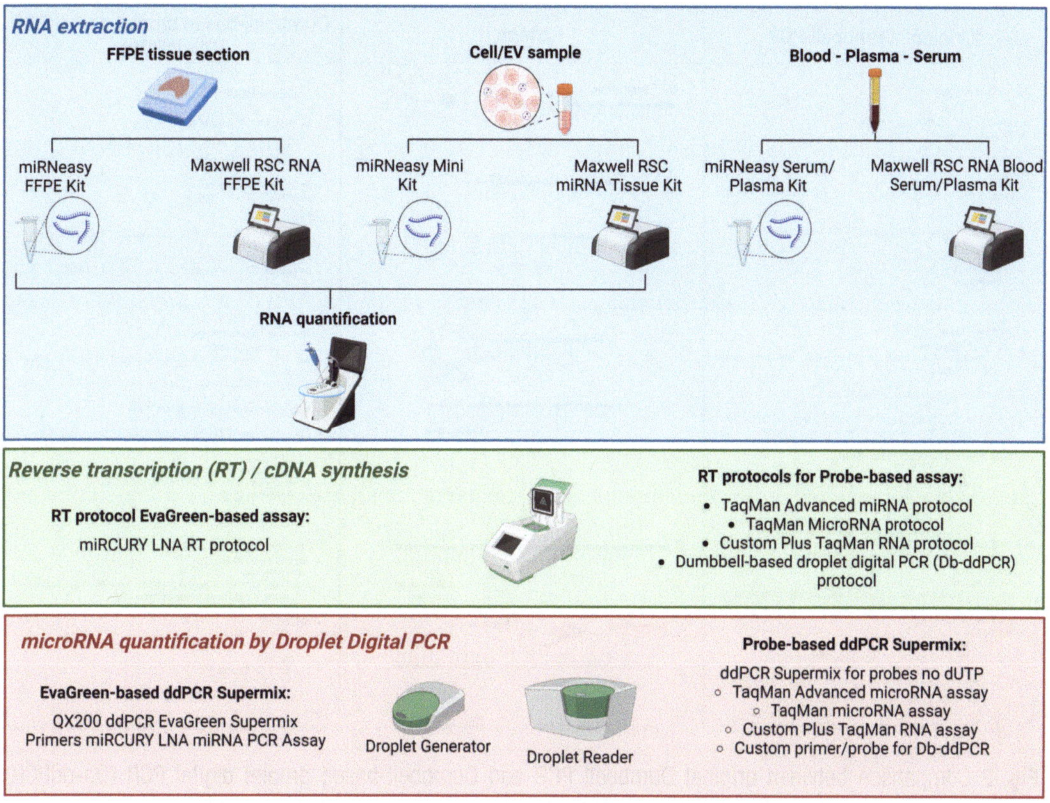

Fig. 1 Workflow for miRNA isolation and quantification in different types of samples with ddPCR technology. Two protocols for RNA extraction are described for FFPE, cell/extracellular vesicles (EV) and plasma/serum (blue box): one that combines phenol/guanidine-based lysis of samples and silica-membrane-based purification of total RNA (Qiagen miRNeasy FFPE Kit, miRNeasy Mini Kit, and miRNeasy Serum/Plasma Kit) and one is an automated system that uses paramagnetic particles (Promega Maxwell RSC miRNA Tissue Kit, Maxwell RSC RNA FFPE Kit, and Maxwell RSC miRNA Plasma and Serum Kit). Reverse transcription (cDNA synthesis) (green box) is performed according to the chosen ddPCR chemistry. For the miRNA experiment based on EvaGreen dye, we provide a protocol based on LNA primers, namely, miRCURY LNA RT Assay (Qiagen). For the probe-based experiment, we provided different protocols (Thermo Fisher Scientific TaqMan Advanced miRNA protocol, TaqMan MicroRNA protocol, Custom Plus TaqMan RNA protocol), including one for isomiR detection: Dumbbell-based droplet digital PCR (Db-ddPCR) protocol. Droplet digital PCR (ddPCR) is performed using Bio-Rad QX200 ddPCR EvaGreen Supermix for EvaGreen-based experiments, and ddPCR Supermix for probes no dUTP for all Probe-based experiments (red box). The figure was created with Biorender.com

TaqMan-based RT-PCR method that distinctively quantifies a specific small RNA variant with single-nucleotide resolution at terminal sequences [20, 21]. For this specific application, we introduced some variations to the Dumbbell PCR that enabled the use of smaller RNA amounts to accommodate the limited RNA yield from, e.g., FFPE samples. We then replaced the last step (qPCR quantification) with ddPCR quantification, to quantify even few copies of isomiRs. To simplify the annotation, we named our optimized method: Dumbbell-based droplet digital PCR assay (Db-ddPCR) (Fig. 2).

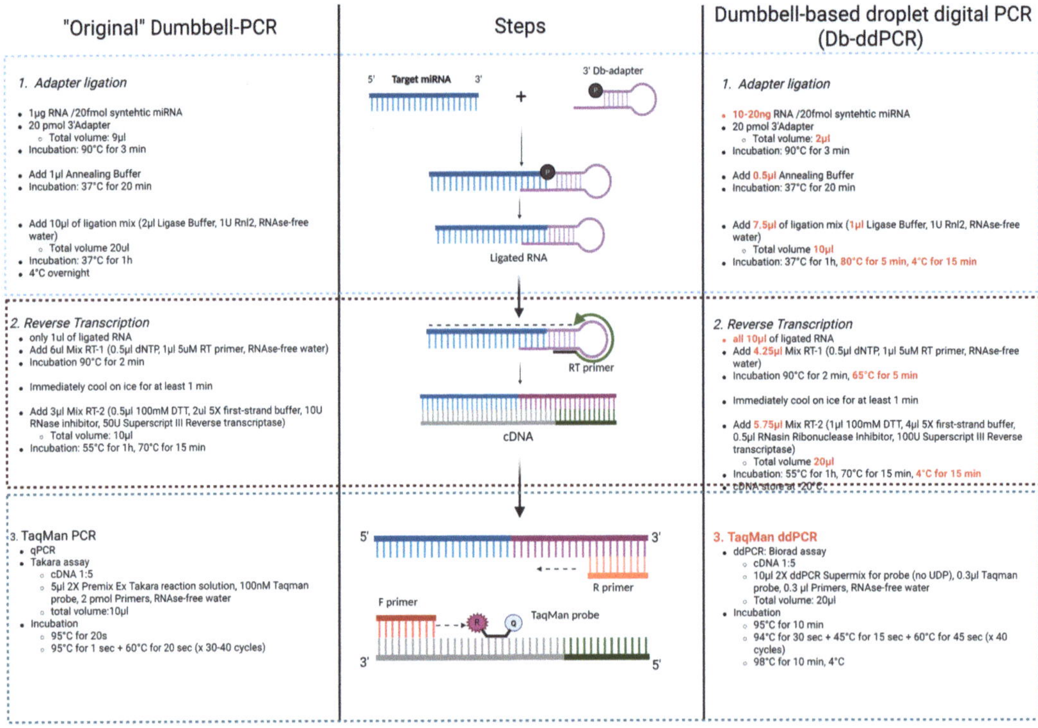

Fig. 2 Comparison between original Dumbbell PCR and Dumbbell-based droplet digital PCR (Db-ddPCR). Dumbbell PCR is a TaqMan RT-PCR-based method that distinctively quantifies a specific small RNA variant with single-nucleotide resolution at terminal sequences [20, 21]. We optimized the method and named it Dumbbell-based droplet digital PCR assay (Db-ddPCR). We introduce variations (marked in red) to be able to use a lower RNA amount, to speed up the protocol, and to have a more sensitive method able to quantify also few copies of the target with droplet digital PCR (ddPCR) quantification. The figure was created with Biorender.com

2 Materials

2.1 Sample Preparation

1. Refrigerated tabletop centrifuge.
2. Refrigerated Ultracentrifuge.
3. 1.5 mL polypropylene microcentrifuge tubes.
4. 15 mL Conical Sterile Polypropylene Centrifuge tubes.
5. 50 mL Conical Sterile Polypropylene Centrifuge tubes.
6. Filtered PBS.
7. Primary blood collection tube(s) without anticoagulants, such as EDTA or citrate, such as BD Vacutainer Venous Blood Collection Tubes: SST Serum Separation Tubes (Fisher Scientific, cat. no. BD 367985), which contain spray-coated silica and a polymer gel for serum separation.

8. Primary blood collection tube(s) containing EDTA as an anti-coagulant for plasma, such as BD Vacutainer Venous Blood Collection Tubes (Fisher Scientific, cat. no. BD 367525) (*see* **Note 1**).

2.2 RNA Extraction

2.2.1 From FFPE Samples

1. miRNeasy FFPE Kit (Qiagen, cat. no. 217504).
   (a) RNase-free needle and syringe.
   (b) RNAse-free water.
   (c) Scalpel and Sterile carbon steel scalpel blades.
   (d) Xylene.
   (e) Ethanol 100%.
   (f) Microcentrifuge (with rotor for 1.5 ml and 2 mL tubes).
   (g) Vortex mixer.
   (h) 1.5 mL polypropylene microcentrifuge tubes.
   (i) Thermoblock for 56 °C incubation.
   (j) Thermoblock for 80 °C incubation.
2. Maxwell RSC RNA FFPE Kit (Promega, cat. no. AS1440).
   (a) Maxwell RSC Instrument (Promega, cat. no. AS4500).
   (b) Thermoblock for 56 °C incubation.
   (c) Thermoblock for 80 °C incubation.
   (d) 1.5 mL polypropylene microcentrifuge tubes.
   (e) DynaMag- Spin Magnet (Thermo Fisher Scientific, cat. no. 12320D).

2.2.2 From Cell/EV Sample

1. miRNeasy Mini Kit (Qiagen, cat. no. 217004).
   (a) Chloroform (without added isoamyl alcohol).
   (b) Ethanol (70% and 96–100%).
   (c) 1.5 mL or 2 mL polypropylene microcentrifuge tubes.
   (d) Vortex mixer.
   (e) Microcentrifuge(s) (with rotor for 1.5 mL and 2 mL tubes) for centrifugation at 4 °C and at room temperature (15–25 °C).
   (f) 1.5 mL polypropylene microcentrifuge tubes.
2. Maxwell RSC miRNA Tissue Kit (Promega, cat. no. AS1460).
   (a) Scalpel and Sterile carbon steel scalpel blades.
   (b) Maxwell RSC Instrument (Promega, cat. no. AS4500).
   (c) 1.5 mL polypropylene microcentrifuge tubes.
   (d) DynaMag- Spin Magnet (Thermo Fisher Scientific, cat. no. 12320D).

2.2.3 From Plasma/Serum Samples

1. miRNeasy Serum/Plasma Kit (Qiagen, cat. no. 217184) (*see* **Note 2**).
   (a) 100 nmol RNA oligo Cel-miR-39-3p (Sequence: UCAC CGGGUGUAAAUCAGCUUG).
   (b) Chloroform (without added isoamyl alcohol).
   (c) Ethanol (70% and 96–100%).
   (d) 1.5 mL or 2 mL polypropylene microcentrifuge tubes.
   (e) Vortex mixer.
   (f) Microcentrifuge(s) (with rotor for 1.5 mL and 2 mL tubes) for centrifugation at 4 °C and at room temperature (15–25 °C).
2. Maxwell RSC miRNA Plasma and Serum Kit (Promega, cat. no. AS1680).
   (a) Maxwell RSC Instrument (Promega, cat. no. AS4500).
   (b) 1.5 mL polypropylene microcentrifuge tubes.
   (c) DynaMag- Spin Magnet (Thermo Fisher Scientific, cat. no. 12320D).
   (d) Thermoblock at 37 °C.

2.3 RNA Quantification

1. NanoDrop One/OneC Microvolume UV-Vis Spectrophotometer (Thermo Fisher Scientific, cat. no. ND-ONE-W).
2. Nuclease-free water.

2.4 Reverse Transcription/ScDNA Synthesis

1. T100 Thermal Cycler (Bio-Rad, cat. no. 1861096).
2. Microcentrifuge (with rotor for 1.5 mL and 2 mL tubes, 0.2 mL tubes).
3. Vortex mixer.
4. 1.5 mL or 2 mL polypropylene microcentrifuge tubes.
5. 0.2 mL PCR tubes.
6. Nuclease-free water.

2.4.1 miRCURY LNA Protocol

1. miRCURY LNA RT Kit (Qiagen, cat number 339340).

2.4.2 TaqMan Advanced miRNA Protocol

1. TaqMan Advanced miRNA cDNA Synthesis Kit (Thermo Fisher Scientific, cat. no. A28007).

2.4.3 TaqMan MicroRNA Protocol

1. TaqMan MicroRNA Reverse Transcription kit (Thermo Fisher Scientific, cat. no. 4366597).
2. TaqMan MicroRNA assay (Thermo Fisher Scientific, cat. no. 4440885) (*see* **Note 3**).

2.4.4 Custom Plus TaqMan RNA Protocol

1. Superscript VILO cDNA synthesis kit (Thermo Fisher Scientific, cat. no. 11754050).

2.4.5 Dumbbell-Based Droplet Digital PCR (Db-ddPCR) Protocol for isomiR

1. Adapter ligation step:
   (a) Thermoblock for 90 °C incubation.
   (b) Thermoblock for 37 °C incubation.
   (c) Specific Adapter ordered by IDT company (100 nm).
   (d) 10× Annealing buffer (50 mM Tris–HCl pH 8, 5 mM EDTA, 100 mM $MgCl_2$).
   (e) T4 RNA Ligase 2 (New England Biolabs, cat. no. M0239S).
   (f) $MgCl_2$ 100 mM.
   (g) 50% PEG8000.
2. Reverse transcription step:
   (a) RT primer ordered by IDT company (100 nm).
   (b) SuperScript III Reverse Transcriptase (Thermo Fisher Scientific, cat. no.18080044).
   (c) dNTP Mix (10 mM each) (Thermo Fisher Scientific, cat. no. R0192).
   (d) RNasin Ribonuclease Inhibitor (Promega, cat. no. N2111).

2.5 cDNA Dilution

1. -Nuclease-free water.

2.6 Droplet Generation and PCR

1. QX200 Droplet Digital PCR System (Bio-Rad): Droplet Generator (Bio-Rad, cat. no. 1864002).
2. T100 Thermal Cycler (Bio-Rad, cat. no. 1861096).
3. PX1 PCR Plate sealer (Bio-Rad, cat. no. 1814000).
4. Microcentrifuge (with rotor for 1.5 mL and 2 mL tubes).
5. Vortex mixer.
6. Microplate Microcentrifuge.
7. 1.5 mL or 2 mL polypropylene microcentrifuge tubes.
8. Eppendorf twin.tec PCR Plate 96, semi-skirted (Eppendorf, cat. no. 0030128605)/ddPCR Plates 96-well Semi-skirted (Bio-Rad, cat. no. 12001925).
9. Rainin Pipet-Lite Multi Pipette L8-20XLS+ (2–20 μL) (Mettler Toledo, cat. no. 17013803).
10. Rainin Pipette Tips SS LTS 20 μL S 960A/5 (Mettler Toledo, cat. No. 17005872).

11. Rainin Pipet-Lite Multi Pipette L8-50XLS+ (5–50 μL) (Mettler Toledo, cat. no. 17013804).
12. Rainin Pipette Tips SS LTS 250 μL S 960A/5 (Mettler Toledo, cat. No. 17005874).
13. Optical Flat 8-Cap Strips (Bio-Rad, cat. no. TCS0803).
14. DG8 Cartridge Holder (Bio-Rad, cat. no. 1863051).
15. DG8 Cartridges (Bio-Rad, cat. no. 1864008).
16. DG8 Gaskets (Bio-Rad, cat. no. 1863009).
17. PCR Plate Heat Seal, foil (Bio-Rad, cat. no. 1814040).
18. Specific EvaGreen-based ddPCR materials:
    (a) QX200 ddPCR EvaGreen Supermix (Bio-Rad, cat. no. 1864033/1864034).
    (b) Primers miRCURY LNA miRNA PCR Assay (Qiagen, cat. no. 339306).
    (c) QX200 Buffer Control for EvaGreen (Bio-Rad, cat. no. 1864052).
    (d) QX200 Droplet Generation Oil for EvaGreen (Bio-Rad, cat. no. 1864005).
19. Specific Probe-based ddPCR.
    (a) ddPCR Supermix for probes no dUTP (Bio-Rad, cat. no. 1863024/1863023).
    (b) ddPCR Buffer Control for Probes (Bio-Rad, cat. no. 1863052).
    (c) Droplet Generation Oil for Probes (Bio-Rad, cat. no. 1863005).
20. Primers and probes (one of these):
    (a) TaqMan Advanced microRNA assay (Thermo Fisher Scientific, cat. no. A25576) if you perform cDNA synthesis with TaqMan Advanced miRNA cDNA Synthesis Kit.
    (b) TaqMan microRNA assay (Thermo Fisher Scientific, cat. no. 4440885) (*see* **Note 3**) if you perform cDNA synthesis with TaqMan MicroRNA Reverse Transcription kit.
    (c) Custom Plus TaqMan RNA assay (Thermo Fisher Scientific, cat. no. 4441118).
    (d) Dumbbell ddPCR primers and probes: PCR primers (forward and reverse) ordered by IDT company (100 nm) or TaqMan MGB Probe (Thermo Fisher Scientific, cat. no. 4316034) [protected from light] or TaqMan Probe ordered by IDT company (100 nm) [protected from light].

2.7 Droplet Reading and Data Analysis

1. QX200 Droplet Digital PCR System (Bio-Rad): Droplet Reader (Bio-Rad, cat. no. 1864003) and a Laptop with QuantaSoft Software (QuantaSoft1.7.4.0917) or QX Manager Software 2.0.
2. ddPCR Droplet Reader Oil (Bio-Rad, cat. no. 1863004).

3 Methods

3.1 Sample Preparation

3.1.1 Cell Preparation

1. Harvest about 1×10^6 cells.
2. Aspirate completely the medium and wash the cells with PBS (*see* **Note 4**).
3. Centrifuge at 300 ×*g* for 5 min.
4. Completely aspirate the supernatant.
5. Aliquots can be stored at −80 °C until use.

3.1.2 EV Preparation from Cell Cultures

1. Before starting, set up the centrifuge at 4 °C and the ultracentrifuge at 4 °C.
2. Collect 5–80 mL of medium (containing EVs) from the cell culture of interest.
3. Centrifuge medium at 2000 ×*g* for 20 min at 4 °C.
4. Ultracentrifuge medium at 20,000 ×*g* for 30 min at 4 °C to isolate microvesicles.
5. Remove supernatant and keep it for further ultracentrifugation.
6. Resuspend the pellet using 500 μL of filtered PBS. Aliquots can be stored at −80 °C until use.
7. Ultracentrifuge residual supernatant at 100,000 ×*g* for 3 h at 4 °C to isolate exosomes.
8. Resuspend the pellet using 500 μL of filtered PBS. Aliquots can be stored at −80 °C until use.

3.1.3 Plasma Preparation

Plasma processing is a critical step in circulating miRNA quantification; as to be comparable, all the samples from the same experiment must be processed using the same workflow.

1. Process the blood within an hour from the isolation to get a better performance.
2. Collect 6 mL of blood in EDTA tubes (BD Vacutainer tubes).
3. Centrifuge samples at 1900 ×*g* for 10 min at 25 °C to remove blood cells.
4. Dispense the supernatant plasma in aliquots.
5. Aliquots can be stored at −80 °C until use.

3.1.4 Serum Preparation

Serum processing is also a critical step in circulating miRNA quantification. To be comparable, all the samples from the same experiment must be processed using the same workflow. Process the blood within 1 h after sampling to get a better performance.

1. Set up the centrifuge at room temperature for serum isolation.
2. Collect 6 mL blood in serum tubes (BD Vacutainer Serum tubes).
3. Keep at room temperature to clot for at least 30 min.
4. Spin at 1100 ×*g* for 20 min at room temperature.
5. Remove the serum and dispense it in aliquots.
6. Aliquots can be stored at −80 °C until use.

3.2 RNA Extraction

3.2.1 From FFPE with miRNeasy FFPE Kit

1. Collect freshly cut 3–5 FFPE tissue sections (each at least 10 μm thick) into 1.5- or 2-mL microcentrifuge tubes. For slide-mounted sections, scrape them off using a scalpel with a clean sterile carbon steel scalpel blade. Centrifuge tubes at maximum speed for 15 s to collect the sample at the tube's bottom.
2. To prepare DNase I, dissolve the lyophilized DNase I in 550 μL RNase-free water. Do not open the vial, inject RNase-free water into the vial using an RNase-free needle and syringe, mix gently by inverting the vial, do not vortex. Prepare aliquots (*see* **Note 5**).
3. To prepare RPE buffer, add 44 mL of 96–100% ethanol to Buffer RPE concentrate (*see* **Note 6**).
4. Preheat a Thermoblock to 56 °C and another Thermoblock to 80 °C.
5. Add 1 mL Xylene. Vortex vigorously for 10 s, and centrifuge at full speed for 2 min.
6. Carefully remove the supernatant by pipetting without disturbing the pellet.
7. Add 1 mL ethanol (96–100%) to the pellet, mix by vortexing, and centrifuge at full speed for 2 min.
8. Carefully remove the supernatant by pipetting without disturbing the pellet. Carefully remove any residual ethanol using a fine pipet tip.
9. Keep the lid open and incubate at room temperature (15–25 °C) for 15 min or until all residual ethanol has evaporated.
10. Add 240 μL Buffer PKD, vortex to mix.
11. Add 10 μL proteinase K. Mix gently by pipetting up and down.

12. Incubate at 56 °C for 15 min, then at 80 °C for 15 min. Ensure that the heating block has reached 80 °C before starting the 15 min incubation (*see* **Note 7**).
13. Incubate on ice for 3 min. Centrifuge for 15 min at 20,000 ×*g*.
14. Transfer the supernatant to a new tube taking care not to disturb the pellet.
15. Add DNase Booster Buffer equivalent to one-tenth of the total sample volume (approximately 25 μL) and 10 μL DNase I stock solution. Mix by inverting the tube. Centrifuge briefly.
16. Let it incubate at room temperature for 15 min.
17. Add 500 μL Buffer RBC and mix the lysate thoroughly.
18. Add 1750 μL ethanol (100%) to the sample and mix well by pipetting. Do not centrifuge.
19. Transfer 700 μL of the sample, including any precipitate, to a RNeasy MinElute spin column placed in a 2 mL collection tube. Close the lid gently and centrifuge for 15 s at ≥8000 ×*g*. Discard the flow-through. Repeat this step until the entire sample has passed through the column.
20. Add 500 μL Buffer RPE to the column. Close the lid gently and centrifuge for 15 s at ≥8000 ×*g*. Discard the flow-through and reuse the previous collection tube.
21. Add 500 μL of Buffer RPE to the RNeasy MinElute spin column. Gently close the lid, and centrifuge for 2 min at ≥8000 ×*g* to thoroughly wash the spin column membrane. Dispose of the collection tube with the flow-through.
22. After centrifugation, carefully remove the spin column from the collection tube to avoid ethanol carryover. Place the RNeasy MinElute spin column in a new 2 mL collection tube. Open the lid of the spin column (*see* **Note 8**) and centrifuge at full speed for 5 min. Discard the collection tube with the flow-through.
23. Place the RNeasy MinElute spin column in a new 1.5 mL collection tube. Add 14–30 μL RNase-free water directly to the spin column membrane (*see* **Note 9**). Let the column sit at room temperature for 5 min with the RNase-free water covering the membrane to enhance elution.
24. Close the lid gently and centrifuge for 1 min at full speed to elute the RNA.
25. Store RNA at −80 °C.

3.2.2 From FFPE with Maxwell RSC RNA FFPE Kit

1. Specimen or reagent spills on any part of the deck tray should be cleaned with a detergent-water solution, followed by a bactericidal spray or wipe, and then water (*see* **Note 10**).
2. Collect freshly cut 3–5 FFPE tissue sections (each at least 10 μm thick) into 1.5- or 2-mL microcentrifuge tubes. For

slide-mounted sections, scrape with a scalpel using a clean, sterile carbon steel scalpel blade. Centrifuge tubes at maximum speed for 15 s to collect the sample at the bottom of the tube.

3. DNase I preparation: Add 275 μL of nuclease-free water to the vial of lyophilized DNase I. Swirl gently to mix, avoiding vortexing. Store reconstituted DNase I at −30 °C to −10 °C after use. It retains activity for up to 10 freeze-thaw cycles.
4. Prepare Maxwell cartridges and elution tubes before starting the extraction steps. Place the cartridges to be used in the deck tray(s) with well #1 (the largest well in the cartridge) facing away from the elution tubes. Press down on the cartridge to snap it into position. Carefully peel back the seal so that all plastic comes off the top of the cartridge. Ensure that all sealing tape and any residual adhesive are removed before placing cartridges in the instrument. Place one plunger into well #8 of each cartridge.
5. Preheat a Thermoblock to 56 °C and a Thermoblock to 80 °C.
6. Add 300 μL of Mineral Oil to the sample tubes and vortex for 10 s.
7. Heat samples at 80 °C for 2 min, then allow them to cool to room temperature while preparing the master mix.
8. Prepare a master mix of Lysis Buffer, Proteinase K Solution, and Blue Dye (Table 1) (*see* **Note 11**). Add 250 μL of master mix to each sample tube and vortex for 5 s.
9. Centrifuge at 10,000 ×*g* for 20 s to separate layers. If a pellet is present in the aqueous layer (lower, blue layer), gently mix the aqueous phase with a pipette.
10. Incubate sample tubes in a 56 °C Thermoblock for 15 min.
11. Move sample tubes to an 80 °C Thermoblock and incubate for 1 h.
12. Remove sample tubes from the Thermoblock and allow them to cool to room temperature for 15 min.
13. Prepare a DNase cocktail containing $MnCl_2$, DNase Buffer, and reconstituted DNase I as instructed in Table 2. If DNase Buffer precipitates, solubilize by heating to 56 °C for 2 min,

Table 1
Lysis Buffer master mix reaction setup for Maxwell RSC RNA FFPE Kit

Reagent	Amount/reaction	Reactions (number to be run + 2)	Total
Lysis buffer	224 μL	$N + 2$	$224 \times (N + 2)$ μL
Proteinase K	25 μL	$N + 2$	$25 \times (N + 2)$ μL
Blue dye	1 μL	$N + 2$	$1 \times (N + 2)$ μL

Table 2
DNase cocktail reaction setup for Maxwell RSC RNA FFPE kit

Reagent	Amount/reaction	Reactions (number to be run + 2)	Total
$MnCl_2$, 0.09 M	26 μL	$N + 2$	$26 \times (N + 2)$ μL
DNase buffer	14 μL	$N + 2$	$14 \times (N + 2)$ μL
DNase I	10 μL	$N + 2$	$10 \times (N + 2)$ μL

followed by brief vortexing (*see* **Note 12**). Add 50 μL of DNase cocktail to the aqueous (blue) phase in each sample tube. Mix by pipetting ten times.

14. Incubate sample tubes at room temperature (15–30 °C) for 15 min.
15. Centrifuge sample tubes at full speed in a microcentrifuge for 2 min.
16. Immediately transfer the blue, aqueous phase to well #1 of a Maxwell FFPE Cartridge, the largest facing away from the elution tubes.
17. Place one plunger into well #8 of each cartridge.
18. Place an empty elution tube into the elution tube position for each cartridge in the deck tray(s).
19. Add 50 μL of nuclease-free water to the bottom of each Maxwell Elution Tube (*see* **Note 13**).
20. If you have a magnet, we recommend removing any remaining black beads from the elution.
21. If sample eluates are not processed immediately, store the eluted RNA at −20 °C or − 80 °C for up to 24 h in the Maxwell Elution Tubes. For longer-term storage, transfer the eluted RNA into labware suitable for long-term storage and store at −80 °C or below.

3.2.3 From Cell/EV Samples with miRNeasy Mini Kit

1. Prepare cell/EV samples or thaw frozen samples on ice.
2. Preparation of RWT and RPE buffers. Before using it for the first time, add the required volumes of ethanol (96%–100%), as indicated on the bottle, to obtain a working solution.
3. Refrigerate a microcentrifuge at 4 °C.
4. Only for cell pellet, aspirate completely any supernatant from the pellet.
5. Disrupt the cells/EVs by adding 700 μL QIAzol Lysis Reagent. Vortex or pipet to mix (*see* **Notes 14** and **15**).

6. To promote dissociation of nucleoprotein complexes, place the tube containing the homogenate on the benchtop at room temperature (15–25 °C) for 5 min.
7. Add 140 μL chloroform to the tube containing the homogenate and cap it securely.
8. Shake the tube vigorously for 15 s. Thorough mixing is important for subsequent phase separation.
9. Place the tube containing the homogenate on the benchtop at room temperature for 2–3 min.
10. Centrifuge for 15 min at 12,000 ×*g* at 4 °C (*see* **Note 16**).
11. After centrifugation, the sample separates into three phases: An upper, colorless, aqueous phase containing RNA; a white interphase; and a lower, red, organic phase. The volume of the aqueous phase should be approximately 350 μL.
12. Transfer the upper aqueous phase to a new collection tube.
13. Add 1.5 volumes (usually 525 μL) of 100% ethanol and mix thoroughly by pipetting up and down several times. Do not centrifuge (*see* **Note 17**).
14. Pipet up to 700 μL of the sample, including any precipitate that may have formed, into a RNeasy Mini spin column in a 2 mL collection tube. Close the lid gently and centrifuge at ≥8000 ×*g* for 15 s at room temperature. Discard the flow-through. Repeat this step using the remainder of the sample. Always reuse the same collection tube.
15. Add 700 μL Buffer RWT to the RNeasy Mini spin column. Close the lid gently and centrifuge for 15 s at ≥8000 ×*g* to wash the column. Discard the flow-through. Reuse the collection tube.
16. Pipet 500 μL Buffer RPE onto the RNeasy Mini spin column. Close the lid gently and centrifuge for 15 s at ≥8000 ×*g* to wash the column. Discard the flowthrough and reuse the collection tube.
17. Add another 500 μL Buffer RPE to the RNeasy Mini spin column. Close the lid gently and centrifuge for 2 min at ≥8000 ×*g* to dry the RNeasy Mini spin column membrane (*see* **Note 18**).
18. Following centrifugation, remove the RNeasy Mini spin column from the collection tube carefully so the column does not contact the flow-through. Otherwise, carryover of ethanol will occur.
19. Place the RNeasy Mini spin column into a new 2 mL collection tube and discard the old collection tube with the flow-through.
20. To eliminate any possible carryover of Buffer RPE, centrifuge in a microcentrifuge at full speed for 1 min.

21. Transfer the RNeasy Mini spin column to a new 1.5 mL collection tube. Pipet 30–50 μL RNase-free water directly onto the RNeasy Mini spin column membrane.
22. Close the lid gently and centrifuge for 1 min at ≥8000 ×*g* to elute the RNA.
23. Store RNA at −80 °C.

3.2.4 From Cell Samples with Maxwell RSC miRNA Tissue Kit

1. Specimen or reagent spills on any part of the deck tray should be cleaned with a detergent-water solution, followed by a bactericidal spray or wipe, and then water (*see* **Note 10**).
2. Prepare cell/EV samples or thaw frozen samples on ice.
3. 1-Thioglycerol/Homogenization Solution preparation: A volume of 200 μL of 1-Thioglycerol/Homogenization Solution is needed for each sample. To prepare a working solution, add 20 μL of 1-Thioglycerol per milliliter of Homogenization Solution. 1-Thioglycerol is viscous, so careful pipetting is required for accurate measurement. We recommend adding 600 μL of 1-Thioglycerol to the 30 mL bottle of Homogenization Solution. Before use, chill the 1-Thioglycerol/Homogenization Solution on ice or at 2–10 °C, where it is stable for up to 30 days.
4. DNase I preparation: Add 275 μL of nuclease-free water to the vial of lyophilized DNase I. Add 5 μL of Blue Dye to the reconstituted DNase I as a visual aid for pipetting. Swirl gently to mix, avoiding vortexing. Store reconstituted DNase I at −30 °C to −10 °C after use. It retains activity for up to ten freeze-thaw cycles.
5. Prepare Maxwell cartridges and elution tubes before starting the extraction process. Place the cartridges to be used in the deck tray(s) with well #1 (the largest well in the cartridge) facing away from the elution tubes. Press down on the cartridge to snap it into position. Carefully peel back the seal so that all plastic comes off the top of the cartridge. Ensure that all sealing tape and any residual adhesive are removed before placing cartridges in the instrument. Place one plunger into well #8 of each cartridge.
6. Only for cell pellet: Aspirate completely any supernatant from the pellet.
7. Work as quickly as possible, homogenize the tissue sample in the chilled 1-Thioglycerol/Homogenization Solution until no visible tissue fragments remain. Homogenize for an additional 15–30 s to ensure complete homogenization. If foaming occurs, let the sample settle on ice. Only 200 μL of homogenate can be processed per cartridge. The final volume of the homogenate to be added to the cartridge should be 200 μL.

Add more Homogenization Solution as needed to bring the sample to a final volume of 200 μL (*see* **Note 19**).

8. Add 200 μL of Lysis Buffer, 200 μL of Lytic Enhancer, and 15 μL of Proteinase K to the cell suspension.
9. Mix by vortexing for 20 s.
10. Incubate at room temperature for 10 min. During this time, prepare the Maxwell RSC Cartridges.
11. Place the cartridges to be used in the deck tray(s) with well #1 (the largest well in the cartridge) facing away from the elution tubes. Press down on the cartridge to snap it into position. Carefully peel back the seal so that all plastic comes off the top of the cartridge. Ensure that all sealing tape and any residual adhesive are removed before placing cartridges in the instrument.
12. Place one plunger into well #8 of each cartridge.
13. Place an empty elution tube into the elution tube position for each cartridge in the deck tray. Add 60 μL of nuclease-free water to the bottom of each elution tube (*see* **Note 9**).
14. Transfer 630 μL of lysate to well #1 (the largest well in the cartridge) of the Maxwell RSC Cartridge.
15. Add 10 μL of blue DNase I Solution to well #4 of the Maxwell RSC Cartridge (well #4 contains yellow reagent). After the blue DNase I Solution is added, the reagent in well #4 will be green.
16. Remove the magnetic beads that may have remained during the extraction process using a magnet.
17. Transfer the eluate to a new 1.5 mL polypropylene tube and put it on ice.
18. Store the RNA at −80 °C.

3.2.5 With miRNeasy Serum/Plasma Kit

1. Buffers RWT and RPE are supplied as concentrates. Before using it for the first time, add the required volumes of ethanol (96–100%), as indicated on the bottle, to obtain a working solution.
2. Refrigerate a microcentrifuge at 4 °C.
3. Prepare serum (200 μL) or plasma (200 μL) or thaw frozen samples on ice (*see* **Note 20**).
4. Add 5 volumes (1 mL) of QIAzol Lysis Reagent to the sample Mix by vortexing or pipetting up and down.
5. Add 2.5 μL of 20 nM cel-miR-39-3p RNA spike-in (*see* **Note 21**).

6. Incubate the homogenate at room temperature (15–25 °C) for 5 min.
7. Add chloroform of an equal volume to the starting sample and cap tube securely (200 μL). Vortex or shake vigorously for 15 s. Thorough mixing is important for subsequent phase separation.
8. Incubate at room temperature for 2–3 min.
9. Centrifuge for 15 min at 12,000 g at 4 °C (*see* **Note 16**). After centrifugation, the sample is separated into three phases: An upper, colorless, aqueous phase containing RNA; a white interphase; and a lower, red, organic phase.
10. Transfer the upper aqueous phase (approximately 700 μL) to a new collection tube. Avoid the transfer of any interphase material.
11. Add 1.5 volumes of 100% ethanol (1000 μL ethanol) and mix thoroughly by inversion.
12. Pipet up to 700 μL sample, including any precipitate, into a RNeasy MinElute spin column in a 2 mL collection tube. Close the lid and centrifuge at ≥8000 ×*g* for 15 s at room temperature. Discard the flow-through. Repeat this step using the remainder of the sample.
13. Add 700 μL Buffer RWT to the RNeasy MinElute spin column. Close the lid, and centrifuge for 15 s at ≥8000 ×*g*. Discard the flow-through.
14. Pipet 500 μL Buffer RPE onto the RNeasy MinElute spin column. Close the lid, and centrifuge for 15 s at ≥8000 ×*g*. Discard the flow-through.
15. Pipet 500 μL of Buffer RPE onto the RNeasy MinElute spin column. Close the lid gently and centrifuge for 2 min at ≥8000 ×*g* to wash the spin column membrane. Discard the collection tube with the flow-through.
16. Place the RNeasy MinElute spin column in a new 2 mL collection tube. Open the lid of the spin column and centrifuge at full speed for 5 min to dry the membrane (*see* **Note 18**). Discard the collection tube with the flow-through.
17. Place the RNeasy MinElute spin column in a new 1.5 mL collection tube. Add 35 μL RNase-free water directly to the center of the spin column membrane. Incubate for 2 min at room temperature.
18. Close the lid gently, and centrifuge for 1 min at full speed to elute the RNA.
19. Discard the column and transfer the tube containing the eluate on ice.
20. Store the RNA at −80 °C.

3.2.6 With Maxwell RSC miRNA Plasma and Serum Kit

1. Specimen or reagent spills on any part of the deck tray should be cleaned with a detergent-water solution, followed by a bactericidal spray or wipe, and then water (*see* **Note 10**).
2. DNase I preparation: Add 275 μL of nuclease-free water to the vial of lyophilized DNase I. Add 5 μL of Blue Dye to the reconstituted DNase I as a visual aid for pipetting. Swirl gently to mix, avoiding vortexing. Store reconstituted DNase I at −30 °C to −10 °C after use. It retains activity for up to ten freeze-thaw cycles.
3. Prepare Maxwell cartridges and elution tubes before starting the extraction process. Place the cartridges to be used in the deck tray(s) with well #1 (the largest well in the cartridge) facing away from the elution tubes. Press down on the cartridge to snap it into position. Carefully peel back the seal so that all plastic comes off the top of the cartridge. Ensure that all sealing tape and any residual adhesive are removed before placing cartridges in the instrument. Place one plunger into well #8 of each cartridge.
4. Preheat a Thermoblock to 37 °C.
5. Use 200 μL of plasma or serum sample (*see* **Note 22**).
6. Add 80 μL of Proteinase K and 230 μL of Lysis Buffer C to the plasma or serum sample. Mix by vortexing for 5 s.
7. Incubate at 37 °C for 15 min.
8. Place an empty elution tube into the elution tube position for each cartridge in the deck tray. Add 50 μL of nuclease-free water to the bottom of each elution tube (*see* **Note 9**).
9. Transfer all of the lysate to well #1 of the Maxwell RSC Cartridge.
10. Add 10 μL of blue DNase I Solution to well #4 of the Maxwell® RSC Cartridge (well #4 contains yellow reagent). After the blue DNase I Solution is added, the reagent in well #4 will turn green.
11. Remove the magnetic beads with a magnet (some of them may have remained during the extraction).
12. Transfer the eluate to a new 1.5 mL polypropylene tube and put it on ice.
13. Store the RNA at −80 °C.

3.3 RNA Quantification

We quantified RNA isolated from FFPE or cell/EV samples, but not serum/plasma samples since the RNA concentration in plasma and serum does not correspond to the real miRNA concentration [22]. Therefore, for plasma and serum samples, we worked with fixed RNA volumes throughout the entire procedure.

1. Mix RNA and spin it to get a homogenized solution.
2. Keep RNA in ice.
3. Use the RNA button to quantify the sample. These applications report nucleic acid concentration and two absorbance ratios (A260/A280 and A260/A230) (*see* **Note 23**).
4. To clean the reading spot, use 2–3 μL of nuclease-free water and wipe it.
5. From the home screen select RNA.
6. Pipette 1 μL of blanking solution (the elution buffer for RNA extraction, namely, nuclease-free water) and lower the arm. Tap blank and wait for the measurement to be completed.
7. Lift the arm and clean both pedals with a new laboratory wipe. Pipette 1 μL of sample solution and lower the arm, then tap "Measure."
8. When you are finished measuring the samples, tap "End Experiment."
9. Lift the arm and clean both pedestals with a new wipe.

3.4 Reverse Transcription (RT)/cDNA Synthesis

The conversion of RNA, including miRNA, to cDNA can be performed using different protocols. For EvaGreen-based droplet digital PCR, we used miRCURY LNA miRNA PCR Assay.

For miRNA detection by probes-based droplet digital PCR, you can choose three different protocols: TaqMan advanced miRNA protocol, TaqMan MicroRNA protocol, or Custom Plus TaqMan RNA protocol (*see* **Note 24**). For the detection of miRNA variants (isomiRs) we use Dumbbell-based droplet digital PCR (Db-ddPCR) protocol.

Since plasma and serum have variable amounts of RNA [22], we use 2 μL of total RNA for each protocol and perform the "absolute" quantification [23]. Dumbbell-based droplet digital PCR (Db-ddPCR) protocol has not been yet tested and optimized for plasma and serum samples.

3.4.1 miRCURY LNA RT Assay

1. Dilute each template RNA sample to 5 ng/μL in 2 μL using nuclease-free water.
2. UniSp6 RNA spike-in preparation (*see* **Note 25**). Resuspend the UniSp6 RNA spike-in by adding 80 μL nuclease-free water to the tube. Mix by vortexing and briefly centrifuge. Leave for 20–30 min on ice to fully dissolve the RNA spike-in. Then mix by vortexing and briefly centrifuge. Store in aliquots at −20 °C.
3. Thaw samples and cDNA synthesis reagents on ice, gently vortex to thoroughly mix, then centrifuge briefly to spin down the contents and eliminate air bubbles.

Table 3
Reverse transcription reaction setup for miRCURY LNA RT assay

Component	For 1 reaction
5× RT SYBR green reaction buffer	2 μL
10× miRCURY RT enzyme mix (*see* **Note 27**)	1 μL
RNA spike-in template	0.5 μL
RNase-free water	4.5 μL
Total RT reaction mix volume	**8 μL**

Table 4
Reverse transcription reaction temperature cycling protocol for miRCURY LNA RT assay

Step	Temperature	Time
Reverse transcription	42 °C	60 min
Stop reaction	95 °C	5 min
Hold	4 °C	Hold

4. In a 1.5 mL microcentrifuge tube, prepare sufficient RT Reaction Mix for the required number of reactions according to Table 3 (*see* **Note 26**).
5. Vortex the RT Reaction Mix to thoroughly mix the contents, then centrifuge briefly to spin down the contents and eliminate air bubbles.
6. Add 2 μL (10 ng of RNA) of sample to each well of a reaction plate or each reaction tube, then transfer 8 μL of RT Reaction Mix to each well or tube.
7. Seal the reaction plate or tubes, then vortex briefly to thoroughly mix the contents.
8. Centrifuge the reaction plate or tubes briefly to spin down the contents and eliminate air bubbles.
9. Place the reaction plate or tubes into a thermal cycler, then incubate using the settings and standard cycling in Table 4.
10. Store the cDNA at 20 °C.

3.4.2 TaqMan Advanced miRNA cDNA Synthesis Protocol

1. Prepare 20 ng of RNA in a volume of 5.1 μL.
2. The 50% PEG 8000 reagent must be at room temperature for the adaptor ligation reaction.
3. Poly(A) polymerase adds a 3′-adenosine tail to the miRNA.

Table 5
Poly(A) reaction mix setup for TaqMan advanced miRNA cDNA synthesis kit

Component	For 1 reaction
10× poly(a) buffer	0.5 μL
ATP	0.5 μL
Poly(A) enzyme	0.3 μL
Total poly(A) reaction mix volume	**1.3 μL**

4. Thaw samples and cDNA synthesis reagents on ice, gently vortex to thoroughly mix, then centrifuge briefly to spin down the contents and eliminate air bubbles.
5. In a 1.5 mL microcentrifuge tube, prepare sufficient Poly (A) Reaction Mix for the required number of reactions according to Table 5 (*see* **Note 26**). Vortex the Poly(A) Reaction Mix to thoroughly mix the contents, then centrifuge briefly to spin down the contents and eliminate air bubbles.
6. Add 5.1 μL (20 ng of RNA) of sample to each well of a reaction plate or each reaction tube, then transfer 1.3 μL of Poly (A) Reaction Mix to each well or tube.
7. Seal the reaction plate or tubes, then vortex briefly to thoroughly mix the contents.
8. Centrifuge the reaction plate or tubes briefly to spin down the contents and eliminate air bubbles.
9. Place the reaction plate or tubes into a thermal cycler, then incubate using the settings and standard cycling in Table 6. Proceed immediately to the adaptor ligation reaction.
10. The miRNA with poly(A) tail undergoes adaptor ligation at the 5′ end.
11. In a 1.5 mL microcentrifuge tube, prepare sufficient Ligation Reaction Mix for the required number of reactions according to Table 7 (*see* **Note 26**). Vortex the Ligation Reaction Mix to thoroughly mix the contents, then centrifuge briefly to spin down the contents and eliminate air bubbles.
12. Transfer 9.6 μL of the Ligation Reaction Mix to each well of the reaction plate or each reaction tube containing the poly (A) tailing reaction product. The total volume should be 16 μL per well or tube.
13. Seal the reaction plate or tubes, then vortex briefly or shake to thoroughly mix the contents.
14. Centrifuge the reaction plate or tubes briefly to spin down the contents.

Table 6
Poly(A) tailing reaction temperature cycling protocol for TaqMan advanced miRNA cDNA synthesis kit

Step	Temperature	Time
Polyadenylation	37 °C	45 min
Stop reaction	65 °C	10 min
Hold	4 °C	Hold

Table 7
Ligation reaction mix setup for TaqMan advanced miRNA cDNA synthesis kit

Component	For 1 reaction
5× DNA ligase buffer	3 μL
50% PEG8000 (*see* **Note 28**)	4.5 μL
25× ligation adaptor	0.6 μL
RNA ligase	1.5 μL
Total ligation reaction mix volume	**9.6 μL**

15. Place the reaction plate or tubes into a thermal cycler, then incubate using the settings and standard cycling in Table 8. Proceed immediately to the reverse transcription (RT) reaction.
16. Universal RT primer binds to the 3'poly(a) tail and miRNA is reverse transcribed.
17. In a 1.5 mL microcentrifuge tube, prepare sufficient RT Reaction Mix for the required number of reactions according to Table 9 (*see* **Note 26**). Vortex the RT Reaction Mix to thoroughly mix the contents, then centrifuge briefly to spin down the contents and eliminate air bubbles.
18. Transfer 14 μL of the RT Reaction Mix to each well of the reaction plate or each reaction tube containing the adaptor ligation reaction product. The total volume should be 30 μL per well or tube.
19. Seal the reaction plate or tubes, then vortex briefly to thoroughly mix the contents.
20. Centrifuge the reaction plate or tubes briefly to spin down the contents.
21. Place the reaction plate or tubes into a thermal cycler, then incubate using the settings and standard cycling in Table 10. Proceed to the miR-Amp reaction.

Table 8
Ligation reaction temperature cycling protocol for TaqMan advanced miRNA cDNA synthesis kit

Step	Temperature	Time
Ligation	16 °C	45 min
Hold	4 °C	Hold

Table 9
Reverse transcription (RT) reaction setup for TaqMan advanced miRNA cDNA synthesis kit

Component	For 1 reaction
5× RT buffer	6 μL
dNTP mix (25 mM each)	1.2 μL
20× universal RT primer	1.5 μL
10× RT enzyme mix	3 μL
RNase-free water	2.3 μL
Total RT reaction mix volume	**14 μL**

Table 10
RT reaction temperature cycling protocol for TaqMan advanced miRNA cDNA synthesis kit

Step	Temperature	Time
Reverse transcription	42 °C	15 min
Stop reaction	85 °C	5 min
Hold	4 °C	Hold

22. Store the RT reaction product at −20 °C for up to 2 months.
23. All miRNAs are amplified for 14 cycles with universal forward and reverse primers, which bind, respectively, adaptor sequence and poly(a) tail.
24. In a 1.5 ml microcentrifuge tube, prepare sufficient miR-Amp Reaction Mix for the required number of reactions according to Table 11 (*see* **Note 26**). Vortex the miR-Amp Reaction Mix to thoroughly mix the contents, then centrifuge briefly to spin down the contents and eliminate air bubbles.
25. Transfer 45 μL of the miR-Amp Reaction Mix to each well of a new reaction plate or reaction tube.

Table 11
miR-Amp reaction mix setup for TaqMan advanced miRNA cDNA synthesis kit

Component	For 1 reaction
2× miR-amp master mix	25 μL
20× miR-amp primer mix	2.5 μL
RNase-free water	17.5 μL
Total miR-amp reaction mix volume	**45 μL**

Table 12
miR-Amp reaction temperature cycling protocol for TaqMan advanced miRNA cDNA synthesis kit

Step	Temperature	Time	Cycles
Enzyme activation	95 °C	5 min	1
Denature	95 °C	3 s	14
Anneal/extend	60 °C	30 s	
Stop reaction	99 °C	10 min	1
Hold	4 °C	Hold	1

26. Add 5 μL of the RT reaction product to each reaction well or each reaction tube. The total volume should be 50 μL per well or tube.
27. Seal the reaction plate or tubes, then vortex briefly to thoroughly mix the contents.
28. Centrifuge the reaction plate or tubes briefly to spin down the contents.
29. Place the reaction plate or tubes into a thermal cycler, then incubate using the settings, MAX ramp speed, and standard cycling in Table 12. Proceed to perform the ddPCR.
30. Store the undiluted miR-Amp reaction product at −20 °C for up to 2 months.

3.4.3 TaqMan Small RNA Assay Protocol

1. Prepare 10 ng of RNA in a volume of 5 μL.
2. Thaw components of the reverse transcription kit on ice.
3. Thaw the RT Primers on ice, vortex briefly then centrifuge briefly to collect the contents at the bottom of the tube.
4. In an appropriately sized microcentrifuge tube, prepare the RT Reaction Mix according to Table 13 (*see* **Note 26**). Invert to mix, then centrifuge briefly to collect the contents at the bottom of the tube.

Table 13
RT reaction mix setup for TaqMan small RNA assays

Component	For 1 reaction
100 mM dNTPs (with dTTP)	0.15 μL
MultiScribe™ reverse transcriptase, 50 U/μL	1 μL
10× reverse transcription buffer	1.5 μL
RNase inhibitor, 20 U/μL	0.19 μL
Nuclease-free water	4.16 μL
Total RT reaction mix volume	**7 μL**

Table 14
RT reaction temperature cycling protocol for TaqMan small RNA assays

Step	Temperature	Time
Reverse transcription	16 °C 42 °C	30 min 15 min
Stop reaction	85 °C	5 min
Hold	4 °C	Hold

5. Place the RT Reaction Mix on ice.
6. Combine 7 μL of RT Reaction Mix and 5 μL of total RNA (10 ng) in a reaction tube or each well of a reaction plate.
7. Mix thoroughly, then centrifuge briefly to collect the contents at the bottom of the tubes or wells.
8. Add 3 μL of 5× RT Primer to each reaction tube or each well of a reaction plate.
9. Seal the tubes or reaction plate. Centrifuge briefly to collect the contents at the bottom of the tubes or wells.
10. Place the reaction plate or tubes into a thermal cycler, then incubate using standard cycling, a reaction volume of 15 μL, and the settings in Table 14. The RT reaction product can be stored at −25 to −15 °C for up to 1 week.

3.4.4 Custom Plus TaqMan RNA Assay Protocol

1. Prepare 20 ng of RNA in a volume of 14 μL.
2. Thaw components of the reverse transcription kit on ice.
3. Thaw the RT Primers on ice, vortex briefly then centrifuge briefly to collect the contents at the bottom of the tube.

Table 15
RT reaction mix setup for custom TaqMan RNA assay

Component	For 1 reaction
5× VILO reaction mix	4 μL
10× Superscript III enzyme blend	2 μL
Total RT reaction mix volume	**6 μL**

Table 16
RT reaction temperature cycling protocol for custom TaqMan small RNA assays

Step	Temperature	Time
Reverse transcription	25 °C	10 min
	42 °C	60 min
Stop reaction	85 °C	5 min
Hold	4 °C	Hold

4. In an appropriately sized microcentrifuge tube, prepare the RT Reaction Mix according to Table 15 (*see* **Note 26**). Invert to mix, then centrifuge briefly to collect the contents at the bottom of the tube.
5. Combine 6 μL of RT Reaction Mix and 14 μL of total RNA (10 ng) in a reaction tube or in each well of a reaction plate.
6. Mix thoroughly, then centrifuge briefly to collect the contents at the bottom of the tubes or wells.
7. Seal the tubes or reaction plate. Centrifuge briefly to collect the contents at the bottom of the tubes or wells.
8. Place the reaction plate or tubes into a thermal cycler, then incubate using standard cycling, a reaction volume of 20 μL, and the settings in Table 16. The RT reaction product can be stored at −20 °C.

3.4.5 Dumbbell-Based ddPCR Assay (Db-ddPCR) for isomiR

1. Design adapters, primers, and probes according to Shigematsu et al. [21].
2. Recipe 10× Annealing buffer: 50 mM Tris–HCl pH 8, 5 mM EDTA, 100 mM $MgCl_2$.
3. 50% PEG 8000 preparation: Transfer 3.5 g of PEG 8000 to a 15 mL falcon and add 5 mL of H_2O. Microwave for 10 s with the unscrewed lid placed loosely on top of the falcon. Add H_2O to the 7 mL line. Prepare 1 mL aliquots. Stored at −20 °C.

4. Prepare 20 ng of RNA in a volume of 1 μL.
5. Preheat a Thermoblock to 90 °C and a Thermoblock to 37 °C.
6. Mix 20 ng cellular total RNA (1 μL) with 0.02 pmol of the 3′-adapter (1 μL volume) in a 1.5 mL centrifuge tube.
7. Incubate at 90 °C for 3 min, and immediately add 0.5 μL of 10× Annealing buffer.
8. Incubate at 37 °C for 20 min to allow annealing and hybridization to occur.
9. In a 1.5 mL microcentrifuge tube, prepare sufficient Ligation Reaction Mix for the required number of reactions according to Table 17 (*see* **Note 26**).
10. Add 7.5 μL of Ligation Reaction Mix to make 10 μL volume solution. Place the reaction tubes into a thermal cycler, then incubate using the settings and standard cycling in Table 18.
11. In a 1.5 mL microcentrifuge tube, prepare sufficient RT Reaction Mix 1 for the required number of reactions according Table 19 (*see* **Note 26**).
12. Add 4 μL of Ligation Reaction Mix to 10 μL of ligated RNA to make 14 μL volume solution.

Table 17
Ligation reaction mix setup for Dumbbell-based ddPCR assay (Db-ddPCR) for isomiR

Component	For 1 reaction
10× ligase buffer	1 μL
T4 RNA ligase 2	0.1 μL
$MgCl_2$ 100 mM	0.5 μL
50% PEG8000 (*see* **Note 28**)	2 μL
RNase-free water	3.9 μL
Total ligation reaction mix volume	**7.5 μL**

Table 18
Ligation reaction temperature cycling protocol for Db-ddPCR for isomiR

Step	Temperature	Time
Ligation	37 °C	60 min
Stop	80 °C	5 min
Hold	4 °C	Hold

Table 19
RT reaction mix setup for Db-ddPCR for isomiR

Component	For 1 reaction
RT primer 5 μM	1 μL
dNTP mix (10 mM each)	0.5 μL
RNase-free water	2.5 μL
Total RT reaction mix 1 volume	**4 μL**

Table 20
RT reaction mix temperature cycling protocol for Db-ddPCR for isomiR

Temperature	Time
90 °C	2 min
65 °C	5 min
Ice	At least 1 min

Table 21
RT reaction mix 2 setup for Db-ddPCR for isomiR

Component	For 1 reaction
5× first strand buffer	4 μL
100 mM DTT	1 μL
RNasin ribonuclease inhibitor	0.5 μL
SuperScript III reverse transcriptase	0.5 μL
Total RT reaction mix 2 volume	**6 μL**

13. Place the reaction tubes into a thermal cycler, then incubate using the settings and standard cycling in Table 20.
14. In a 1.5 mL microcentrifuge tube, prepare sufficient RT Reaction Mix 2 for the required number of reactions according to Table 21 (*see* **Note 26**).
15. Add 6 μL of Ligation Reaction Mix to RT product to make 20 μL volume solution.
16. Place the reaction tubes into a thermal cycler, then incubate using the settings and standard cycling in Table 22.
17. cDNA can be stored at −20 °C.

Table 22
RT reaction mix 2 temperature cycling protocol for Dumbbell-based ddPCR assay

Step	Temperature	Time
RT	55 °C	60 min
Stop	70 °C	15 min
Hold	4 °C	Hold

3.5 cDNA Dilution

1. Dilute the amount of cDNA template needed for the planned ddPCR reactions in nuclease-free water (*see* **Note 29**).
2. Store the diluted cDNA at 20 °C. The diluted cDNA proved to be stable for at least 3 months at 20 °C.

3.6 Droplet Generation and PCR

3.6.1 Sample Preparation

1. Decide the desired concentration of cDNA. Depending on the miRNA target, cDNA can be used directly or diluted to prevent saturation (*see* **Note 29**).
2. If you expect a low concentration of miRNA, then you can use undiluted cDNA. Conversely, if the expected quantity is high, then the sample must be diluted. Tests must be made to determine a quantity that avoids saturation (all positive droplets).
3. NTC, no-template control (add water instead of sample) must be added for each miRNA Reaction Mix (*see* **Notes 30** and **31**).
4. Samples must be loaded along the columns (A1, B1, C1, etc.). Make sure that the column has all wells filled. If there are empty wells, add 20 μL of ddPCR Buffer Control.

3.6.2 Reaction Setup

1. Thaw all components to room temperature. Mix thoroughly by vortexing each tube to ensure homogeneity, then centrifuge briefly to collect contents at the bottom of the tubes.
2. Prepare cDNA samples at the desired concentration before setting up the reaction mix.
3. In a 1.5 mL microcentrifuge tube, prepare sufficient Reaction Mix for the required number of reactions according to Tables 23, 24, and 25 (*see* **Note 26**). Assemble all required components except the sample.
4. Dispense equal aliquots into each well of the plate. Add the samples as the final step Seal the reaction plate with optical flat caps, then vortex briefly to thoroughly mix the contents.
5. Centrifuge the plate briefly to spin down the contents.
6. Allow the plate to equilibrate at room temperature for about 3 min.

Table 23
ddPCR reaction mix setup for EvaGreen Supermix

Component for EvaGreen-based ddPCR	For 1 reaction
QX200 ddPCR EvaGreen Supermix	10 μL
miRCURY LNA miRNA PCR assay	1 μL (*see* **Note 32**)
RNase-free water	Variable
cDNA sample	Variable
Total volume	11 μL

Table 24
ddPCR reaction mix setup for Supermix for probes (No dUTP)

Component for probe-based ddPCR	For 1 reaction
2× ddPCR Supermix for probes (no dUTP)	10 μL
20× target primers/probes (FAM)	1 μL
RNase-free water	Variable
cDNA sample	Variable
Total volume	**11 μL**

Table 25
ddPCR reaction mix setup for Dumbbell-based ddPCR isomiR detection

Component for isomiR detection (Db-ddPCR)	For 1 reaction
2× ddPCR Supermix for probes (no dUTP)	10 μL
TaqMan MGB probe 10 μM	0.3 μL
Forward primer 10 μM	0.3 μL
Reverse primer 10 μM	0.3 μL
RNase-free water	Variable
cDNA sample	Variable
Total volume	**20 μL**

3.6.3 Droplet Generation

1. Place a DG8 cartridge into the DG8 Cartridge Holder.
2. With a multichannel pipette (2–20 μL), transfer 20 μL of PCR mixture into the middle row of wells (labeled sample) of the DG8 cartridge (Transfer first column of wells) (*see* **Note 33**).
3. Add 70 μL of droplet generation oil in all wells into the left row (labeled oil).

4. Place a new rubber gasket over the top. Plug in the droplet generator (cord in the back) to turn on. Press the green button to open the droplet generator. Place the prepared cassette into the droplet generator and press the button to close and begin droplet generation.
5. Following droplet generation (about 2–3 min), remove the cartridge and, using a multichannel pipette (5–50 μL), pipet 45 μL the droplets (placed in the right row of the cartridge) into a new plate (*see* **Note 34**).
6. Be sure to cover the plate with optical flat caps to minimize evaporation as components are volatile.
7. Continue making droplets (other columns) until complete.
8. Once the droplet generation is completed, seal the top of the plate with a foil seal with the plate sealer (5.0 s at 180 °C) (*see* **Note 35**).

3.6.4 Thermocycling Condition (PCR)

1. Place sealed plate into Thermal Cycler and run PCR.
2. Use a heated lid set to 105 °C and set the sample volume to 40 μL.
3. Thermocycling protocol changes according to assay (see Tables 26, 27, and 28).

3.7 Droplet Reading and Data Analysis

3.7.1 Droplet Reading

1. After PCR, the plate should be left in the thermal cycler for at least 15 min at 4 °C before droplet reading. Otherwise, the plate can be left in the thermal cycler overnight or stored for a few days at 4 °C before droplet reading. Check if the volume of the solution is consistent across the wells.
2. Power on the QX200 droplet reader and let it warm up for at least 30 min.
3. Prime the system the first time you use the Reader or if you just change Droplet Reader Oil.

Table 26
ddPCR temperature cycling protocol for EvaGreen Supermix

Step for EvaGreen-based ddPCR	Temperature (*see* Note 36)	Time	Cycles
Enzyme activation	95 °C	5 min.	1
Denaturation	95 °C	30 s	40
Annealing/extension	58 °C (*see* **Note 37**)	1 min	
Signal stabilization	4 °C	5 min	1
Enzyme deactivation	90 °C	5 min	1
Hold	4 °C	Hold	1

Table 27
ddPCR temperature cycling protocol for Supermix for probes (No dUTP)

Step for probe-based ddPCR	Temperature (*see* Note 36)	Time	Cycles
Enzyme activation	95 °C	10 min	1
Denaturation	94 °C	30 s	40
Annealing/extension	60 °C	1 min	
Enzyme deactivation	98 °C	10 min	1
Hold	4 °C	Hold	1

Table 28
ddPCR temperature cycling protocol for Dumbbell-based ddPCR isomiR detection

Step for Db-ddPCR for isomiR	Temperature	Time	Cycles
Enzyme activation	95 °C	10 min	1
Denaturation	94 °C	30 s	40
Annealing/extension	45 °C 60 °C	15 s 45 s	
Enzyme deactivation	98 °C	10 min	1
Hold	4 °C	Hold	1

4. Flush the system if the instrument has been unused for longer than a week.
5. Make certain that there is enough ddPCR Droplet Reader Oil in the reader compartment.
6. Prepare plate template: Start the QuantaSoft software from the system PC and click Template > New to create a new template.
   (a) To open the well editor, double-click on the well(s) you wish to edit.
   (b) In the Sample panel, enter the sample Name of the sample and select the Experiment from the drop-down menu: ABS (for absolute quantification).
   (c) Select the Supermix from the drop-down menu (QX200 ddPCR EvaGreen Supermix for EvaGreen experiment, 2X ddPCR Supermix for Probes (no dUTP) for Probe experiment).
   (d) Define Target 1 (channel 1, the FAM channel) and Target 2 (channel 2, the VIC or HEX channel). Also, if testing only a target for well (as in the EvaGreen experiment), you have to select the type of target 2. Target 1 à Name:

miRNA target; Type: Ch1 Unknown; Target 2 à Name: second miRNA target or Name empty (if you test only a single miRNA in the well), Type: Ch2 unknown.

(e) Settings appear in the Applied Well Settings box as you enter them. Then, click Apply or OK to save the information.

7. Carefully move the plate from the thermal cycler to the droplet reader.
8. Place the 96-well PCR plate containing the post-PCR droplets into the base of the plate holder.
9. Place the top of the plate holder on the PCR plate. Firmly press both release tabs down to secure the PCR plate in the holder.
10. Start the QuantaSoft software from the system PC.
11. In QuantaSoft software, click Setup in the left navigation bar to define your experiment and open a saved template for a plate map (settings only, no data), click Template > Load and select the file.
12. In QuantaSoft software, click Run to start the droplet reading defining your experiment: select columns and the Supermix name (EvaGreen or Probe).
13. When droplet reading is complete, click the Analyze button to open and analyze the data.

3.7.2 DdPCR Data Analysis

1. Events: you can see total droplets. The number of total droplets for each well must be greater than 10,000. If you have some wells with not enough droplets, repeat the experiment for these samples.
2. 2D Amplitude: Use NTC to define negative droplets. Define the threshold for all samples based on NTC, the blank. All NCT droplets should be negative droplets. You can select the positive droplets with the lasso tool. In your samples, you should detect two clouds of droplets: negative droplets in gray and positive droplets in blue. If you do not have negative droplets that means you have saturation. Repeat the experiment with less cDNA. If you do not have positive droplets, namely, the sample and NTC have the same droplet cloud, that means the miRNA is not detected in the sample.
3. Concentration: You can see the concentration of your sample. Concentration is determined using the Poisson distribution and it is given as the number of target copies/μl.
4. Once the positive droplets are selected, export the miRNA concentration values by selecting all the wells and using the export .csv option from the Concentration tab.

3.7.3 miRNA Quantification

Absolute quantification of circulating miRNA can be achieved by maintaining constant volumes in all steps (RNA extraction, cDNA conversion, and ddPCR), making it possible to standardize the data and compare it between samples.

To obtain a faithful quantification of circulating miRNA levels, it is necessary to standardize all the pre-analytical variables that challenge the accurate quantification of miRNA levels such as: source type, blood tube used, tissue centrifugation protocol, and conservation.

Reverse-transcription reactions can introduce a certain degree of variability in cDNA generation [24, 25]. Therefore, this factor must be considered in the absolute quantification of miRNAs. After selecting the most efficient reverse transcription protocol, the absolute number of miRNA copies in 1 μL of plasma or serum is calculated by multiplying the concentration obtained from the ddPCR reaction by the dilution factor [23].

Relative quantification should be used for tissue miRNAs. It is possible to normalize the miRNA expression data, and compare miRNA expression between samples, by using and quantifying one or more invariant miRNA(s)/smallRNA(s) (*see* **Note 38**) by ddPCR, to use as normalizer(s).

Calculate the ratio between the target miRNA concentration obtained from ddPCR (n° copies/μL) and the invariant gene concentration (n° copies/μL) for each sample.

4 Notes

1. Do not use heparin-containing blood collection tubes as this anticoagulant can interfere with downstream assays, such as RT-PCR.
2. Reagents, such as buffers and MinElute spin columns, in miRNeasy Serum/Plasma Kit, and miRNeasy Mini Kit are the same. For RNA isolation from plasma and serum, miRNeasy Mini Kit with miRNeasy Serum/Plasma protocol can be used.
3. TaqMan MicroRNA Assays are delivered in two tubes: one containing the RT primer and the other containing the specific preformulated TaqMan Assay (TaqMan probe and PCR primer set).
4. Incomplete removal of cell-culture medium will inhibit lysis and dilute the lysate, affecting the conditions for binding of RNA to the RNeasy Mini spin column membrane. Both effects may reduce RNA yield.
5. For long-term storage of DNase I, transfer the stock solution into single-use aliquots and store at −30 °C to −15 °C for up to 9 months. Thawed aliquots can be kept at 2–8 °C for up to 6 weeks. Avoid refreezing aliquots after thawing.

6. Ethanol addition may cause precipitates, which will not interfere with the procedure.
7. The 80 °C incubation partially reverses the formaldehyde modification of nucleic acids. Longer or higher temperature incubations may increase RNA fragmentation but might also slightly decrease cycle threshold (Ct) values in downstream applications like real-time RT-PCR.
8. It is important to dry the spin column membrane since residual ethanol may interfere with downstream reactions. Centrifugation with the lids open ensures that no residual ethanol is carried over during RNA elution. To avoid damage to their lids, place the spin columns into the centrifuge with at least one empty position between columns. Orient the lids so that they point in a direction opposite to the rotation of the rotor (e.g., if the rotor rotates clockwise, orient the lids counterclockwise).
9. Eluting with smaller volumes of RNase-free water increases total RNA concentrations but lowers RNA yields. This process results in about 2 μL of lost eluate: using 14 μL RNase-free water yields a 12 μL eluate.
10. . Do not use bleach on any instrument parts. Use only the 0.5 mL Elution Tubes provided in the kit, as other tubes may be incompatible with the Maxwell Instrument.
11. Use the master mix within 1 h of preparation; it cannot be stored for later use. For fewer than six samples, prepare enough master mix for $n + 1$ samples.
12. Add DNase cocktail reagents in the specified order, ensuring thorough mixing. Store remaining reconstituted DNase I at −30 °C to −10 °C.
13. During the elution process, approximately 15 μL may be lost during this process.
14. Incomplete loosening of the cell pellet may lead to inefficient lysis and reduced RNA yields.
15. Homogenized cell lysates can be stored at −90 °C to −65 °C for several months.
16. After centrifugation, heat the centrifuge up to room temperature if the same centrifuge will be used for the next centrifugation steps.
17. A precipitate may form after the addition of ethanol, but this will not affect the procedure.
18. The long centrifugation dries the spin column membrane, ensuring that no ethanol is carried over during RNA elution. Residual ethanol may interfere with downstream reactions.

19. After homogenization, samples may be stored frozen at −80 °C for later processing. Thaw frozen homogenates on ice or at 2–10 °C to avoid RNA degradation.
20. Since RNA concentration cannot be determined accurately, we suggest using fixed volumes as a measure of the input amount.
21. One of the primary purposes of the RNA spike-ins and the matching primer pairs for their detection is to provide controls for the quality of the RNA isolation, since RNA isolations may vary in yield, purity, and integrity. To control differences in efficiencies during RNA isolation from plasma and serum samples, we suggest adding known RNA spike-ins to the sample prior to isolation. Then, detect RNA spike-in levels in ddPCR to check RNA isolation quality: in all samples we expect the same amount of the RNA spike-in. If not, repeat RNA isolation.
22. For plasma or serum samples less than 100 μL, add 100 μL of nuclease-free water to the sample.
23. A260/A280 and A260/A230 purity ratios should be ~2 for RNA.
24. We provide the following guidance to suggest the choice of reverse transcription kit for miRNA quantification experiments based on the use of probes:
    (a) TaqMan Advanced miRNA protocol: Use this protocol for cDNA synthesis when you need to test multiple miRNAs from a single amplified sample. In fact, it is a universal reverse transcription (RT), one RT step for all TaqMan Advanced miRNA Assays.
    (b) TaqMan Advanced miRNA protocol: Use this protocol for cDNA synthesis for miRNA-specific RT. TaqMan MicroRNA Assay employs a target-specific stem–loop primer during cDNA synthesis to produce a template for real-time PCR or ddPCR.
    (c) Custom plus TaqMan RNA protocol: Use this protocol if you have custom plus TaqMan probe/primers (not only miRNAs, but also long no coding RNA).
25. Another purpose of the RNA spike-ins and the matching primer pairs for their detection is to provide controls for the quality of cDNA synthesis reaction. Some sample types may contain compounds that inhibit the cDNA synthesis, even if the RNA was purified using the best standard procedures. This may result in different efficiencies of the reverse transcription or PCR between compared samples. One way to control differences in efficiencies during cDNA synthesis is by adding known RNA spike-ins to the sample prior to cDNA synthesis. We recommend UniSp6 RNA spike-in provided with the

miRCURY LNA RT Kit. Then, detect RNA spike-in level in ddPCR to check cDNA conversion quality: in all samples, we expected the same amount of the RNA spike-in. If not, repeat cDNA conversion.

26. If you are setting up multiple reactions, prepare a master mix with a volume 10% greater than required for the total number of reactions.
27. Immediately before using it, remove the 10x miRCURY RT Enzyme from the freezer, mix by flicking the tube, and then place it on ice. Briefly centrifuge it to collect residual liquid from the sides of the tubes, and then keep it on ice.
28. For accurate pipetting of 50% PEG 8000: Use 50% PEG 8000 at room temperature, aspirate and dispense solution slowly.
29. We recommend diluting the cDNA reaction between 1:50 and 1:500 in water, depending on the target miRNA abundance. The amount of each miRNA varies greatly, and some miRNAs are more abundant than others. Using a 1:50 diluted cDNA in a ddPCR reaction is usually sufficient to obtain an adequate number of positive and negative droplets. In case of positive droplet saturation (i.e., no negative droplets), a further dilution of the cDNA sample is necessary.
30. The NTC (No Template Control) sample needs to be run in every plate to verify the expected fluorescent signal of the individual droplets. There is always the possibility that due to the short length of the miRNAs, nonspecific products are amplified. Since the size of targeted miRNA amplicons and that of primer dimers are relatively close in size, it may be difficult to discriminate between them using fluorescence amplitude only. Analysis using 2D plots adds a second fluorescent dimension to the droplets and sometimes allows for separation of two distinct populations. If the "cloud" of nonspecific target amplification is not overlapping with the true signal, the miRNA can still be quantified by selecting only the true positive droplets. In this case, the NTC sample is essential for droplet selection. In the case of overlapping true- and false-positive droplet distributions, the assay cannot be used with the ddPCR system. A different assay, from a different supplier, should be considered.
31. We advise using A1 well for NTC; in fact, sometimes the number of droplets analyzed from the first well is lower compared to the others.
32. Adjust the primer concentration (range 0.5–2 μL) according to the degree of separation between positive and negative droplets [25].

33. Avoid bubbles at this point as they will disturb droplet formation.
34. At this point, it is important to pipet the droplet mixture very gently. Fast pipetting or mixing can shear (damage) the droplets and ruin your run.
35. Leave the plate sealer tray ejected during warm up time to avoid accidentally heating your samples.
36. Check/adjust ramp rate settings to −2 °C/s.
37. The miRCURY miRNA assay usually works at 58 °C in EvaGreen ddPCR mixture, but an initial optimization could be required (range 56–60 °C).
38. When measuring miRNA expression, we suggest using small-nucleolar RNAs (snoRNAs) as reference genes, such as *SNORD44* or *SNORD48*, or stably expressed housekeeping miRNAs that do not show variability of expression in your sample cohort (e.g., invariant miRNAs in microarray/small RNA sequencing experiments).

References

1. Nemeth K, Bayraktar R, Ferracin M, Calin GA (2024) Non-coding RNAs in disease: from mechanisms to therapeutics. Nat Rev Genet 25(3):211–232. https://doi.org/10.1038/s41576-023-00662-1
2. Ferracin M, Negrini M (2015) Micromarkers 2.0: an update on the role of microRNAs in cancer diagnosis and prognosis. Expert Rev Mol Diagn 15(10):1369–1381. https://doi.org/10.1586/14737159.2015.1081058
3. Finnegan EF, Pasquinelli AE (2013) MicroRNA biogenesis: regulating the regulators. Crit Rev Biochem Mol Biol 48(1):51–68. https://doi.org/10.3109/10409238.2012.738643
4. Moutinho C, Esteller M (2017) MicroRNAs and epigenetics. Adv Cancer Res 135:189–220. https://doi.org/10.1016/bs.acr.2017.06.003
5. Griffiths-Jones S, Grocock RJ, van Dongen S, Bateman A, Enright AJ (2006) miRBase: microRNA sequences, targets and gene nomenclature. Nucleic Acids Res 34(Database issue):D140–D144. https://doi.org/10.1093/nar/gkj112
6. Cloonan N, Wani S, Xu Q, Gu J, Lea K, Heater S et al (2011) MicroRNAs and their isomiRs function cooperatively to target common biological pathways. Genome Biol 12(12):R126. https://doi.org/10.1186/gb-2011-12-12-r126
7. Liu CG, Calin GA, Meloon B, Gamliel N, Sevignani C, Ferracin M et al (2004) An oligonucleotide microchip for genome-wide microRNA profiling in human and mouse tissues. Proc Natl Acad Sci USA 101(26):9740–9744. https://doi.org/10.1073/pnas.0403293101
8. Volinia S, Calin GA, Liu CG, Ambs S, Cimmino A, Petrocca F et al (2006) A microRNA expression signature of human solid tumors defines cancer gene targets. Proc Natl Acad Sci USA 103(7):2257–2261. https://doi.org/10.1073/pnas.0510565103
9. Loher P, Londin ER, Rigoutsos I (2014) IsomiR expression profiles in human lymphoblastoid cell lines exhibit population and gender dependencies. Oncotarget 5(18):8790–8802. https://doi.org/10.18632/oncotarget.2405
10. Telonis AG, Magee R, Loher P, Chervoneva I, Londin E, Rigoutsos I (2017) Knowledge about the presence or absence of miRNA isoforms (isomiRs) can successfully discriminate amongst 32 TCGA cancer types. Nucleic Acids Res 45(6):2973–2985. https://doi.org/10.1093/nar/gkx082
11. Kakimoto Y, Tanaka M, Kamiguchi H, Ochiai E, Osawa M (2016) MicroRNA stability in FFPE tissue samples: dependence on GC content. PLoS One 11(9):e0163125. https://doi.org/10.1371/journal.pone.0163125
12. Boisen MK, Dehlendorff C, Linnemann D, Schultz NA, Jensen BV, Hogdall EV et al

(2015) MicroRNA expression in formalin-fixed paraffin-embedded cancer tissue: identifying reference MicroRNAs and variability. BMC Cancer 15:1024. https://doi.org/10.1186/s12885-015-2030-2

13. Ferracin M, Pedriali M, Veronese A, Zagatti B, Gafa R, Magri E et al (2011) MicroRNA profiling for the identification of cancers with unknown primary tissue-of-origin. J Pathol 225(1):43–53. https://doi.org/10.1002/path.2915
14. Cortez MA, Bueso-Ramos C, Ferdin J, Lopez-Berestein G, Sood AK, Calin GA (2011) MicroRNAs in body fluids – the mix of hormones and biomarkers. Nat Rev Clin Oncol 8(8):467–477. https://doi.org/10.1038/nrclinonc.2011.76
15. Cocks A, Martinez-Rodriguez V, Del Vecchio F, Schukking M, Broseghini E, Giannakopoulos S et al (2021) Diverse roles of EV-RNA in cancer progression. Semin Cancer Biol 75:127–135. https://doi.org/10.1016/j.semcancer.2020.11.022
16. Liu T, Zhang Q, Zhang J, Li C, Miao YR, Lei Q et al (2019) EVmiRNA: a database of miRNA profiling in extracellular vesicles. Nucleic Acids Res 47(D1):D89–D93. https://doi.org/10.1093/nar/gky985
17. Landgraf P, Rusu M, Sheridan R, Sewer A, Iovino N, Aravin A et al (2007) A mammalian microRNA expression atlas based on small RNA library sequencing. Cell 129(7): 1401–1414. https://doi.org/10.1016/j.cell.2007.04.040
18. Schmittgen TD, Lee EJ, Jiang J, Sarkar A, Yang L, Elton TS et al (2008) Real-time PCR quantification of precursor and mature micro-RNA. Methods 44(1):31–38. https://doi.org/10.1016/j.ymeth.2007.09.006
19. Ferracin M, Negrini M (2018) Quantification of circulating MicroRNAs by droplet digital PCR. Methods Mol Biol 1768:445–457. https://doi.org/10.1007/978-1-4939-7778-9_25
20. Honda S, Kirino Y (2015) Dumbbell-PCR: a method to quantify specific small RNA variants with a single nucleotide resolution at terminal sequences. Nucleic Acids Res 43(12):e77. https://doi.org/10.1093/nar/gkv218
21. Shigematsu M, Honda S, Kirino Y (2018) Dumbbell-PCR for discriminative quantification of a small RNA variant. Methods Mol Biol 1680:65–73. https://doi.org/10.1007/978-1-4939-7339-2_4
22. Wang K, Yuan Y, Cho JH, McClarty S, Baxter D, Galas DJ (2012) Comparing the MicroRNA spectrum between serum and plasma. PLoS One 7(7):e41561. https://doi.org/10.1371/journal.pone.0041561
23. Ferracin M, Lupini L, Salamon I, Saccenti E, Zanzi MV, Rocchi A et al (2015) Absolute quantification of cell-free microRNAs in cancer patients. Oncotarget 6(16):14545–14555. https://doi.org/10.18632/oncotarget.3859
24. Lindner L, Cayrou P, Jacquot S, Birling MC, Herault Y, Pavlovic G (2021) Reliable and robust droplet digital PCR (ddPCR) and RT-ddPCR protocols for mouse studies. Methods 191:95–106. https://doi.org/10.1016/j.ymeth.2020.07.004
25. Miotto E, Saccenti E, Lupini L, Callegari E, Negrini M, Ferracin M (2014) Quantification of circulating miRNAs by droplet digital PCR: comparison of EvaGreen- and TaqMan-based chemistries. Cancer Epidemiol Biomarkers Prev 23(12):2638–2642. https://doi.org/10.1158/1055-9965.EPI-14-0503

Chapter 14

Digital Droplet PCR (ddPCR) for Absolute Quantification of 16S rRNA Copy Number in Metagenomic Data

Claudia Leoni, Marinella Marzano, Ermes Filomena, and Anna Maria D'Erchia

Abstract

Digital Droplet PCR (ddPCR) is a quantitative PCR method that offers high sensitivity and accuracy in measuring the amount of nucleic acid in a sample, without the need of a standard curve. In ddPCR, a single sample is partitioned into up to 20,000 droplets, using the water–oil emulsion technology, and the amplification reaction occurs within each droplet using a fluorescent hydrolysis probe (Taqman) or a DNA-binding fluorescent dye. Following PCR, the emitted signals are individually measured in each droplet. Here, we describe a ddPCR optimized protocol for accurately quantifying the total copy number of the 16S rRNA gene in a metagenomic DNA sample. The protocol utilizes a primer pair, targeting the 16S V5-V6 hypervariable regions, in combination with a double-strand DNA-binding fluorescent dye.

Key words Metagenomics, 16SrDNA, Absolute quantification, ddPCR, Microbiome

1 Introduction

Digital Droplet PCR (ddPCR) is a third-generation PCR technology, which offers an accurate absolute quantification of a specific nucleic acid target in a sample. In recent years, this technology has found a wide range of applications, from research to diagnostics, including metagenomics [1, 2]. In the field of metagenomics, ddPCR can be used to assess the bacterial abundance in various metagenomic DNA samples (e.g., stool, tissues, environment, etc.), typically by quantifying the total number of copies of the prokaryotic 16S rRNA gene. This gene comprises multiple conserved regions flanking nine hypervariable (V1-V9) ones, which vary across bacterial species. Due to these characteristics, the 16S rRNA is used as a reliable molecular marker and can be targeted using different primer pairs, designed to amplify specific regions

Claudia Leoni and Marinella Marzano contributed equally with all other contributors.

Guillaume Pavlovic (ed.), *Digital PCR: Methods and Protocols*, Methods in Molecular Biology, vol. 2969,
https://doi.org/10.1007/978-1-0716-4767-7_14,

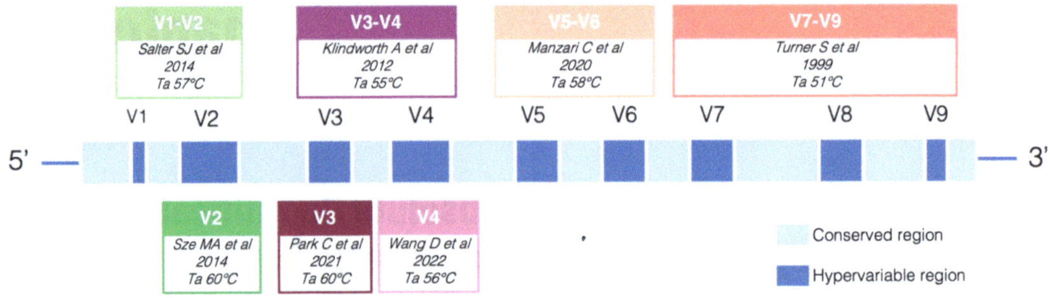

Fig. 1 Graphical representation of the 16S rRNA gene, a conserved prokaryotic gene consisting of multiple conserved regions (light blue), flanking nine hypervariable ones (V1-V9, dark blue). Boxes, up and down the structure of the gene, indicate the hypervariable regions targeted in ddPCR studies, along with their references and the annealing Temperature (Ta) used

[3–9] (Fig. 1). ddPCR can be also applied to integrate taxonomic results obtained from the microbiome sequencing, through the metabarcoding approach, using Next-Generation Sequencing (NGS) technologies, which provide information about the relative abundance of the different identified taxa in a sample. Relative abundance represents the proportion or percentage of a specific taxon in the microbial population present in a sample, while quantitative abundance indicates the absolute number of total microbial cells or gene copies or can measure the abundance of specific taxa or species, within a defined volume or mass of a sample. For instance, relative abundance can highlight taxa that are over-represented, whereas quantitative data helps determine whether these taxa are more numerous or if their apparent dominance is due to a decline in other taxa [10]. Combining relative abundance (derived from NGS sequencing) and quantitative abundance data (measured via ddPCR) provides a comprehensive understanding of microbial communities. The integrated data could be crucial in various contexts and for addressing specific biological questions, such as assessing the abundance of a particular pathogen within the host [11–13], confirming pathogen detection [14], or evaluating the impact of a specific treatment on a targeted microbial population [15].

Here, we describe the complete workflow for quantifying the total copy number of the 16S rRNA gene in a metagenomic DNA sample, using an optimized ddPCR protocol, targeting the 16S V5-V6 hypervariable regions, that we have previously successfully applied [6].

2 Materials

2.1 Kit, Reagents and Consumables

1. FastDNA Spin kit for soil (MP Biomedicals, 6560200) for fecal sample (*see* **Note 1**).
2. Sterile Nuclease-Free Water (Ambion AM9932).

3. Forward and Reverse primer (10 μM).
4. QX200 ddPCR 2X EvaGreen Supermix (Bio-Rad code 1864034).
5. DG8 Cartridges for QX200 Droplet Generator (Bio-Rad 1864008).
6. Droplet Generation Oil for EvaGreen (Bio-Rad 1864006).
7. Droplet Generator DG8 Gasket (Bio-Rad- 1863009).
8. QX200 Buffer Control Kit for EvaGreen (Bio-Rad 1864052).
9. ddPCR Plates 96-well, Semi-Skirted (Bio-Rad code 12001925).
10. Pierceable Foil Heat Seal (Bio-Rad code 1814040).
11. ddPCR Droplet Reader oil (Bio-Rad 1863004).
12. P2-P20-P200 Pipettes and 8-channel pipette (200 μL).
13. Filtered pipette tips.
14. V5/V6 primers (*see* **Note 5**).

2.2 Equipment

1. Droplet Generator DG8 Cartridge Holder (Bio-Rad 1863051).
2. QX200 Droplet Generator (Bio-Rad 1864002).
3. PX1 PCR Plate Sealer (Bio-Rad 1814000).
4. QX200 Droplet Reader (Bio-Rad 1864003).
5. QuantaSoft Analysis Software (Bio-Rad).
6. FastPrep-24 (MP Biomedicals, recommended).
7. 2100 Bioanalyzer (Agilent, recommended).
8. Tape Station 2200 (Agilent, recommended).
9. Qubit Fluorometer (Thermo Fisher Scientific, recommended).

3 Methods

3.1 DNA Extraction, Qualitative and Quantitative Analyses

In metagenomic analysis, the extraction of high-quality DNA from complex matrices is often a challenging task, and this, in turn, may affect its accurate quantification [6]. Thus, before starting ddPCR quantification of 16S copies, it is essential to evaluate the quality and the yield of the metagenomic DNA to guarantee the reliability of observed differential abundances among samples.

1. Carry out the metagenomic DNA isolation and purification using the appropriate kit in accordance with the typology of the sample (*see* **Note 1**). A negative control (no sample in the vial) must be processed together with the samples to monitor possible contaminations during the experimental procedure.

2. Perform qualitative and quantitative evaluations of the extracted metagenomic DNA, by electrophoretic analysis and fluorometric dosage, respectively (*see* **Notes 2** and **3**).

3.2 Selection of 16S rRNA Gene Primers and PCR Set up

As for any PCR, also for ddPCR, the primer pair design is essential to obtain accurate quantification. For metagenomic DNA, the primer pair should be capable of pairing with a conserved region of the target gene for the specific kingdom (*see* **Note 4**). For the 16S rRNA gene, several primer pairs have already been used in ddPCR for bacterial quantification [3–9] (Fig. 1). The ddPCR guidelines suggest choosing a primer pair able to produce an amplicon of a length ranging from 80 to 250 bp. In our recent paper, we demonstrated the reliability of a primer pair targeting the 16S V5–V6 regions, producing a 287 bp amplicon (*see* **Note 5**). In different DNA samples with a known number of 16S copies, using this primer pair, a perfect concordance was observed between the expected number of 16S copies and that calculated by ddPCR, both in DNA from single bacterial strains and from bacterial communities [6]. Here, we describe the ddPCR workflow using the V5-V6 primer pair with a dsDNA-binding dye (Eva green) (*see* **Note 6**).

Before starting with the ddPCR protocol, it is advisable to perform a classic PCR in order to: (i) define the optimal amount of metagenomic DNA to be used as template for ddPCR (*see* **Note 7**); (ii) verify the correctness of the amplicon produced and the absence of non-specific products (*see* **Note 8**).

3.3 Droplet Generation and PCR

Unless otherwise specified, carry out all procedures on ice. Clean the workbench with ethanol, sterilize pipettes to UV, and use filtered pipette tips. Before use, aliquot the enzyme, primers, and all reagents.

1. Prepare the PCR, mixing the metagenomic DNA (*see* **Note 9**) with 11 μL of 2X EvaGreen Supermix, 100–180 μM forward and reverse primers (*see* **Note 10**), nuclease-free water to a final volume of 22 μL (*see* **Note 11**). Each sample must be analyzed at least in duplicate, preparing independent reactions. Also prepare a negative control (no template) and a positive control, consisting of a mock template of known bacterial DNAs (*see* **Note 12**).
2. Vortex the tubes containing the reaction in short pulse, centrifuge briefly at maximum speed and incubate for 3 min at room temperature (*see* **Note 13**).
3. To generate the droplet, place the DG8 cartridge into the cartridge holder and transfer 20 μL of each sample into the middle wells of the cartridge (*see* **Note 14**), then dispense 70 μL of droplet oil into the bottom wells (*see* **Notes 15** and **16**). Close the top of the DG8 cartridge with the gasket, place the cartridge holder in the QX200 droplet generator and press the button to start the emulsion production.

4. Once droplet generation is completed, open the QX200 droplet generator and remove the cartridge holder; then remove the gasket from the cartridge and slowly aspirate 40 μL of droplets from each well of the top row of the cartridge and slowly transfer the droplets into the wells of a ddPCR plate (*see* **Note 17**).
5. Cover the ddPCR plate with a perforable aluminum seal, compatible with the needles of the droplet reader, and place it in the PX1 PCR plate Sealer. Set the plate sealer at 180 °C for 5 s.
6. Insert the sealed ddPCR plate in a thermocycler and carry out the PCR according to the following thermal cycling conditions: 95 °C for 5 min; 40 cycles at 95 °C for 30 s and at 58 °C for 1 min; 1 cycle at 4 °C for 5 min, 1 cycle at 90 °C for 5 min and final hold at 4 °C (*see* **Notes 18** and **19**).

3.4 Droplet Reading

1. Before starting the droplet reading in the QX200 Droplet reader, it is advisable to perform a wash of the microfluidic system:
   (a) Open the QuantaSoft Analysis Software on the computer.
   (b) Flush and prime the microfluidic system of the QX200 plate reader with the buttons in the top right corner of the QuantaSoft Analysis Software (*see* **Note 20**).
2. Design the layout of the ddPCR plate according to the experimental design and select the experiment type, with the QuantaSoft Analysis software in setup mode.
3. Once the droplet PCR is complete, insert the ddPCR plate into the QX200 Droplet Reader and proceed with the plate reading step, by clicking the 'Run' button on the designed layout (*see* **Note 21**).

3.5 Data Analysis

For each sample, PCR-positive and PCR-negative droplets are counted for the absolute quantification of the target by the QuantaSoft Analysis Software.

1. At the end of the reading of the samples, the software provides a 1-D amplitude plot that, for each read droplet (event), reports the magnitude of its fluorescent signal (*see* **Note 22**). The plot presents an upper cloud of "positive events" and lower cloud of "negative events", separated by a threshold line (Fig. 2). The threshold line is automatically determined for each sample by the software, although it may be manually chosen to only consider the target-specific dense clouds of positive events and exclude the so-called "rain". For this, use the threshold button (*see* **Notes 23** and **24**).

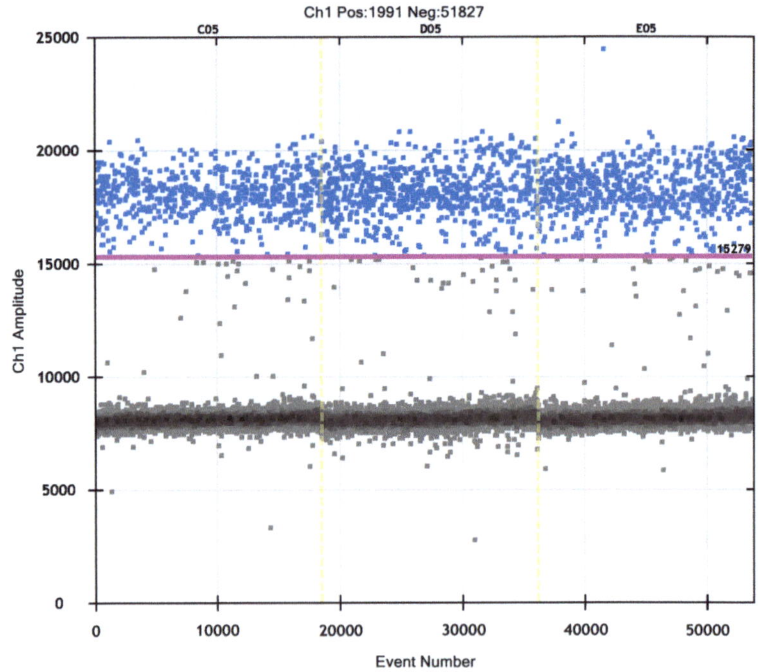

Fig. 2 1D-amplitude plot of a successful ddPCR experiment. Three ddPCRs with various DNA template (C05, D05, E05) are divided by the vertical yellow lines. The ordinate scale indicates fluorescent amplitude. The horizontal magenta line is the threshold, above which are positive droplets (blue) and below which are negative droplets (gray). The distinct separation between positive and negative events facilitates the quantification of the target

2. The software provides the quantification of the 16S in terms of copies/μL of reaction, available in a table that can be exported as a csv file, by using the "Export csv" button.
3. Calculate the 16S copies/ng of input DNA, considering: the reaction volume (22 μL); the amount of DNA used in the reaction; the dilution factor (if the DNA template was diluted), as reported below:

   16S copies· ng^{-1} = [16S copies · μL^{-1} · reaction volume (22 μL) · dilution factor]/ng input DNA (*see* **Notes 25** and **26**).

4 Notes

1. For fecal samples, we suggest adopting the FastDNA SPIN kit for soil (MP Biomedicals, Santa Ana, CA, USA), following the manufacturer's procedure and performing the bead-beating step in the FastPrep Instrument (BIO 101, Carlsbad, Canada), at a 6.0 m/s speed for 40 s [16]. This kit has been

successfully employed on biopsies and bakery products, too [17–20]. Elute the DNA in sterile and nuclease-free water.

2. For the quality control of the extracted DNA, use electrophoresis to estimate the level of degradation of the samples. We suggest using an automated electrophoresis Agilent system (Agilent, Santa Clara, CA, USA), such as Tape Station 2000, which can evaluate the integrity of a DNA sample, providing an integrity value (DNA Integrity Number, DIN).
3. For the quantitative analysis of the extracted DNA, assess the DNA concentration by fluorimetry, using selective dyes that emit fluorescence only when bound to DNA, to obtain a more accurate quantification. Indeed, traditional UV spectrophotometers can overestimate sample concentrations due to contaminants such as salts, solvents, proteins, and free nucleotides. For a fast and accurate quantification, we suggest using a Qubit Fluorometer (Qubit 4 or Qubit Flex, depending on low or mid-throughput processing) (Thermo Fisher Scientific, Waltham, MA, USA), with the dsDNA HS or BR assays, depending on DNA concentration of the samples.
4. Characteristics of a good molecular marker are: (i) universality across the taxonomic range under examination (e.g., 16S rDNA for bacteria, Internal Transcribed Spacers of the 18S rRNA (ITS) for fungi, 18S rDNA for metazoa); (ii) high discriminatory power up to the species level; (iii) presence of hypervariable regions, flanked by highly conserved regions, within the taxonomic range under examination; (iv) size compatible with the application of interest (e.g., ddPCR and/or NGS technologies).
5. The universal 16S primer pair (B-V5: 5′- ATTAGATACC CYGGTAGTCC -3′, A-V6: 5′- ACGAGCTGACGACARC CATG -3′) amplifies the 16S V5-V6 regions, producing an amplicon 287 bp long [6].
6. Since the 16S amplicon contains variable regions that may differ between strains and species [3–9], making it difficult to design a hydrolysis probe, ddPCR should be carried out using a dsDNA-binding dye (EvaGreen for Biorad).
7. Considering the nature of the sample, in terms of biodiversity or for the predominant presence of the host DNA that could inhibit the reaction, it is necessary to test different DNA input quantities. If the sample is highly biodiverse (e.g., mouse or human feces), we suggest testing quantities ranging from 0.5 ng to 10 ng. Conversely, for samples with low biodiversity and/or a high presence of the DNA host (e.g., mouse or human tissues), the DNA input quantities to be tested should range from 50 ng up to 100 ng [20].

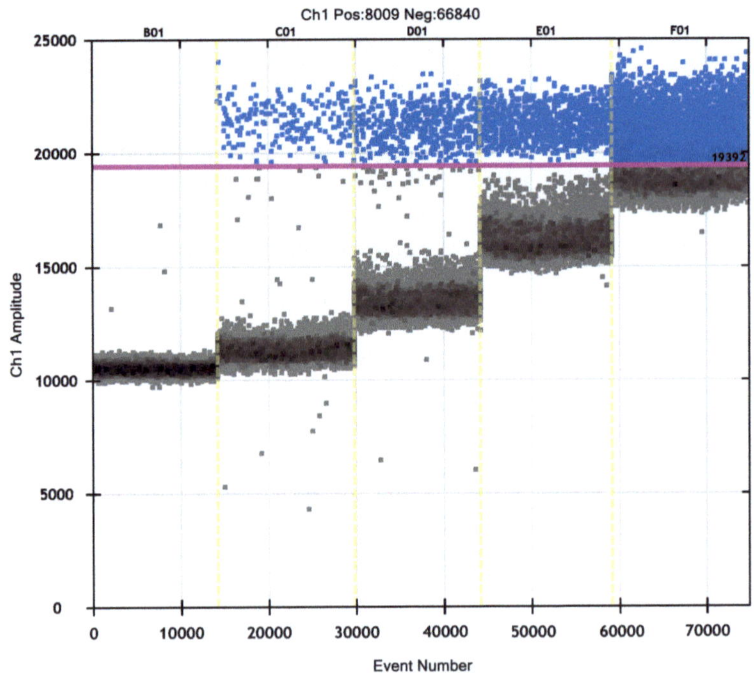

Fig. 3 Effect of the amount of fecal DNA template on V5-V6 ddPCR. Increasing the amount of DNA template shortens the distance between positive (in blue) and negative events (in gray) in the 1D-amplitude plot. High amounts of DNA template hinder the quantitation of V5-V6 target (lane F01). The horizontal magenta line is the threshold. Lane B01: negative control; lane C01: dilution 1:500 of 1 ng of DNA template; lane D01: dilution 1:200 of 1 ng of DNA template; lane E01: dilution 1:100 of 1 ng of DNA template; lane F01: 1 ng of DNA template. Lanes are divided by the vertical yellow lines

8. Verify the correct amplicon size produced by PCR using a mock template of known bacterial DNAs and confirm it by Sanger sequencing. Include a negative control (no template, only sterile nuclease-free water) for quality assurance of the PCR experiments to monitor possible reagent contaminations.
9. According to the ddPCR manufacturer's guidelines, the template amount in ddPCR should not exceed 100 ng. Because of the preliminary analysis described in Note 7, if the DNA metagenomic sample exhibits high biodiversity and yield, it should be diluted through serial dilutions up to 1:500–1:800 (Fig. 3).
10. Primer concentrations can influence the ddPCR efficiency and results. It is recommended to use a primer concentration ranging between 100 and 180 μM to minimize the occurrence of the so-called "rain," through a clear separation of the positive and negative events in the amplitude plots.
11. Vortex all components of the ddPCR, especially the EvaGreen Supermix, which is very thick, and centrifuge briefly before

dispensing. The reaction volume to prepare is more than 20 μL to ensure that 20 μL of the mix is then transferred in the cartridge. Use aerosol-barrier tips for all the procedures. To process multiple samples, start preparing the reagent mix for the total number of samples; then dispense the mix into the vials and in each add the necessary volume of metagenomic DNA.

12. The negative control must be prepared by adding a volume of water corresponding to the μl of DNA added in the reactions. The mock DNA template must be dosed by fluorometric assay (Qubit) and diluted to have 0.05 pg in the reaction.

13. For greater sample homogeneity, vortex all samples for the same time.

14. Avoid air bubbles in the well by pipetting the sample very slowly into the cartridge, positioning the pipette tip at a 15° angle near the bottom of the well. If any bubbles form during loading, they must be burst with the tip. All sample wells must be filled with sample, diluted supermix, or 1× control sample buffer.

15. The droplet generation oil for EvaGreen is stored at room temperature and, before use, check that it is transparent. Do not use oil if it appears opaque.

16. Use an 8-channel pipette to fill each well of the bottom row with 70 μL of droplet generator oil.

17. Aspirate droplets with a 200 μL tip, individually from each well or use an 8-channel pipette. Droplets are stable for 30 min after emulsion. Aspirate the sample very slowly to avoid air bubbles inside the sample. After transferring the droplets to the ddPCR plate, the presence of two different phases can be observed and this indicates that the droplets did not undergo alterations during the transfer. Once the droplets have been generated, no contamination of the sample can occur in the subsequent phases and the ddPCR plate can remain uncovered until it is filled with all samples.

18. The annealing/extension step is carried out at 58 °C, that is the optimal annealing temperature of the primer pair targeting the 16S V5-V6 regions. To reduce the "rain" effect of events, it is possible to increase the annealing temperature from 58 °C to 60 °C (Fig. 4) and/or to increase the cycle number up to 50 (Fig. 5). The final 90 °C and 4 °C steps help to stabilize the emulsion beads.

19. Keep the ddPCR plate for at least 30 min at 4 °C at the end of the ddPCR before reading. The plate can be left in the thermal cycler overnight at 4 °C or stored at 4 °C for 3 days.

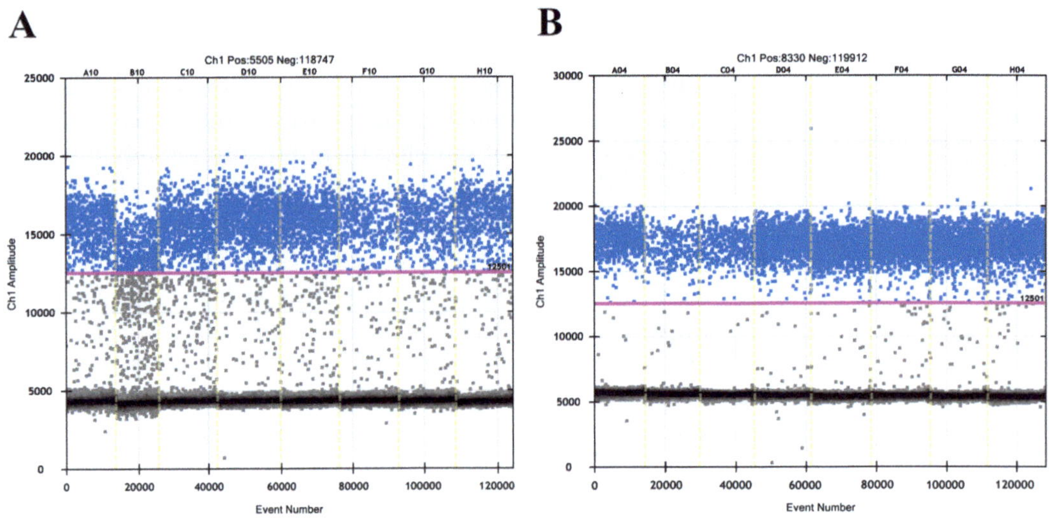

Fig. 4 Effect of the annealing temperature on V5-V6 ddPCR. Increasing the annealing temperature reduces the amount of the so-called "rain" in the 1D-amplitude plot. Four fecal DNA samples were analyzed as template (dilution 1:100 of 1 ng each) in duplicate, at two annealing temperatures (58 °C in panel A and 60 °C in panel B). Sample 1: lanes A10 and B10 (panel A) and A04 and B04 (panel B); Sample 2: lanes C10 and D10 (panel A), C04 and D04 (panel B); Sample 3: lanes E10 and F10 (panel A), E04 and F04 (panel B); Sample 4: lanes G10 and H10 (panel A), G04 and H04 (panel B). Lanes are divided by the vertical yellow lines

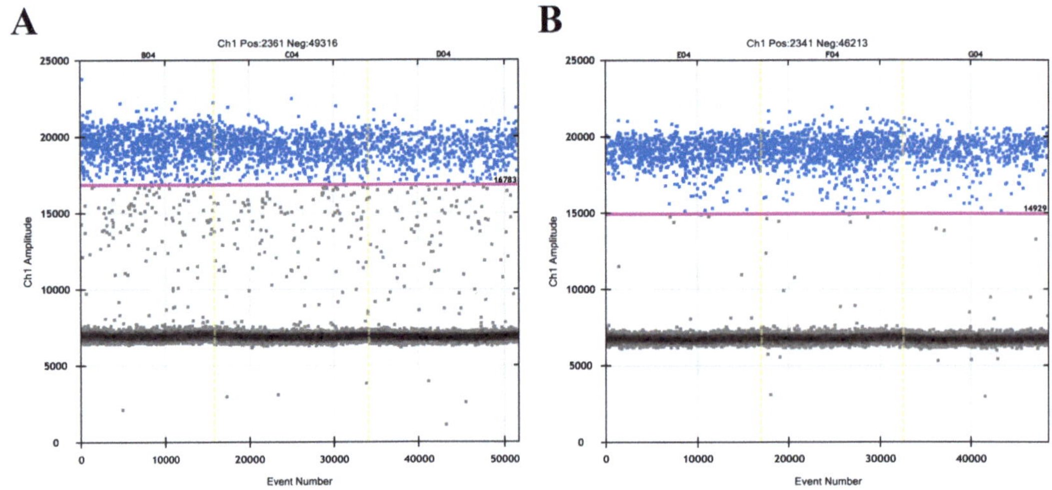

Fig. 5 Effect of the cycle number on V5-V6 ddPCR. Increasing the cycles number reduces the amount of the so-called "rain" in the 1D-amplitude plot. Three fecal DNA samples were used as template (dilution 1:100 of 1 ng each) for ddPCR, at different cycles number (40 cycles in panel A, 45 cycles in panel B). Sample 1: lane B04 (panel A) and lane E04 (panel B); Sample 2: lane C04 (panel A) and lane F04 (panel B); Sample 3: lane D04 (panel A) and lane G04 (panel B). Lanes are divided by the vertical yellow lines

20. If the instrument has not been used for a long time (e.g., more than 1 month), four flush and priming wash cycles must be performed, otherwise one cycle of each is sufficient.
21. Verify the status of the oil and waste tanks of the Droplet Reader beforehand, although the reader can automatically detect if enough oil is available to perform the reading for all samples. About 90 s are needed to process each sample.
22. The robustness of the statistical quantification method underlying the ddPCR depends on the number of *total* events per sample, which should at least be $1 \cdot 10^4$ [21]. This may be controlled via the dedicated plot the software renders by pressing the "Events" button above the amplitude plot and checking the *total* box on the right. Do not consider the results, if less than $1 \cdot 10^4$ events are read and repeat the quantification.
23. The threshold line is unique for all samples and can be set manually.
24. The negative control must have a maximum number of positive events equal to 3; anyway, if the number of the positive events of the control is lower than the number of the positive events of the samples, it is possible to manually set the threshold line to eliminate them.
25. It is possible to estimate the number of bacterial genomes present in 1 ng of DNA used as template, considering the average copy number of 16S rRNA gene per bacterial genome (four copies) [22]: *Bacterial genome* $\cdot$ *ng* $^{-1}$ *of template DNA* = *(16S copies* $\cdot$ *ng*$^{-1}$*)/4.*
26. It is possible to estimate the amount of bacterial DNA present in 1 ng of DNA used as template, considering: (i) the average length of a bacterial genome (5 Mb) [23]; (ii) the average mass of a base pair ($1.1 \cdot 10^{-12}$ ng, obtained starting from the 650 g/mole as average molar mass of 1 bp); (iii) the average copy number of 16S rRNA gene per bacterial genome (four copies) [22], as below: *ng bacterial DNA$\cdot$ng*$^{-1}$ *of template DNA* = *(5x10e6 bp* $\cdot$ *1.1x10e-12 (ng/bp))/4* = *1.4x10e-6 ng.*

Acknowledgments

This work was supported by: PRIN2017 project (MUR, prot. 2017M9L3L8_003) and INNOVA project (Italian Health Minister, code PNC-E3-2022-23683266 PNC-HLS-DA).

References

1. Verhaegen B, De Reu K, De Zutter L, Verstraete K, Heyndrickx M, Van Coillie E (2016) Comparison of droplet digital PCR and qPCR for the quantification of Shiga toxin-producing Escherichia coli in bovine feces. Toxins 8:157. https://doi.org/10.3390/toxins8050157
2. Li F, Liu J, Maldonado-Gómez MX, Frese SA, Gänzle MG, Walter J (2024) Highly accurate and sensitive absolute quantification of bacterial strains in human fecal samples. Microbiome 12:168. https://doi.org/10.1186/s40168-024-01881-2
3. Park C, Kim SB, Choi SH, Kim S (2021) Comparison of 16S rRNA gene based microbial profiling using five next-generation sequencers and various primers. Front Microbiol 12: 715500. https://doi.org/10.3389/fmicb.2021.715500
4. Salter SJ, Cox MJ, Turek EM, Calus ST, Cookson WO, Moffatt MF, Turner P, Parkhill J, Loman NJ, Walker AW (2014) Reagent and laboratory contamination can critically impact sequence-based microbiome analyses. BMC Biol 12:87. https://doi.org/10.1186/s12915-014-0087-z
5. Klindworth A, Pruesse E, Schweer T, Peplies J, Quast C, Horn M, Glöckner FO (2013) Evaluation of general 16S ribosomal RNA gene PCR primers for classical and next-generation sequencing-based diversity studies. Nucleic Acids Res 41:e1–e1. https://doi.org/10.1093/nar/gks808
6. Manzari C, Oranger A, Fosso B, Piancone E, Pesole G, D'Erchia AM (2020) Accurate quantification of bacterial abundance in metagenomic DNAs accounting for variable DNA integrity levels. Microb Genomics 6. https://doi.org/10.1099/mgen.0.000417
7. Turner S, Pryer KM, Miao VPW, Palmer JD (1999) Investigating deep phylogenetic relationships among cyanobacteria and plastids by small subunit rRNA sequence Analysis1. J Eukaryot Microbiol 46:327–338. https://doi.org/10.1111/j.1550-7408.1999.tb04612.x
8. Sze MA, Abbasi M, Hogg JC, Sin DD (2014) A comparison between droplet digital and quantitative PCR in the analysis of bacterial 16S load in lung tissue samples from control and COPD GOLD 2. PLoS One 9:e110351. https://doi.org/10.1371/journal.pone.0110351
9. Wang D, Wang S, Du X, He Q, Liu Y, Wang Z, Feng K, Li Y, Deng Y (2022) DDPCR surpasses classical QPCR technology in quantitating bacteria and fungi in the environment. Mol Ecol Resour 22:2587–2598. https://doi.org/10.1111/1755-0998.13644
10. Vandeputte D, Kathagen G, D'hoe K, Vieira-Silva S, Valles-Colomer M, Sabino J, Wang J, Tito RY, De Commer L, Darzi Y, Vermeire S, Falony G, Raes J (2017) Quantitative microbiome profiling links gut community variation to microbial load. Nature 551:507–511. https://doi.org/10.1038/nature24460
11. Abram TJ, Cherukury H, Ou C-Y, Vu T, Toledano M, Li Y, Grunwald JT, Toosky MN, Tifrea DF, Slepenkin A, Chong J, Kong L, Del Pozo DV, La KT, Labanieh L, Zimak J, Shen B, Huang SS, Gratton E, Peterson EM, Zhao W (2020) Rapid bacterial detection and antibiotic susceptibility testing in whole blood using one-step, high throughput blood digital PCR. Lab Chip 20:477–489. https://doi.org/10.1039/C9LC01212E
12. Li H, Bai R, Zhao Z, Tao L, Ma M, Ji Z, Jian M, Ding Z, Dai X, Bao F, Liu A (2018) Application of droplet digital PCR to detect the pathogens of infectious diseases. Biosci Rep 38:BSR20181170. https://doi.org/10.1042/BSR20181170
13. Wouters Y, Dalloyaux D, Christenhusz A, Roelofs HMJ, Wertheim HF, Bleeker-Rovers CP, Te Morsche RH, Wanten GJA (2020) Droplet digital polymerase chain reaction for rapid broad-spectrum detection of bloodstream infections. Microb Biotechnol 13:657–668. https://doi.org/10.1111/1751-7915.13491
14. Hu B, Tao Y, Shao Z, Zheng Y, Zhang R, Yang X, Liu J, Li X, Sun R (2021) A comparison of blood pathogen detection among droplet digital PCR, metagenomic next-generation sequencing, and blood culture in critically ill patients with suspected bloodstream infections. Front Microbiol 12:641202. https://doi.org/10.3389/fmicb.2021.641202
15. Shao Z, Zhu J, Wei Y, Jin J, Zheng Y, Liu J, Zhang R, Sun R, Hu B (2022) Pathogen load and species monitored by droplet digital PCR in patients with bloodstream infections: a prospective case series study. BMC Infect Dis 22: 771. https://doi.org/10.1186/s12879-022-07751-2
16. Piancone E, Fosso B, Marzano M, De Robertis M, Notario E, Oranger A, Manzari C, Bruno S, Visci G, Defazio G, D'Erchia AM, Filomena E, Maio D, Minelli M, Vergallo I, Minelli M, Pesole G (2022) Natural and after colon washing fecal samples: the two sides of the coin for investigating the human gut microbiome. Sci Rep 12:

17909. https://doi.org/10.1038/s41598-022-20888-z

17. Leoni C, Ceci O, Manzari C, Fosso B, Volpicella M, Ferrari A, Fiorella P, Pesole G, Cicinelli E, Ceci LR (2019) Human endometrial microbiota at term of normal pregnancies. Genes 10:971. https://doi.org/10.3390/genes10120971
18. Marzano M, Calasso M, Caponio GR, Celano G, Fosso B, De Palma D, Vacca M, Notario E, Pesole G, De Leo F, De Angelis M (2022) Extension of the shelf-life of fresh pasta using modified atmosphere packaging and bio-protective cultures. Front Microbiol 13: 1003437. https://doi.org/10.3389/fmicb.2022.1003437
19. Calasso M, Marzano M, Caponio GR, Celano G, Fosso B, Calabrese FM, De Palma D, Vacca M, Notario E, Pesole G, De Angelis M, De Leo F (2023) Shelf-life extension of leavened bakery products by using bio-protective cultures and type-III sourdough. LWT 177:114587. https://doi.org/10.1016/j.lwt.2023.114587
20. Leoni C, Vinci L, Marzano M, D'Erchia AM, Dellino M, Cox SN, Vitagliano A, Visci G, Notario E, Filomena E, Cicinelli E, Pesole G, Ceci LR (2024) Endometrial cancer: a pilot study of the tissue microbiota. Microorganisms 12:1090. https://doi.org/10.3390/microorganisms12061090
21. Pinheiro LB, Coleman VA, Hindson CM, Herrmann J, Hindson BJ, Bhat S, Emslie KR (2012) Evaluation of a droplet digital polymerase chain reaction format for DNA copy number quantification. Anal Chem 84:1003–1011. https://doi.org/10.1021/ac202578x
22. Větrovský T, Baldrian P (2013) The variability of the 16S rRNA gene in bacterial genomes and its consequences for bacterial community analyses. PLoS One 8:e57923. https://doi.org/10.1371/journal.pone.0057923
23. Land M, Hauser L, Jun S-R, Nookaew I, Leuze MR, Ahn T-H, Karpinets T, Lund O, Kora G, Wassenaar T, Poudel S, Ussery DW (2015) Insights from 20 years of bacterial genome sequencing. Funct Integr Genomics 15:141–161. https://doi.org/10.1007/s10142-015-0433-4

Chapter 15

Extending the Potential of Digital Droplet PCR by Expressing Amplified Genes and Assaying Their Phenotype

Janine Kehrli, Radi Khodr, and Michael Ryckelynck

Abstract

Besides its key role in genetic information storage, DNA is also highly relevant to be used as an amplifiable and readable biomarker for plenty of applications including diagnostics, sterility, and quality controls. To this end, several molecular analytical methods emerged over the past decades to specifically detect and quantify DNA sequences from different sample origins. Significant gain in sensitivity was reached with the advent of digital droplet PCR (ddPCR), while also overcoming many important limitations of quantitative PCR. In ddPCR, DNA templates are typically isolated in water-in-oil droplets containing a PCR amplification mixture supplemented with a fluorogenic reporter. Whereas the analytical process usually ends with the counting of the fraction of fluorescent droplets in which PCR amplification took place, in some cases these droplets can be the starting point of more complex analytical pipelines. This is typically the case of the microfluidic-assisted in vitro compartmentalization technology introduced by our team and routinely used to perform in vitro functional screenings and directed evolution experiments. In this chapter, we present this technology in which DNA templates are first amplified by ddPCR, prior to being in vitro expressed as RNA or protein within these droplets. Finally, the phenotype of each gene is analyzed using a fluorescence-based assay and the droplets sorted accordingly.

Key words Droplet digital PCR, DNA quantification, In vitro gene expression, Directed evolution, Droplet microfluidics, Droplet fusion, Droplet sorting

1 Introduction

Deoxyribonucleic Acid (DNA) serves diverse roles in living organisms. It not only acts as the repository of genetic information, but it can also be the target of numerous epigenetic modifications involved in gene expression regulation. Specific detection and quantification of DNA sequences present in biological or environmental samples is frequently used to diagnose and/or monitor disease (e.g., cancer) progression [1], identify pathogens (e.g., bacteria, viruses) [2], profile microbiota composition or in forensic science, just as a few examples. To this end, highly sensitive amplification-based molecular techniques have been developed

Guillaume Pavlovic (ed.), *Digital PCR: Methods and Protocols*, Methods in Molecular Biology, vol. 2969,
https://doi.org/10.1007/978-1-0716-4767-7_15,

[3, 4], Polymerase Chain Reaction (PCR) and its derivatives being by far the most widely used. Based on iterative rounds of oligonucleotide primers annealing and extension by a DNA polymerase, PCR allows the exponential accumulation of DNA copies with a single original DNA template being amplified up to several million times in a very short time. PCR can also be turned quantitative by including dedicated fluorogenic probes (i.e., sequence-specific dually labeled oligonucleotides or intercalating dyes) in the reaction mixture [5–7]. In this approach known as quantitative PCR (qPCR), the interaction between the probe and the DNA (templates and copies) leads to a fluorescence emission whose recording enables monitoring of the overall amplification process in real time. Determining the number of amplification cycles required to reach a fluorescence exceeding a threshold allows a threshold cycle (Ct) to be computed (the lower the initial amount of template DNA, the higher the Ct). Finally, comparing the Ct of a target to that of an internal reference or to a calibration curve allows the relative initial quantity of the target DNA to be calculated. Though highly sensitive and widely used, qPCR also suffers significant limitations. Indeed, as the quantification is relative to a standard (such quantification being said analogue), any variation linked to sample composition (e.g., presence of polymerase inhibitors in the assayed sample but absent in the calibration samples) or the use of non-optimized primer sets (leading to unequal amplification efficiency of the target and its reference) may significantly challenge quantification accuracy. Moreover, detecting underrepresented point mutants diluted within a large excess of wildtype molecules (a scenario typically encountered in cancer diagnostics) may also be particularly challenging due to the competition between templates for the same set of primers, which can limit qPCR to sensitivity [8]. Many limitations of qPCR are overcome by digitalizing the process, i.e., by individualizing each template DNA molecule into a compartment prior to amplifying it in the presence of fluorogenic probes [9, 10]. In the so-called digital PCR (dPCR), simply counting the fraction of fluorescent compartments (i.e., those initially occupied by a DNA template) enables the absolute concentration of the target DNA to be computed with only a marginal impact of the presence of some polymerase inhibitor (provided a fraction of enzyme remains functional) or suboptimal primer design. Moreover, since templates are individualized, they do not compete with each other. Therefore, even very rare variants can be identified and quantified; the limit of detection of the method being only dictated by the number of compartments that can be generated and analyzed. Whereas dPCR was initially performed in microtiter plates [9], the advent of microfluidics rapidly led to its miniaturization down to nanoliter-volume microchambers using large-scale integration device [11], and more recently down to a few picoliters with the introduction of droplet-based microfluidics [12]. Briefly,

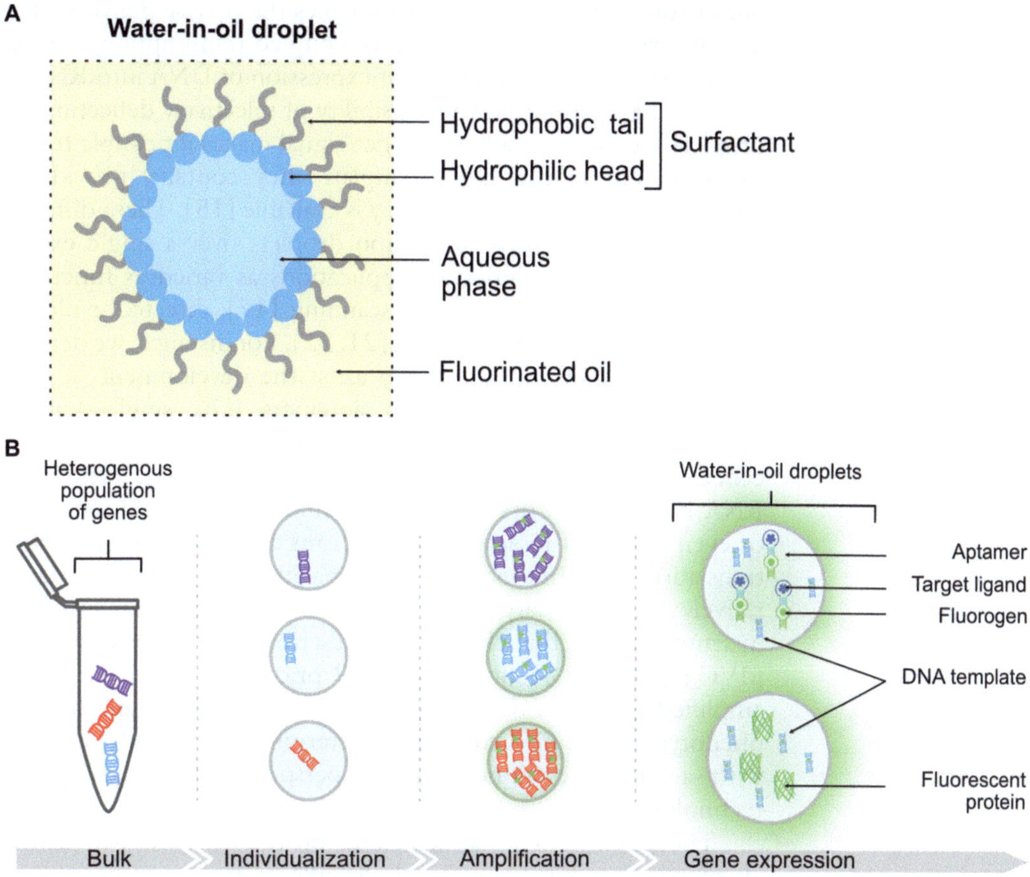

Fig. 1 Microfluidic-assisted in vitro compartmentalization. (**a**) Organization of a water-in-oil droplet. (**b**) Analytical pipeline. DNA templates are individualized in picoliter droplets where they are amplified in the presence of an intercalating dye. Upon amplification, occupied droplets become fluorescent. Next, an in vitro gene expression mixture is delivered to each droplet and amplified DNA is expressed as RNA and possibly protein

in digital droplet PCR (ddPCR), the mixture of DNA is diluted into qPCR mixture and later dispersed in water-in-oil droplets carried by an oil phase supplemented in a surfactant that stabilizes the droplets and prevents their uncontrolled coalescence (Fig. 1a). This allows the production and the analysis of thousand to million droplets providing unprecedented sensitivity and accuracy. In most ddPCR applications, the counting of fluorescent droplets is the last step of the analytical pipeline, leaving the information carried by the amplified DNA unexploited. On the contrary, ddPCR is instead the first step of the microfluidic-assisted in vitro compartmentalization (μIVC) pipeline that we introduced [13], and in which the serial use of microfluidic devices enables the precise control of droplets while preserving the confinement of the genetic information (Fig. 1b). Indeed, the possibility of fusing each ddPCR droplet one-to-one with a larger droplet enables the precise delivery of an in vitro expression (transcription or coupled transcription/translation) mixture and the subsequent expression of DNA into RNA or protein [14]. Moreover, the possibility of selectively deflecting the droplets according to a fluorescence signal makes it possible to sort them (based on the DNA template they contain) provided a fluorescence-based reporting assay is available [15]. These different steps, being performed on million droplets over a single experiment, unlock possibilities for applications as varied as functional screening [13, 16], mutational scanning [17], directed evolution [18–20], or even drug screening [21, 22]. For instance, we demonstrated that μIVC can efficiently assist the development of optimized enzymes [20], light-up aptamers (i.e., single-stranded oligonucleotides able to specifically bind and activate the fluorescence of a small molecule called fluorogen) [18, 23], fluorogenic RNA-based biosensors of small molecules [24, 25] and ions [26], or even to dissect biological mechanisms as complex as eukaryotic translation [27]. In every case, the capacity of precisely and quantitatively analyzing large mutant libraries expressed in vitro was instrumental. In this chapter, we describe the main steps of the μIVC pipeline, starting from ddPCR prior to explaining how the amplified DNA can be expressed in RNA or protein and how the phenotype of the molecule can be determined. We finally conclude with a short description of the perspectives opened by this technology.

The μIVC functional screening procedure described in this chapter is aimed at being highly flexible by making it possible to in vitro express genes coding for proteins and nucleic acids (in the example depicted here RNA, but it is easily adaptable to DNA and other chemistries). Moreover, the use of independent microfluidic devices provides great flexibility in terms of incubation time between each step. An important advantageous feature of μIVC is the preservation of the template DNA and the possibility to recover it at the end of the process. This makes it possible, for instance, to start a new round of μIVC screening directly from the DNA molecules isolated at the end of a previous round. Besides, the pool of recovered DNA molecules can be analyzed in depth using high-throughput sequencing and bioinformatics as this is done in the μIVC-seq approach [19].

This pipeline is well suited for plenty of applications. For instance, one can prepare mutant libraries of a known gene and screen them for mutants with improved or reprogrammed functions. We have already illustrated several times how such a directed evolution strategy can be used to assist the development of new nucleic acid-based biosensors (illustrated on Fig. 7), light-up aptamers [18, 28–30] or even enzymes [13]. The great resolution and reproducibility offered by μIVC allows the precise selection and recovery of genes encoding a phenotype of interest. Finally, the completely in vitro nature of our pipeline allows to extend its use for the screening of gene libraries coding for protein, and also for nonnatural polymers like nucleic acids made of xenobiotic building blocks that would be difficult to handle in living cells.

2 Materials

2.1 General Material

2.1.1 Microfluidic Chip

1. A mold prepared by photolithography [31].
2. A vacuum jar connected to a pump.
3. A plasma cleaner (e.g., Femto, Diener).
4. An oven or an incubator set to 65 °C.
5. A source of compressed air.
6. 75 × 50 mm glass slides (Corning).
7. 1 mL plastic syringes (Terumo).
8. 23G needles (Terumo).
9. 0.45 μm PTFE filters for 1 mL syringes (Millipore).
10. Tweezers.
11. 1 mm diameter biopsy punch (Harris Uni-core).
12. Scotch tape (3M).
13. Sylgard 184 PDMS kit (Dow Corning).
14. 1H,1H,2H,2H-Perfluorodecyltrichlorosilane 97% stabilized with copper (ABCR).
15. Novec 7500 fluorinated oil (3M).

2.1.2 Collection Tubes Preparation

1. 0.2 mL or 0.5 mL tubes (Thermo Scientific).
2. 1 mL syringes (Terumo).
3. 1 mm, 6 mm, and 7 mm diameter biopsy punches (Harris Uni-core).
4. 23 G needles (Terumo).
5. I.D. 0.76 mm, O.D. 1.07 mm PTFE tubing (Thermo Scientific).
6. Tweezers.
7. An 8 mm thick slab of cured PDMS.
8. Novec 7500 fluorinated oil (3M).

2.2 DNA Amplification and Characterization

2.2.1 ddPCR Mixture

1. 0.5 μM of each primer (forward and reverse).
2. 0.2 mM of each deoxyribonucleotide (dNTP).
3. 0.1% of pluronic F68 (Sigma Aldrich).

4. A DNA polymerase and its corresponding buffer at the recommended concentration (e.g., Q5 enzyme, New England Biolabs).
5. 10 μM sulfo-cyanine 5 (Cy5, Lumiprobe).
6. 1X of EvaGreen (Genbiotech).
7. DEPC-treated dideionized water (Carl Roth).
8. DNA template diluted in 200 μg/mL total yeast RNA (Thermo Scientific).

2.2.2 ddPCR Production

1. A state-of-the-art microfluidic workstation equipped with pressure-driven pumps (Fluigent) (*see* **Note 1**) and optics as described in [19].
2. A thermocycler (e.g., T100 thermocycler, Bio-Rad).
3. 0.2 mL collection tube (Fig. 2) filled with surfactant-free Novec 7500.
4. A droplet maker device (Fig. 3).
5. 100 μL of PCR mixture (Subheading 2.2.1).
6. 2% Fluorosurfactant prepared in Novec 7500 fluorinated oil (*see* **Note 2**).

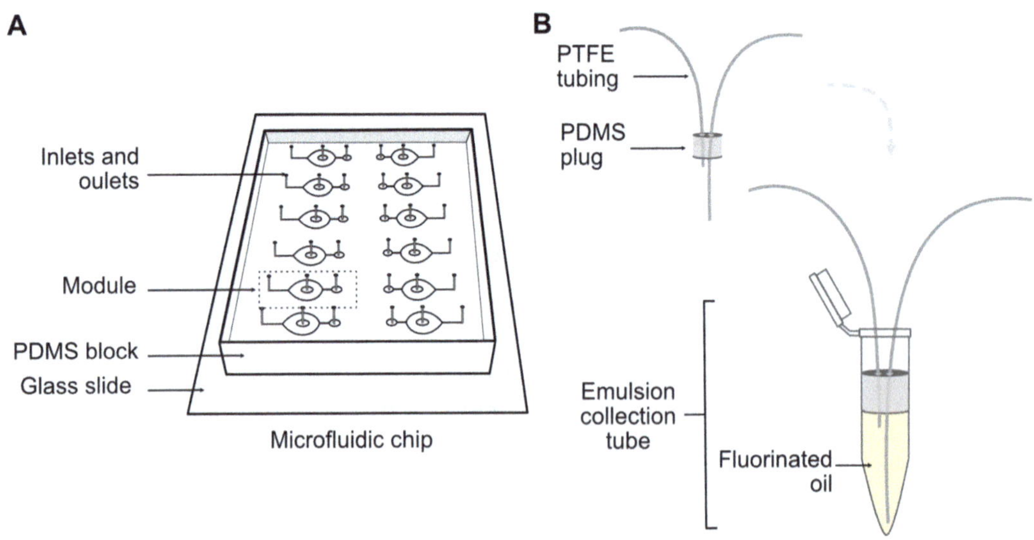

Fig. 2 Material to prepare for droplet production, manipulation, collection, and analysis. (**a**) Typical microfluidic chip consisting of a PDMS block with channels imprinted and bound to a glass slide. Upon molding, inlets and outlets are punched in the PDMS block to enable tubing connections for liquids injection/collection. (**b**) Preparation of droplet collection tubes. Two lengths of tubing are inserted through a PDMS plug, ensuring a tight seal. The assembly is then inserted into a collection tube by inserting. Finally, the tube is filled with fluorinated oil and the system is ready to collect emulsions

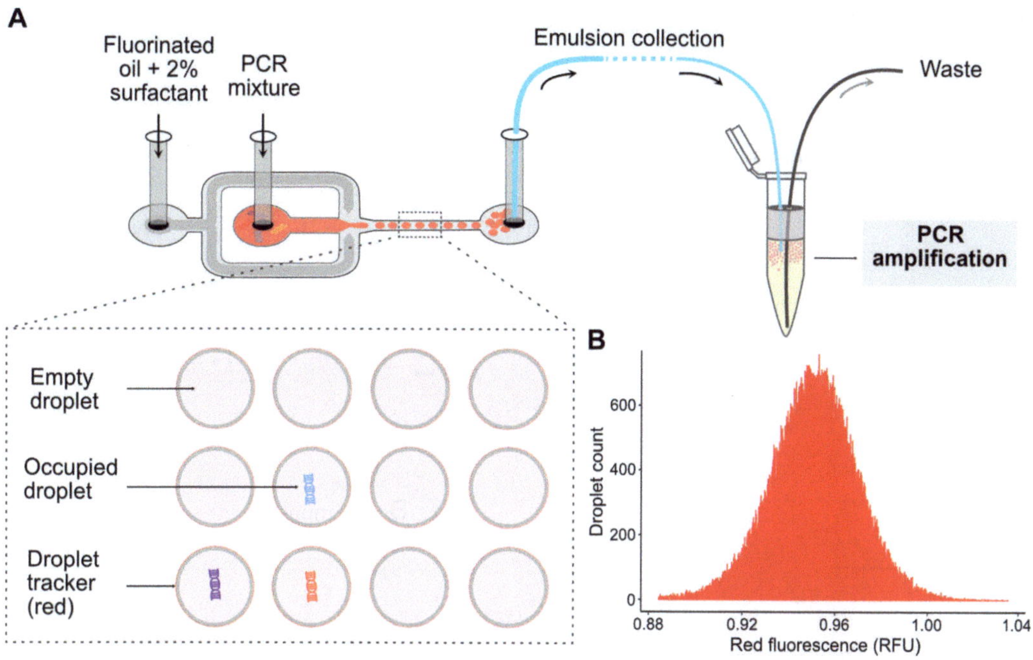

Fig. 3 Digital droplet PCR droplet production. (**a**) Experimental setup. The aqueous phase made of DNA templates diluted in a PCR mixture (red) is infused into a droplet generator microfluidic chip. A stream of HFE 7500 supplemented with 2% of surfactant (gray) is infused through a second inlet. An emulsion of 2.5 pL droplets is then generated and collected off the chip into a collection tube. (**b**) Dispersity of the produced emulsion. The monitoring of droplet fluorescence using the tracker (Cy5) contained in the PCR mixture allows to confirm emulsion monodispersity (<5% of size variation) as confirmed by the tight gaussian distribution profile

2.2.3 Reinjection and Fluorescence Tracking

1. A state-of-the-art microfluidic workstation equipped with pressure-driven pumps and optics as described in [19].
2. A droplet reinjection/analysis device (Fig. 4a).
3. I.D. 0.76 mm, O.D. 1.07 mm PTFE tubing (Thermo Scientific).
4. Novec 7500 fluorinated oil (3M)

2.3 Fusion of ddPCR and dIVT/dIVTT: RNA or Protein Synthesis

2.3.1 In Vitro Transcription Reaction Mixture

1. 2 mM of each triphosphate ribonucleoside (rNTPs, Larova).
2. 25 mM $MgCl_2$.
3. 44 mM Tris–HCl pH 8.0 (at 25 °C).
4. 5 mM DiThioThreitol (DTT) solution (Carl Roth).
5. 1 mM spermidine (Carl Roth).
6. 0.1% Pluronic F68 (Sigma Aldrich).
7. 5 μg/mL of Inorganic PyroPhosphatase (IPP) from yeast solution (Roche).

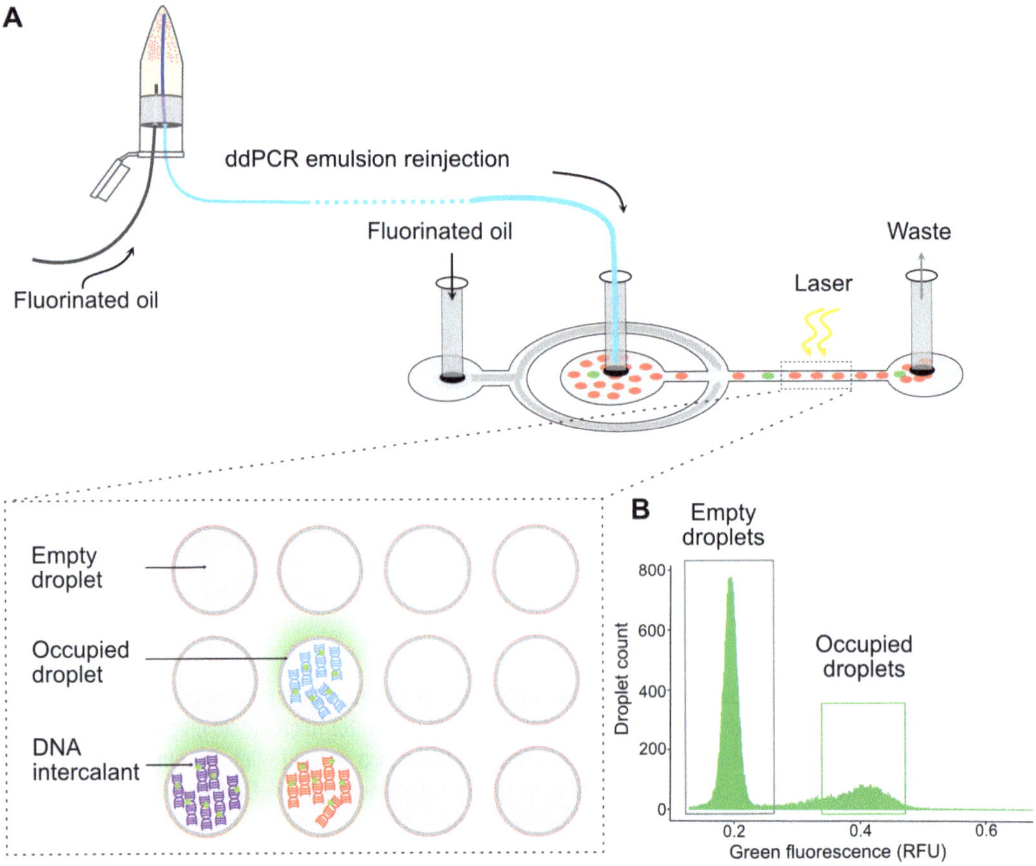

Fig. 4 ddPCR droplet fluorescence analysis. (**a**) Experimental setup. The tube containing the thermocycled emulsion is placed upside down and connected to the reinjection inlet of the analysis device. A fluorinated oil stream is infused through the short tubing to push the droplet into the long tubing that is connected to the chip. The droplets are then spaced by a stream of surfactant-free fluorinated oil infused through a second inlet and those droplets in which a template DNA was amplified (green circles) can be discriminated from those left empty (red circles) by measuring their fluorescence. (**b**) Green fluorescent profile of the droplets. Occupied droplets can easily be distinguished from empty ones by their high and low green fluorescence, respectively

8. 10 μM sulfo-cyanine 5 (Cy5, Lumiprobe).
9. T7 RNA polymerase (prepared in the laboratory) (*see* **Note 3**).
10. DEPC-treated dideionized water (Carl Roth).
11. Optional: a solution of fluorogen when the studied gene sequence codes for a light-up RNA aptamer (*see* **Note 4**).

2.3.2 Coupled In Vitro Transcription and Translation

1. PURExpress In Vitro Protein Synthesis Kit (New England Biolabs), with 40% Solution "A" and 30% Solution "B" of the total volume (*see* **Note 5**).
2. 20 units of RNase Inhibitor, Murine (New England Biolabs).
3. 0.1% Pluronic F68 (Sigma Aldrich).

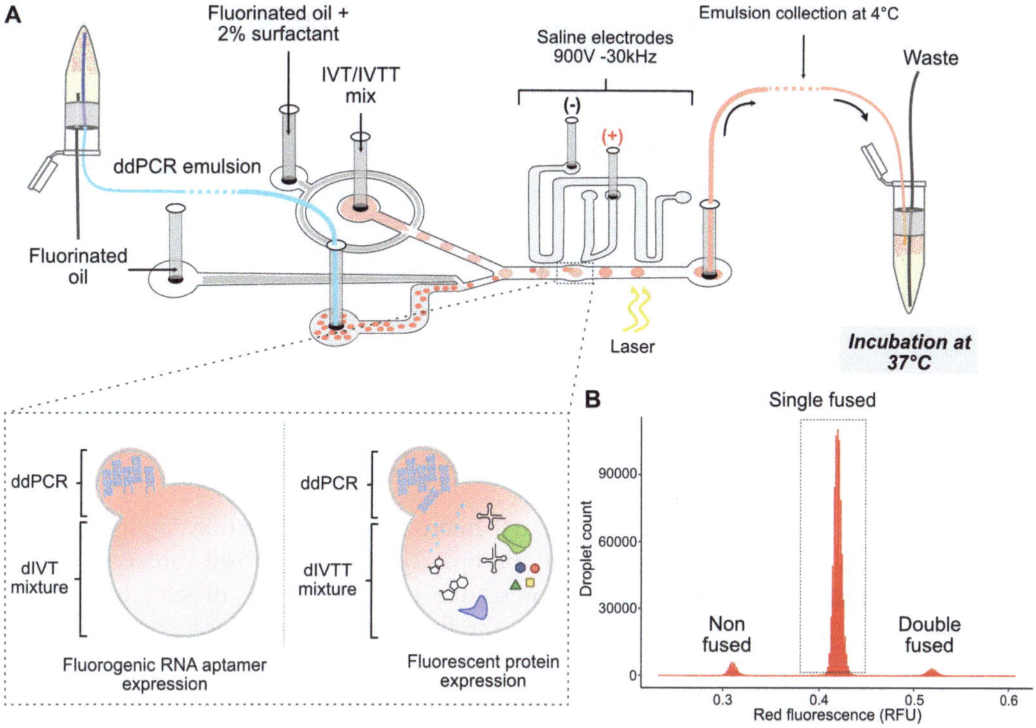

Fig. 5 Droplet fusion for gene expression. (**a**) Experimental setup. The tube containing the thermocycled ddPCR emulsion is placed upside down and connected to the reinjection inlet of the analysis device. A fluorinated oil stream is infused through the short tubing to push the droplet (dark red circles) into the long tubing that is connected to the chip. The droplets are then spaced by a stream of surfactant-free fluorinated oil (gray) infused through a second inlet. A second set of larger 17.5 pL droplets is produced on-chip by infusing an in vitro infusion expression mixture (IVT/IVTT, light-red circles) and dispersing it using a stream of fluorinated oil supplemented with 2% of surfactant infused through another inlet. 2.5 pL and 17.5 pL are paired one-to-one and fused while passing between two saline electrodes positioned near the channel. The resulting 20 pL droplets are collected and incubated at 37 °C to allow transcription/translation reaction to take place. (**b**) Monitoring of the fusion process. The use of different concentrations of cyanine 5 in ddPCR (10 μM Cy5) and IVT/IVTT (1 μM Cy5) droplets enables the process to be monitored. Indeed, low fluorescent unfused IVT/IVTT droplets (1 μM cyanine 5) can easily be distinguished from 1 ddPCR/1 IVT single fused droplets (expected to contain 2 μM Cy5) and 2 ddPCR/1 IVT single fused droplets (expected to contain 3 μM Cy5)

4. 10 μM sulfo-cyanine 5 (Cy5, Lumiprobe).
5. DEPC-treated dideionized water (Carl Roth).

2.3.3 Droplet Fusion

1. A droplet fusion device (Fig. 5a).
2. A Peltier device (e.g., cooling element CP-031, TE Technology, Inc) set at 4 °C.
3. An incubator set at 37 °C (e.g., Thermomixer, Eppendorf).
4. ddPCR thermocycled emulsion stored in a 0.2 mL collection tube (Subheading 2.1.2).
5. An ice box filled with blue ice.

6. Lengths of I.D. 0.76 mm, O.D. 1.07 mm PTFE tubing.
7. Tweezers.
8. 1 mL syringes (Terumo).
9. 23G needles (Terumo).
10. Saline solution of 5 M NaCl (Carl Roth) prepared in demineralized water.
11. Novec 7500 fluorinated oil (3M).
12. 2% Fluorosurfactant in Novec 7500.
13. A 0.5 mL collection tube filled with surfactant-free Novec 7500 fluorinated oil.
14. 200 μL of dIVT (Subheading 2.3.1, **item 1**) or of dIVTT mixture (Subheading 2.3.1, **item 2**).

2.4 Droplet Reinjection: Fluorescence Profiling or Sorting

1. A state-of-the-art microfluidic workstation equipped with pressure-driven pumps, optics, and a software enabling to detect the fluorescence emitted by the droplet tracker, the DNA interaclant, the fluorogen, or the fluorescent protein as described in [19].
2. A droplet analysis and sorting device (Fig. 6).
3. A Peltier device (e.g., cooling element CP-031, TE Technology, Inc) set at 4 °C.
4. Lengths of I.D. 0.76 mm, O.D. 1.07 mm PTFE tubing.

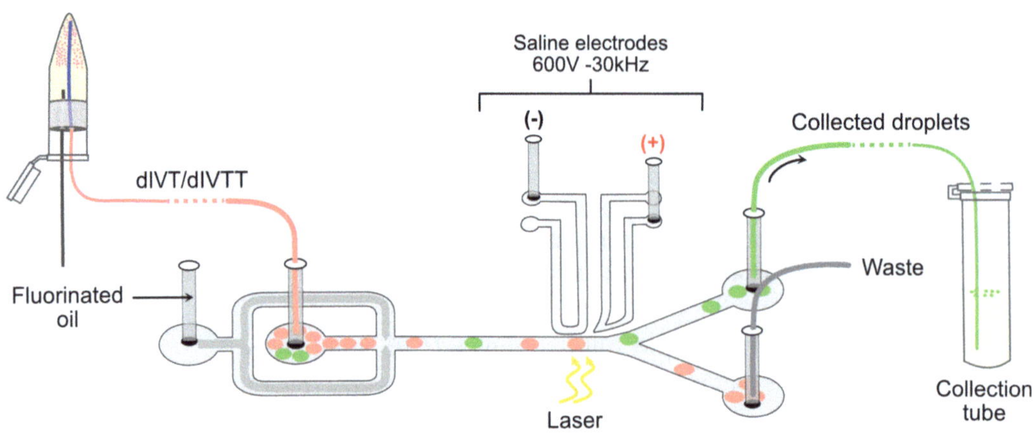

Fig. 6 Droplet fluorescence analysis and sorting. The tube containing the emulsion in which genes were in vitro expressed is placed upside down and connected to the reinjection inlet of the analysis and sorting device. A fluorinated oil stream is infused through the short tubing, inserted into the emulsion tube, to push the droplets into the long tubing that is connected to the chip (light red tubing). The droplets are then spaced by a stream of surfactant-free fluorinated oil infused through a second inlet. The fluorescence of each droplet is measured with a laser and an optical setup. Droplets are then sorted according to this fluorescence signal and droplets of interest (green circles) are collected into a dedicated tube

5. 1 mL syringes (Terumo).
6. 23G needles (Terumo).
7. Novec 7500 fluorinated oil (3M).
8. Thermocycled ddPCR droplets or incubated dIVT or dIVTT droplets stored in a collection tube (prepared in Subheading 3.3).

3 Methods

Prior to being able to perform a μIVC screening, microfluidic chips (Fig. 2a) and collection tubes (Fig. 2b) have to be prepared.

3.1 General Material Preparation

3.1.1 Microfluidic Chip

1. A mold (i.e., a silicon wafer onto which the design of the chip has been patterned) is placed into a 9 cm Petri dish and covered by 30 grams of a thoroughly mixed 10:1 PDMS/curing agent solution.
2. The mixture is degassed under vacuum until no bubble forms prior to being baked for at least 1 h 30 min at 65 °C.
3. The PDMS replicate is cut and peeled off the mold.
4. Inlets and outlets are punched into the PDMS using a 1 mm biopsy punch for both fluidic and electrode inlets.
5. Dusts and pieces of PDMS are blown off using pressurized air.
6. A glass slide and the PDMS slab are activated by an oxygen plasma for 1 min at a 0.2 power and 0.3 mbar of oxygen (*see* **Note 6**).
7. Activated glass and PDMS faces are then put into physical contact prior to being placed onto a hot plate for 15 min at 65 °C.
8. Finally, the surface of the channels is passivated using a 1% solution of 1H,1H,2H,2H-Perfluorodecyltrichlorosilane (ABCR) diluted in Novec 7500, filtered and injected into channels. The excess of liquid is finally flushed with pressurized air.
9. When required (i.e., droplet fusion and sorting), saline electrodes (*see* **Note 7**) are prepared by injecting a 5 M NaCl solution into the "electrode" channels. To do so, the solution is loaded into a syringe connected to the chip via a PTFE tubing mounted onto a needle. Electrical connection is then established using two alligator clips clamped on the needle and connected to a high voltage amplifier.

3.1.2 Collection Tubes

1. An 8 mm thick slab of cured 10:1 PDMS/curing agent is prepared.
2. 6 mm (for 0.2 mL tube used with ddPCR) or 7 mm (for 0.5 mL tube used with dIVT and dIVTT) PDMS plugs are prepared by punching the slab of PDMS with biopsy punch.
3. Two 1 mm holes are punched into each plug.
4. A length of PTFE tubing is inserted in each hole (ensure that you cut the tubing with a beveled edge). The first tubing is inserted up to 1 mm below the PDMS and the second 1.5 cm away from the PDMS (Fig. 2a). Insert the plug/tubing ensemble into the corresponding tube (0.2 mL or 0.5 mL).
5. The assembly is filled with Novec 7500 fluorinated oil using a syringe through the short tubing (shown in blue on Fig. 3a) while air leaves the system via the long tubing with the tube kept upside down.
6. Once the system is filled with oil, another syringe is filled with oil (surfactant-free Novec 7500 fluorinated oil and is connected to the other tubing.

3.2 Droplet Digital PCR

For each template to amplify/quantify, a set of primers should be designed (*see* **Note 8**) according to general guidelines used to design primers for conventional and quantitative PCR. Depending on the experiment conducted, the template DNA can either originate from a biological sample (e.g, cells) or can be obtained by solid-phase chemical synthesis (*see* **Note 9**). We recommend diluting the template in yeast total RNA in order to limit its loss by adsorption to the surface of tubes and tubings.

3.2.1 ddPCR Production

DNA template molecules are diluted in PCR reaction mixture prior to being individualized into water-in-oil droplets, using a dedicated microfluidic device (Fig. 3a).

1. DNA template (*see* **Note 10**) is diluted in a 200 μg/mL solution of yeast total RNA down to a concentration at which not all the droplets are occupied (*see* **Note 11**).
2. 1 μL of this dilution is mixed with 100 μL of PCR mixture containing all the components (Subheading 2.2.1), especially the set of specific primers, a droplet tracker (10 μM Cy5, *see* **Note 12**), pluronic F68 (*see* **Note 13**), and 1× EvaGreen.
3. The PCR mixture is loaded into a 50 cm-long piece of PTFE tubing that is then connected to the pump (in our case the flowmeter of a MFCS device) on one side and to the aqueous inlet of the droplet generator chip on the other side.
4. The PCR mixture is injected through a first inlet of the chip (Fig. 3a) together with Novec 7500 fluorinated oil supplemented with 2% of fluorosurfactant injected through the second

inlet. The flow rates are adjusted to produce 2.5 pL droplets at a frequency of ~20,000 droplets/sec (*see* **Note 14**), and to ensure the production of a monodisperse emulsion (Fig. 3b). There should be less than 5% of size variation.

5. The resulting emulsion is collected in a 0.2 mL collection tube (prepared in Subheading 3.1.2) connected to the chip outlet (Fig. 3a).
6. At the end of the production, tubings are cut, one is removed, and the other one is inserted into the second hole to close the system and avoid evaporation during the thermocycling.
7. The tube is placed into a thermocycler while paying attention that all the emulsion is entirely accommodated in the Peltier element.
8. Thermocycle the emulsion to amplify DNA using the program (*see* **Note 15**): 2 min at 98 °C (initial denaturation step) and 30 cycles of:
   (a) 10 s at 98 °C (denaturation)
   (b) 30 s at 50–72 °C (primer annealing)
   (c) 30 s to 2 min at 72 °C (elongation)
   (d) 2 min at 72 °C (final elongation step)

3.2.2 ddPCR Reinjection and Analysis

Upon thermocycling, the fluorescence of each droplet is analyzed by reinjecting the emulsion in a microfluidic device in which droplets are spaced with a fluorinated oil stream (*see* **Note 16**). The fluorescence of each droplet is then recorded and used to determine its occupied/unoccupied status.

1. The PCR tube (0.2 mL) containing the emulsion is recovered from the thermocycler and the closing tubing is removed.
2. A first new PTFE tubing (10 cm) is then inserted in one hole of the PDMS plug and pushed all the way down the bottom of the tube, while a second length (10 cm) PTFE tubing connected to a syringe filled with Novec 7500 fluorinated oil is inserted in the PDMS plug and pushed only 1 mm below the plug.
3. The PCR tube is turned upside down to allow the emulsion to cream toward the narrowest part of the tube.
4. The PCR tube is then connected in the microfluidic chips via the longest tubing, while the shortest is connected to the pump (Fig. 4a).
5. Surfactant-free Novec 7500 fluorinated oil is infused into the PCR tube and pushes the emulsion into the microfluidic device.
6. Surfactant-free Novec 7500 fluorinated oil is also infused into the chip and used to space the droplets.

7. Droplets are identified by their red fluorescence (Cy5 droplet tracker) and flow rates are adjusted to reach a reinjection rate of 500 droplets/second.
8. Both red (Cy5 droplet tracker) and green (EvaGreen intercalated into amplified DNA) fluorescence are recorded for each droplet. Whereas all the droplets are expected to display the same red fluorescence, those containing an amplified DNA are expected to display a green fluorescence much higher than the empty ones (Fig. 4b).

3.2.3 Absolute Quantification of the Target DNA (Optional)

Several studies established that DNA molecules are individualized in droplets following a Poisson law described by Eq. (1):

$$P_{(X=k)} = \frac{e^{-\lambda}}{k!}\lambda^k \tag{1}$$

where the probability of having exactly k element per compartment ($P_{(X=k)}$) is directly linked to the average number of molecules per compartment (λ). From this equation, one can derive Eq. 2 that allows to compute the absolute concentration of the target DNA in a sample (in molecule per compartment, so λ) directly from droplet occupancy.

$$\lambda = -\mathrm{Ln}\,(1 - \mathrm{Occupancy}) \tag{2}$$

3.3 Digital Gene Expression

Upon verification of the proper DNA amplification and ddPCR droplets occupancy, an aliquot of in vitro expression mixture should be added to each ddPCR droplet, while not compromising the confinement of genetic information and maintaining each phenotype with its encoding genotype. To do so, ddPCR droplets are reinjected in a droplet fusion device (Fig. 5a) equipped with saline electrodes (*see* Subheading 2.3.2) where they are pair-wised and fused one-to-one with larger 17.5 pL droplets produced on-chip and containing the in vitro expression mixture.

1. The tubing is connected to the corresponding aqueous inlet of the droplet fusion chip (Fig. 5a) and the other side of the tubing is connected to the pump.
2. 100 μL of IVT or IVTT mixture (*see* Subheading 2.3.1) are loaded into a length of PTFE tubing kept on ice during all the experiment.
3. 17.5 pL droplets are produced at a rate of 2500 droplets per second by pinching the aqueous stream by a stream of Novec 7500 fluorinated oil supplemented with 2% of fluorosurfactant infused through the corresponding inlet (Fig. 5a). Including 1 μM Cy5 as droplet tracker allows adjusting the flow rates to form droplets of expected volume.
4. ddPCR droplets are reinjected as described in (Subheading 3.2.2) through the corresponding inlet (Fig. 5a). They are

spaced by a surfactant-free Novec 7500 fluorinated oil stream prior to being synchronized one-to-one with dIVT/dIVTT droplets produced on chip.

5. Pair-wised droplets are then fused when passing in between electrodes to which an electrical field (900 V, 30 kHz) is applied. The differential concentration of Cy5 tracker in ddPCR and dIVT/dIVTT droplets (10 and 1 μM, respectively) allows to monitor the overall process and adjust the infusion rate of ddPCR droplets to maximize situations in which one ddPCR droplet is fused with one dIVT/dIVTT while limiting unfused or multiple-fused droplets (Fig. 5b).
6. The emulsion is collected into a 0.5 mL collection tube (prepared in Subheading 3.1.2) kept at 4 °C via a length of PTFE tubing.
7. At the end of the process, collection tubings are cut and disconnected from the chip and the holes are closed as described in Subheading 3.2.1.
8. Droplets are incubated for 45 min to 2 h at 37 °C for in vitro transcription, or for 2–4 h at 37 °C for coupled in vitro transcription/translation.

3.4 Droplet Fluorescence Analysis and Sorting

Upon the incubation step enabling in vitro gene expression to take place, the phenotype of each gene is evaluated by measuring the fluorescence of each dIVT/dIVTT droplet. To do so, droplets are reinjected into a droplet analysis and sorting device (Fig. 6) equipped with saline electrodes (*see* Subheading 2.3.2) and those droplets displaying the fluorescence profile of interest are deflected into a sort channel.

1. Proceed as before for the analysis of ddPCR droplets fluorescence by handling the collection tube containing dIVT/dIVTT droplets in a similar manner (Subheading 3.2.2, **steps 1–6**).
2. 2 mL collection tubes are connected to each outlet via a length of PTFE tubing.
3. As before, droplets are spaced by a stream of surfactant-free fluorinated oil and identified by their red fluorescence (Cy5 droplet tracker), and flow rates are adjusted to reach a reinjection rate of 800 droplets/second in order to limit false positive events (i.e., negative droplets passively flowing toward the sort channel).
4. The fluorescence of each droplet is then analyzed in real time using a specific software developed by the team. Beside the red fluorescence (Cy5 used as droplet tracker), up to three additional fluorescence emissions can be simultaneously recorded with our current setup. In general, the green fluorescence channel is used for the phenotypic assay. Note that the signal

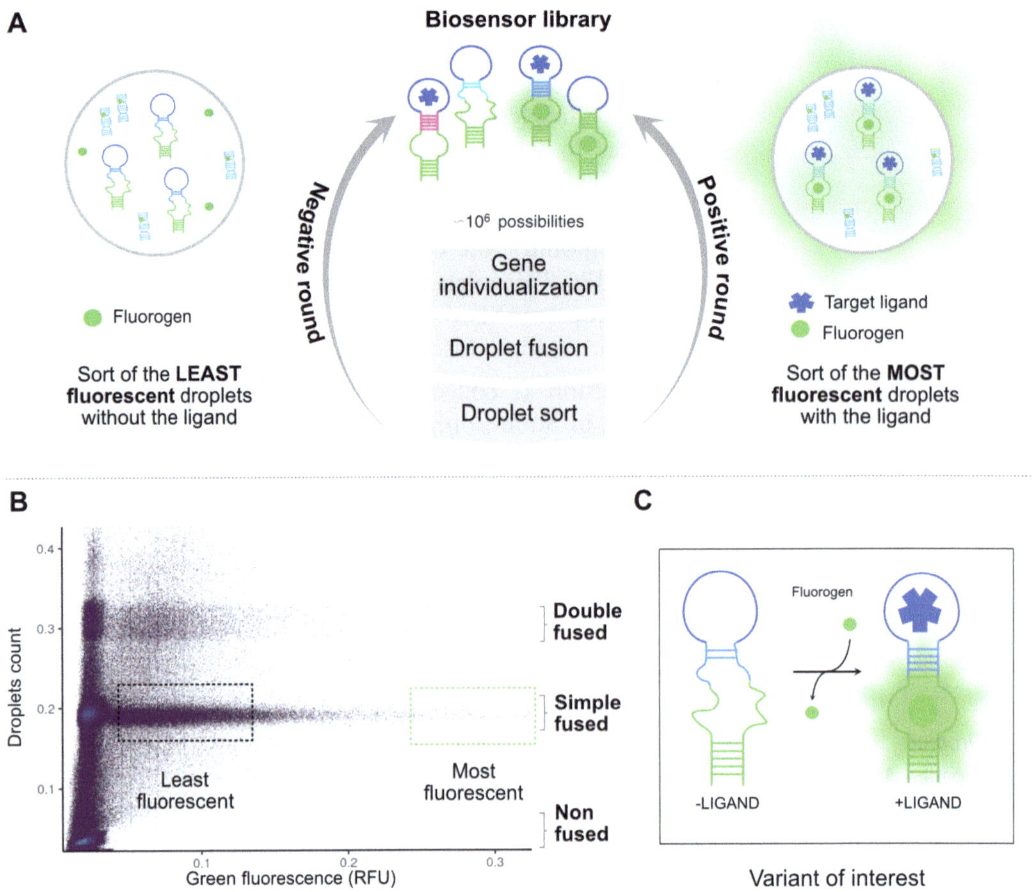

Fig. 7 Application case: selection of fluorogenic RNA-based biosensors. (**a**) Principle of fluorogenic RNA-based biosensor selection. A library consisting of variants that can fluoresce in the presence and/or absence of the target, as well as variants that remain nonfluorescent is prepared. Using the μIVC pipeline, rounds of positive (right) and negative (left) selections are performed to enrich the library in variants emitting fluorescence only in the presence of the ligand. (**b**) Fluorescent profile of the sorted droplets. Depending on the type of selection (positive or negative), the least (negative) or the most (positive) fluorescent droplets are gated and sorted. These populations correspond to those already identified on Fig. 5b. (**c**) Typical variant of interest with a sequence adapted for specific target recognition and fluorescence emission conditional on this recognition after molecular restructuring

contributed by the EvaGreen brought by ddPCR droplets is marginal after the 1/10 dilution that occurred after droplet fusion (Fig. 7).

5. Droplets displaying the fluorescence of interest are deflected into the sort channel by applying an electrical field of 600 V and 30 kHz, while ungated droplets passively flow into the "waste" channel (Fig. 6).

6. Sorted droplets are flushed from the collection tubing into the collection tube using 100 μL of surfactant-free Novec 7500 fluorinated oil.
7. The recovered droplets are broken using 100 μL of 1H, 1H, 2H, 2H-perfluoro-1-octanol followed by the same volume of 200 μg/mL yeast total RNA solution sequentially flushed into the collection tubing.
8. The tube is closed and vortexed for 30 s prior to being centrifuged for 30 s at 10,000 g.
9. The DNA-containing aqueous phase is then transferred into a new 1,5 mL tube and ready for further analyses and characterizations.

4 Notes

1. In our lab, we use a 7-bar MFCS pressure-driven flow controller equipped with S Flowmeters and driven by MAESFLO software (Fluigent). However, any other microfluidic flow controller can be used: either pressure-driven flow control (e.g., from ElveFlow, Dolomite or Parker) or a syringe pump (e.g., from Harvard apparatus or Cetoni).
2. The fluorinated surfactant we use is synthesized in the laboratory but several fluorosurfactants are commercially available: Pico-Surf (Sphere Fluidics), 008-F (Ran biotechnologies), dSurf (Fluigent), or FluoSurf-S (Emulseo). We recommend these surfactants to be prepared in Novec 7500 fluorinated oil.
3. We prepare our own T7 RNA polymerase in the lab, but it can also be obtained from several commercial suppliers (e.g., New England Biolabs, Promega, ThermoFisher).
4. The fluorogens (e.g., DFHBI, DFHO, DFHBI-1 T, TO1-Biotin, TO3-Biotin) are chosen depending on the light-up aptamer used for the experiment. Many fluorogens are commercially available (Lucerna, ABM, Merck, Jena Bioscience).
5. We use a commercial reconstituted IVTT mixture following the manufacturer's instructions, but home-made or commercial (e.g., from New England Biolabs, Promega, ThermoFisher or Arbor Biosciences) bacterial extracts can be used as well.
6. The settings described here are adapted to our material, but adaptations may be required when using other devices.
7. A saline solution is used to conduct the electrical signal into the microfluidic chip, but ionic liquids can be used as well.
8. Should the amplified DNA be later expressed, care should be taken that transcription promoter and translation initiation

sites are included either in one of the primer or be present in the amplified sequence.

9. The DNA templates are obtained from Integrated DNA technology (IDT) but any other company can be used as long as the purity of the molecule meets all the requirements to avoid contamination during the process.
10. Prior to the experiments, the DNA is amplified in bulk and purified using a PCR product cleaning kit such as SPRISelect (Beckman Coulter) or Monarch PCR & DNA Cleanup Kit (New England Biolabs) to allow exact quantification of the DNA by using a Nanodrop device (ThermoFisher Scientific).
11. To optimize the encapsulation process, it is advised to target a droplet occupancy of around 20% to maximize the number of usable droplets while minimizing the risk of multiple encapsulation events. At 20% occupancy, the latter should be ~2% according to Poisson statistics.
12. Any other droplet tracker can be used, as long as the wavelengths do not interfere with the intercalating dye or the probe.
13. Pluronic F68 is used to avoid nonspecific adsorption of hydrophobic compounds to PTFE tubings.
14. Droplet production frequency is monitored in real time using the droplet tracker (here, Cyanine 5) included in droplet as well as an optical setup and a software developed by the team [19]. Using this frequency and knowing the infusing rate of the aqueous solution allows droplet volume to be simply computed.
15. (a) All PCR mixes are prepared according to the protocol recommended by the brand of the enzyme being used and the choice of the enzyme is determined by the DNA template to be amplified and can be adapted. (b) The annealing temperature must be adjusted according to the primer, size, and nucleotide composition. (c) Elongation time should be adapted to the length of the DNA template and the enzyme used and could be slightly higher in the emulsion setup compared to a classic PCR mix. (d) Care should be taken to ensure that the PDMS plug is pushed so that the whole emulsion is inside the thermal cycling block.
16. If the background noise is too high, making it difficult to distinguish between the two populations, the analysis chip can be mounted onto a thermoplate set at 40 °C. Warming the droplets this way destabilizes weak primer–primer nonspecific interactions and intramolecular folding that otherwise lead to dye intercalation and background fluorescence.

Acknowledgments

This work received financial support from Agence Nationale de la Recherche (Microfluor project ANR-21-CE04-0010 and Amplisense project ANR-21-CE42-0019). Performed within Interdisciplinary Thematic Institute "IMCBio, as part of the ITI 2021-2028 program of the University of Strasbourg, CNRS and Inserm, this work was supported by IdEx Unistra (ANR-10-IDEX-0002), and by SFRI-STRAT'US project and EUR IMCBio (ANR-17-EURE-0023) under the framework of the French Investments for the Future Program. This work was also supported by the "Centre National de la Recherche Scientifique" (CNRS) and the "Université de Strasbourg" whom it received support from its Initiative of Excellence (IdEx).

References

1. Campos-Carrillo A, Weitzel JN, Sahoo P et al (2020) Circulating tumor DNA as an early cancer detection tool. Pharmacol Ther 207: 107458
2. Sciuto EL, Leonardi AA, Calabrese G et al (2021) Nucleic acids analytical methods for viral infection diagnosis: state-of-the-art and future perspectives. Biomolecules 11:1585
3. Petralia S, Conoci S (2017) PCR technologies for point of care testing: progress and perspectives. ACS Sens 2:876–891
4. Gill P, Ghaemi A (2008) Nucleic acid isothermal amplification technologies—a review. Nucleosides Nucleotides Nucleic Acids 27: 224–243
5. Higuchi R, Fockler C, Dollinger G et al (1993) Kinetic PCR analysis: real-time monitoring of DNA amplification reactions. Nat Biotechnol 11:1026–1030
6. Ginzinger DG (2002) Gene quantification using real-time quantitative PCR: an emerging technology hits the mainstream. Exp Hematol 30:503
7. Ji H, Chang L, Yan Y et al (2023) Development and validation of a duplex real-time PCR for the rapid detection and quantitation of HTLV-1. Virol J 20:9
8. Lima A, França A, Muzny CA et al (2022) DNA extraction leads to bias in bacterial quantification by qPCR. Appl Microbiol Biotechnol 106:7993–8006
9. Vogelstein B, Kinzler KW (1999) Digital PCR. Proc Natl Acad Sci 96:9236–9241
10. Quan P-L, Sauzade M, Brouzes E (2018) dPCR: a technology review. Sensors 18:1271
11. Ottesen EA, Hong JW, Quake SR et al (2006) Microfluidic digital PCR enables multigene analysis of individual environmental bacteria. Science 314:1464–1467
12. Pekin D, Skhiri Y, Baret J-C et al (2011) Quantitative and sensitive detection of rare mutations using droplet-based microfluidics. Lab Chip 11:2156
13. Ryckelynck M, Baudrey S, Rick C et al (2015) Using droplet-based microfluidics to improve the catalytic properties of RNA under multiple-turnover conditions. RNA 21:458–469
14. Mazutis L, Araghi AF, Miller OJ et al (2009) Droplet-based microfluidic systems for high-throughput single DNA molecule isothermal amplification and analysis. Anal Chem 81: 4813–4821
15. Baret J-C, Miller OJ, Taly V et al (2009) Fluorescence-activated droplet sorting (FADS): efficient microfluidic cell sorting based on enzymatic activity. Lab Chip 9:1850
16. Fallah-Araghi A, Baret J-C, Ryckelynck M et al (2012) A completely in vitro ultrahigh-throughput droplet-based microfluidic screening system for protein engineering and directed evolution. Lab Chip 12:882
17. Romero PA, Tran TM, Abate AR (2015) Dissecting enzyme function with microfluidic-based deep mutational scanning. Proc Natl Acad Sci 112:7159–7164
18. Autour A, Westhof E, Ryckelynck M (2016) iSpinach: a fluorogenic RNA aptamer optimized for *in vitro* applications. Nucleic Acids Res 44:2491–2500
19. Bouhedda F, Cubi R, Baudrey S et al (2021) μIVC-Seq: a method for ultrahigh-throughput

development and functional characterization of small RNAs. In: Rederstorff M (ed) Small non-coding RNAs. Springer, New York, pp 203–237

20. Gantz M, Neun S, Medcalf EJ et al (2023) Ultrahigh-throughput enzyme engineering and discovery in *in vitro* compartments. Chem Rev 123:5571–5611
21. Liu Y, Sun L, Zhang H et al (2021) Microfluidics for drug development: from synthesis to evaluation. Chem Rev 121:7468–7529
22. Trinh TND, Do HDK, Nam NN et al (2023) Droplet-based microfluidics: applications in pharmaceuticals. Pharmaceuticals 16:937
23. Bouhedda F, Autour A, Ryckelynck M (2017) Light-up RNA aptamers and their cognate fluorogens: from their development to their applications. Int J Mol Sci 19:44
24. Autour A, Bouhedda F, Cubi R et al (2019) Optimization of fluorogenic RNA-based biosensors using droplet-based microfluidic ultrahigh-throughput screening. Methods 161:46–53
25. Husser C, Baudrey S, Ryckelynck M (2023) High-throughput development and optimization of RNA-based fluorogenic biosensors of small molecules using droplet-based microfluidics. Methods Mol Biol 2570:243–269
26. Husser C, Vuilleumier S, Ryckelynck M (2023) FluorMango, an RNA-based fluorogenic biosensor for the direct and specific detection of fluoride. Small 19:2205232
27. Pernod K, Schaeffer L, Chicher J et al (2020) The nature of the purine at position 34 in tRNAs of 4-codon boxes is correlated with nucleotides at positions 32 and 38 to maintain decoding fidelity. Nucleic Acids Res 48:6170–6183
28. Trachman RJ, Truong L, Ferré-D'Amaré AR (2017) Structural principles of fluorescent RNA Aptamers. Trends Pharmacol Sci 38:928–939
29. Autour A, Jeng SCY, Cawte AD, Abdolahzadeh A, Galli A, Panchapakesan SS, Rueda D, Ryckelynck M, Unrau PJ et al (2018) Fluorogenic RNA mango aptamers for imaging small non-coding RNAs in mammalian cells. Nat Commun 9:656
30. Bouhedda F, Fam KT, Collot M et al (2020) A dimerization-based fluorogenic dye-aptamer module for RNA imaging in live cells. Nat Chem Biol 16:69–76
31. Baudrey S, Cubi R, Ryckelynck M (2022) Droplet-based microfluidic chip design, fabrication, and use for ultrahigh-throughput DNA analysis and quantification. In: Caballero D, Kundu SC, Reis RL (eds) Microfluidics and biosensors in cancer research. Springer International Publishing, Cham, pp 445–460

Chapter 16

20-Plex Gene Expression Analysis Using Hyperplex PCR (hpPCR) Toward the Development of Salt-Tolerant Wheat

Esteri Viitanen, Samuel Narmack, Margarita Psallida, Johan Banér, Umear Naseem, Ruben R. G. Soares, and Henrik Aronsson

Abstract

Climate change exacerbates the challenge of soil salinity, directly threatening the agricultural yields of staple crops such as wheat. Salinity currently affects at least 20% of total arable land and this figure is expected to worsen depending on future climate change trends. Aiming at improving the salt tolerance of wheat and ensuring future food security, researchers use, for example, multi-omics tools and genome-wide association studies (GWAS) to understand plant stress tolerance, with transcription factors playing a crucial role in regulating stress responses. Here, we describe the use of a next-generation PCR method, Hyperplex PCR (hpPCR), to analyze a 20-plex transcription factor panel focusing on three underexplored wheat transcription factor families—bHLH, C3H, and RAV—to identify their involvement in salt stress. All measured transcription factors, except one of the *RAV* (related to *ABI3* and *VP1*) targets in shoot tissue, were observed to be downregulated in the presence of 100 mM NaCl. Compared to 2-step RT-qPCR, hpPCR measured the same upregulation and downregulation trends for all targets with up to 18 fewer PCR cycles.

Key words Hyperplex PCR, hpPCR, 20-plex, Transcription factor, Gene expression, Salt tolerance, Crop, qPCR, Quantification

1 Introduction

Ongoing climate change increases the risk of severe negative effects on agricultural yields due to factors such as drought, salinity, and heat, driven by fluctuations in temperature, humidity, and rainfall. Salinity, a major abiotic stressor, significantly reduces agricultural production [11]. Soil salinity arises from natural primary salinization, including ocean spray deposition, and rock weathering, but is also largely driven by human-induced secondary salinization, such as excessive irrigation without proper drainage and the use of saline water for irrigation. As a result, climate change and human activity have led to salt stress affecting 20% of the

Guillaume Pavlovic (ed.), *Digital PCR: Methods and Protocols*, Methods in Molecular Biology, vol. 2969,
https://doi.org/10.1007/978-1-0716-4767-7_16,

world's arable land, with numbers continuing to rise [3]. Developing stress-tolerant crops is thus increasingly crucial for modern agriculture, particularly in regions vulnerable to salinity.

Triticum aestivum (common wheat) is one of the most important staple crops worldwide, playing a crucial role in global human consumption [6]. However, its cultivation is increasingly threatened by rising soil and water salinity [10]. The ability of plants, including wheat, to tolerate environmental stress is a complex quantitative trait [1, 7]. Researchers use, for example, genome-wide association studies (GWAS) and multi-omics to determine the genetic basis of such traits ([2, 19, 5]). Numerous studies have identified salt stress-related genes and transcription factors that can enhance plant tolerance [4, 17, 18, 20].

Transcription factors play a key role in regulating plant growth and stress responses [4, 17, 18]. They serve as key regulatory elements that function as molecular switches, activating or suppressing gene expression in response to environmental stressors. By inducing stress-responsive genes, transcription factors alter cellular metabolism, enabling adaptation to saline conditions. Several transcription factor families, including WRKY, NAC, AP2/EREBP, MYB, and bZIP, have been studied for their role in plant salt stress responses [8, 14–17]. Understanding their molecular mechanisms can support targeted genetic modifications to enhance stress tolerance in wheat and other crops. In our study, we focus on three less-characterized transcription factor families in wheat that are related to salt stress: bHLH, C3H, and RAV. By analyzing these transcription factors, we aim to improve our understanding in their potential regulatory roles during salt stress, and subsequently improve our understanding of their downstream target genes. Identifying these regulatory genes could facilitate their use as molecular markers in breeding programs, contributing to the development of salt-tolerant wheat varieties and supporting sustainable agriculture in saline environments.

To enable a more comprehensive analysis of these markers, we approached the use of the novel Hyperplex PCR (hpPCR) method as it allows to simultaneously measure, as a 20-plex panel, 17 specific transcription factors and 3 internal control genes for each sample. In brief, hpPCR combines (1) RT-PCR amplification, (2) padlock probes (PLP) ligation, (3) rolling circle amplification (RCA), and (4) labeling with optically encoded probes—Nanopixels—enabling precise identification and counting of up to 100 different target sequences per sample. Compared to standard RT-qPCR, this alternative method enables significantly improved efficiency in quantification of multiple markers routinely without sample splitting, facilitating high frequency analysis of a larger number of samples and/or post-stress time-points. Furthermore, the method described in this chapter aims also at cross-validating hpPCR against standard SYBR Green-based RT-qPCR concerning measured upregulation and downregulation, as well as detection sensitivity.

2 Materials

This section describes the tested samples, RNA extraction procedures, the reagents, equipment, and consumables required to carry out the hpPCR assay (1-step reverse transcription), as well as the cDNA conversion and subsequent qPCR used for cross-validation.

2.1 Wheat Samples and Extraction Procedure

2.1.1 Reagents

1. Nelson Garden Hydroponic Nutrition™ (Nelson Garden, Tingsryd, Sweden), an alternative is to use Murashige-Skoog (MS) media (Duchefa, Netherlands).
2. HCl to adjust the pH to 6.0.
3. 100 mM NaCl.
4. Root and shoot tissue material from BARI Gom-25 (*See* **Note 1**).
5. Liquid nitrogen.
6. NucleoSpin RNA Plant kit (Macherey-Nagel, Düren, Germany).
7. Dithiothreitol (DTT) as reducing agent.

2.1.2 Equipment and Accessories

1. Filter paper (Munktell-filter paper—A1-100-80™).
2. Petri dishes (9 cm diameter).
3. Greenhouse or phytotron exhibiting optimal growing temperature for wheat at 20 °C, light/dark cycle of 14:10 h, light intensity of 400 μmol/m^2/s, and relative humidity of 60%.
4. Eppendorf tubes: 1.5 mL.
5. (Black) buckets (1 liter) with perforated lids to fit the diameter of 1.5 mL Eppendorf tubes.
6. A pair of scissors.
7. Liquid nitrogen container.
8. 15 mL or 50 mL Falcon tubes depending on volume of tissue material collected.
9. Optional; tissuelyser such as Mixer Mill MM 301 (Retsch GmBH, Haan, Germany).
10. Glass beads 4 mm diameter.
11. Vortex.
12. Flake ice machine or cooling block (+4 °C only).
13. Pipettes (1–200 μL) are preferred for some steps.

14. NanoDrop™ One (Thermo Fisher Scientific Inc., Waltham, USA).

2.2 20-Plex hpPCR Custom Gene Expression Assay

2.2.1 Reagents

1. Custom hpPCR assay kit.
   (a) RT-PCR MasterMix.
   (b) 20-plex Primer Panel.
   (c) Dilution Factor.
   (d) Ligation Buffer.
   (e) 20-plex PLP Panel.
   (f) hpPCR Ligase.
   (g) RCA Buffer.
   (h) hpPCR Polymerase.
   (i) Nanopixel™ Labeling Buffer.
   (j) Mounting medium.
   (k) Control Sample.
   (l) hpPCR Capture Slides 5-pack.
   (m) Slide Frame.
   (n) Cover slips.
   (o) Reconstitution Buffer.
   (p) Binding and Wash buffer.

2.2.2 Equipment and Accessories

1. Bio-rad T100 thermal cycler or similar system.
2. VWR INCU-Line IL 10 Incubator oven or similar system.
3. 100–200 μL PCR tubes or PCR plates (with suitable lids. Referred to as "PCR tubes" below).
4. Eppendorf tubes: 1.5 mL up to 5 mL, depending on number of samples assayed.
5. Vortex and table centrifuge with adapter for PCR tubes or plates.
6. Flake ice machine or cooling block (+4 °C only).
7. Dark cover to protect slides from light and dust (*see* **Note 2** below).
8. Optional: dispensers (20–200 μL) or multichannel pipettes (1–200 μL) are preferred for some steps.
9. Olympus APX100 Imaging System equipped with the following accessories:
   (a) UPLXAPO 20X NA. 0.80/0.17 mm objective.
   (b) Hamamatsu ORCA-Fusion Digital sCMOS camera (C14440-20UP).

(c) X-Cite XYLIS II Broad Spectrum LED Illumination System.
(d) Chroma Technology filter cube 1: ET395/25×; T412lpxt; ET431/28 m.
(e) Chroma Technology filter cube 2: ZET436/20×; ZT442rdc; ET480/40 m.
(f) Chroma Technology filter cube 3: ET470/40×; T495lpxr; ET525/50 m.
(g) Chroma Technology filter cube 4: ET545/25×; T565lpxr; ET605/52 m.
(h) Chroma Technology filter cube 5: ET640/30×; T660lpxr; ET690/50 m.
(i) Chroma Technology filter cube 6: ET735/75×; T760lpxr; CT785/25 m.

2.3 Reverse Transcription and qPCR

2.3.1 Reagents

1. RNA extracted from plant tissue of interest.
2. iScript cDNA Synthesis Kit (Bio-Rad, Hercules, CA, USA) or qScript Ultra SuperMix (Quantabio).
3. SsoAdvanced Universal SYBR Green Supermix (Bio-Rad).
4. Primers of specific gene(s) of interest at a working concentration of 5 μM. Primers need to be designed to avoid primer dimer formation. Sequences according to Table 2.

2.3.2 Equipment and Accessories

1. MyCycler™ (Bio-Rad Laboratories AB, California, USA).
2. 50–100 μL PCR tubes or PCR plates (with suitable lids. Referred to as "PCR tubes" below).
3. Vortex and table centrifuge with adapter for PCR tubes or plates.
4. Flake ice machine or cooling block (+4 °C only).
5. Multichannel or single pipettes (1–200 μL) are preferred for some steps.
6. Bio-Rad CFX96 Real-Time PCR Detection System.

3 Methods

3.1 Plant Growth and RNA Extraction

3.1.1 Step

1. Soak 40 seeds of BARI Gom-25 in water and stratify them in a fridge at 6–8 °C overnight to synchronize germination (*See* **Note 3**).
2. Cover the surface of a petri dish with a filter paper (Munktell-filter paper—A1-100-80TM).
3. Soak the filter paper with 4 mL of tap water.

4. Add five seeds to each wet filter paper and put on the lid of the petri dish.
5. Leave the seeds to germinate for 4 days.
6. Prepare 1-liter (black) buckets to be used for hydroponic growth. Make holes into the lids to fit the width of 1.5 mL Eppendorf tubes. Ensure a distance of at least 3 cm between each hole.
7. Use a pair of scissors to cut open the bottom of 1.5 mL Eppendorf tubes, ~5 mm from the bottom. The opening needs to be large enough to get the roots through but small enough to keep the seed within the tube.
8. Place the cut Eppendorf tubes into the holes of the bucket lid.
9. Fill up the bucket with water mixed with Nelson hydroponic nutrition (2 mL/L).
10. Adjust the pH to 6.0 with 1 M HCl.
11. Place the hydroponic bucket system into a greenhouse or phytotron. Ensure the system reaches the same temperature as the surrounding room temperature before adding the wheat seedlings. Set the growth conditions to the following parameters: temperature at 20 °C, light/dark cycle of 14:10 h, light intensity of 400 μmol/m^2/s, and relative humidity of 60%.
12. Place evenly germinated seeds into the cut Eppendorf tubes mounted onto the lid. Ensure the roots are in contact with the hydroponic solution and that the seed does not fall into the solution.
13. After 7 days of growth, transfer half of the seedlings to new buckets filled with the same solution as previously grown (in control treatment). Transfer the remaining half to buckets mixed with Nelson Garden Hydroponic Nutrition™ (2 mL/L) and 100 mM NaCl (salt treatment).
14. After 7 days of hydroponic growth in each condition, collect shoot tissue from four plants into a 15 mL Falcon tube (one biological shoot replicate), and the corresponding root tissue from the same four plants into another 15 mL Falcon tube (one biological root replicate). Repeat until at least four biological replicates are collected from both treatments.
15. Snap freeze the collected samples by placing the Falcon tubes into a container filled with liquid nitrogen.
16. Store at −80 °C until further use or use directly.
17. Add 4–8 glass beads to each Falcon tube and vortex with strong vibration to pulverize the samples. Alternatively, use a tissuelyser. Ensure the samples do not thaw during the procedure.

18. Extract RNA by using the NucleoSpin RNA Plant kit (Macherey-Nagel, Düren, Germany) (or similar) to the pulverized tissue in accordance with the manufacturer's protocol.
19. Add dithiothreitol (DTT) as reducing agent.
20. Keep the tubes on ice or cooling block (+ 4 °C only) when indicated during the different extraction steps.
21. Measure the RNA concentration with a NanoDrop.
22. Store the RNA at −80 °C until further use.

3.2 20-Plex hpPCR Assay Protocol

The 20-plex hpPCR assay comprises four main steps described in detail below, namely (1) RT-PCR amplification of RNA; (2) Ligation of padlock probes; (3) Rolling circle amplification and Nanopixel labeling and; (4) Fluorescence microscopy imaging. The primer sequences and padlock probe (PLP) target sequences are listed in Tables 1 and 2.

3.2.1 RT-PCR Amplification of RNA Samples

1. Arrange and label required PCR tubes on ice. Also on ice, prepare the following RT-PCR mix: 2.8 μL primer panel and 2.8 μL RT-PCR Master Mix per reaction.
2. Add 5 μL RT-PCR mix to each PCR tube.
3. Transfer 5 μL sample to corresponding PCR tube. Mix by gentle vortex and spin down.
4. Place all tubes in a thermocycler. Make sure lid is set to 105 °C and incubate at: 25 °C for 2 min, 50 °C for 15 min, 95 °C for 2 min; 18 cycles of [95 °C for 3 s, 60 °C for 1 min], followed by 4 °C until next step.
5. Move all PCR tubes to wet ice, then add 10 μL of reconstituted Dilution Factor to each PCR tube.
6. Place all tubes in a thermocycler. Incubate at: 50 °C for 10 min, 95 °C for 10 min, followed by 4 °C until next step. The assay can be paused at this step. Keep samples at 4 °C or −20 °C.

3.2.2 Ligation of Padlock Probes

1. Label new PCR tubes and prepare the following ligation mix on ice: 13.2 μL PLP panel, 4.4 μL ligation buffer, and 2.2 μL ligase per reaction.
2. Aliquot 18 μL ligation mix to the new PCR tubes.
3. Retrieve samples from **step 3.2.1.6** and transfer 2 μL of each diluted RT-PCR sample to the corresponding new tubes. Mix by gentle vortex and spin down. Keep tibes on ice during this step. The remaining RT-PCR dilution can be stored at +4 °C until further use.
4. Label a new PCR tube on ice and prepare the following control ligation mix: 14 μL of control sample, 4 μL ligation buffer, and 2 μL ligase.

Table 1
List of sequences targeted by each of the 20 padlock probes (PLPs)

Target ID	Target RNA	PLP target sequence (5′-3′)[a]	Nanopixel code
1	*C3H1*	GGAGCAGAAGACCAAGCA AGACGCGGCGGGGGAGCT	1 (C1-C2) + C3
2	*C3H3*	CGGTGAACGTCGGCTCGC AGCGCGTGTTCGTGGGCG	5 (C1-C2) + C3
3	*C3H5*	CGCCACGCCGGCCGCATA CAACCGCAAGCCGAGCTC	6 (C1-C2) + C3
4	*C3H7*	CCGGTGAACGTCGGCTCG CGCGTGTTCGTGGGCGTG	7 (C1-C2) + C3
5	*ACTIN*	GGTACATGGTTAGTGTTC TGTAGAAGATGTGTACCC	3 (C1-C2) + C3
6	*RAV1*	GAGTCACGCGCCCGAGCA CGAGGACTGCAGCATGGT	4 (C1-C2) + C3
7	*RAV2*	CACCTCACAGTGCCCGAG GCCAAGAGATCAACTAAA	2 (C1-C2) + C3
8	*RAV3*	AGCAACCAGCTCTTCATC GACTGCAAGCTCCGGTCC	1 (C1-C2) + C4
9	*RAV4*	GTGCTCGGCGTACGGCCA GGAGAAGCAGTTCTTCAT	5 (C1-C2) + C4
10	*RAV5*	AAGACCACGACGGTGAAC AGCGGCGCGTCGGCGTCA	6 (C1-C2) + C4
11	*RAV6*	CGTGCTCGCTGTACGAAC AGGAGAAGCAGTTCTTCA	2 (C1-C2) + C4
12	*RAV7*	TCTCGTGCTCGGCGTACG GTCAAGAGAAGCAGTTCT	3 (C1-C2) + C4
13	*RAV8*	ACGTCAAGAGATCAACTA GAAGGAATACAAGCAGCA	4 (C1-C2) + C4
14	*bHLH0100*	ATGAAGGAGAAGCTATAC GAGCTCAGGGCTCTCGTC	7 (C1-C2) + C4
15	*bHLH6900*	GCCACAACCGCCGCAGAT CCATCACTTCGAGAACGG	1 (C1-C2) + C5
16	*bHLH9500*	CTCGCGGGTGATCAAGTA CATCCCGGAGCTGCAGAA	5 (C1-C2) + C5
17	*bHLH5400*	GATCACACCGAAAGATGA GTCACAACGCGTACGAGC	6 (C1-C2) + C5
18	*GAPDH*	AAGTCTGACATCGACATT GTCTCCAACGCTAGCTGC	2 (C1-C2) + C5
19	*AP5.1*	CATCCTAGAACAACGCAG CGAGTAAATTCATCATTT	3 (C1-C2) + C5
20	*bHLH8000*	TTCGCGTCCAACGCATCC GTCACGGTGACCCCGCCT	7 (C1-C2) + C5

[a]The space indicates the 5′-3′ arm ligation site of each probe. *RAV* (related to *ABI3* and *VP1*) gene numbers relate to internal identifiers

Table 2

List of primer sequences for each target RNA used for qPCR (after reverse transcription to cDNA) or hpPCR (direct RNA amplification as 1-step RT) and respective final concentrations in the PCR master mix for hpPCR. *RAV* (related to *ABI3* and *VP1*) gene numbers relate to internal identifiers

Target name	Forward primer (5′-3′)	Reverse primer (5′-3′)	Final concentration (nM) for hpPCR (F/R)
C3H1	AAAGGCGAGGAGCAGAAGAC	TCTCCTCCCACTTG TTGCAC	600/800
C3H3	GGCTACCTCAAGATGGACAGG	TACACCTCGAAC TGCAGCC	600/800
C3H5	ACCTCAAGATGAACCAGGCG	ACATGCCCTGGTTG TACACG	600/800
C3H7	ACCTGAAGATGGACAGGAACG	GCTGTACACC TCGAACTGC	600/800
ACTIN	TCTATTTTGGCCTCTCTTAGCAC	TTTCCTGTACCCC TTATTCCTC	150/200
RAV1	ACCGTGCGACTCTTTGG	TCTTGTATGTC TTGGGCACC	150/200
RAV2	ACCTACAGACATGCCACCAC	ACACCCTGCTGC TTGTACTC	150/200
RAV3	GGTTCTACCGCTCTGCCTCA	TCATCGTCGTGGTC TTGGAC	150/200
RAV4	ACTCGTATTGGAACAGCAGC	GGTCGTGTTCTTC TTGCAGT	150/200
RAV5	TCGTGTTCTCCTGCTCAGAGT	GCTTGTTCTCC TTTGCCCGT	150/200
RAV6	TCCTATTCTCGTGCTCGCTG	GCCTCCGTTCATGC TAGTGT	150/200
RAV7	GCGATTCCATCGTGTTCTCG	CGTCTTGTTCTTC TTGCAGTCG	75/100
RAV8	TATAGTCATGCCACCACCTCAC	CGAATGGTGACAAG TCTGC	150/200
bHLH0100	TGTCGGAGCGCAAGAGGA	GGCGATGA TGGAAGCCTTGT	150/200
bHLH6900	TCCGTCGAGATCTCCGTCTC	ACTGGGGTTCTCCG TGAGC	75/100
bHLH9500	CCTCCTCCCCGATGACGAT	CCTTCTTCTTC TCCAGGCCG	150/200
bHLH5400	GGATAACGACATACCGTCAGCA	GCGTTGTGACTCATC TTTCG	150/200
GAPDH	GGTGCCAAGAAGGTCATCAT	TGGTCATCAAACCC TCAACA	18,75/25

(continued)

Table 2
(continued)

Target name	Forward primer (5′-3′)	Reverse primer (5′-3′)	Final concentration (nM) for hpPCR (F/R)
AP5.1	GCCTCAGTGGTAGGAGCATT	TTCAGCAAATGCGG TGGTTG	150/200
bHLH8000	TTTCCGCCGGGAGCTCAAT	CCAGACTCGGGC TTGAAGAA	150/200

5. Place all tubes in a thermocycler. Incubate at: 95 °C for 20 s; 60 °C for 30 min; followed by 4 °C. The assay can be paused at this step. Keep samples at −20 °C.

3.2.3 Capture and Amplification by RCA

1. Place all tubes on ice and add 60 μL of binding and wash buffer to each tube. Mix by gentle vortex and spin down.
2. Retrieve a slide and assemble the slide frame complex (*see* **Note 4**).
3. Transfer 80 μL of each sample to separate wells.
4. Cover the slide frame assembly to protect from evaporation and place it in the incubator oven at 38 °C for 30 min.
5. Place the labeling buffer at room temperature to thaw.
6. While keeping Polymerase on ice, prepare the following RCA mix on ice: 86 μL RCA buffer and 2.2 μL hpPCR polymerase.
7. Retrieve slide assembly and discard all samples (be careful not to scratch the slide surface).
8. Add 80 μL RCA mix to each well.
9. Cover the slide frame assembly to protect from evaporation and place it in the incubator oven at 38 °C for 2 h.
10. Retrieve slide assembly and discard all samples (be careful not to scratch the slide surface).
11. Vortex labeling buffer for 20 s, then add 45 μL to the side of each well.
12. Cover the assembly to protect it from evaporation and light. Place assembly in incubator oven at 38 °C for 1 h.
13. Discard all samples (be careful not to scratch the slide surface).
14. Add 50 μL binding and wash buffer to each well and subsequently discard the entire volume.
15. Repeat the previous step three times (i.e., total of four washes) (*see* **Note 5**).
16. Place a cover slip on the bench and dispense 35 μL mounting medium evenly along the center line of the entire cover slip.

17. Disassemble the slide frame complex and carefully place the slide with samples facing the cover slip so that Mounting Medium spreads evenly across the samples (*See* **Note 6**).
18. Incubate the slide at room temperature protected from light for at least 2 h.

3.2.4 Fluorescence Imaging

1. Start the computer, software, camera, light source, and APX100 imager.
2. Launch cellSens APEX software.
3. Select the sample holder for either slides or plates based on the sample type and insert the sample.
4. Open the microscope lid, place your sample holder into the APX100, and proceed to the software.
5. Insert the sample holder into the imaging position by pressing "Insert/Eject" and load well configuration for either slides or plates.
6. Select the 20× objective, click on select pattern, select multiple image alignment, and define number of tiles acquired per well (5 × 5). Mark the wells containing samples in the well navigator tab.
7. Calibrate well configuration and define initial Z-heights of the predefined focus points for each slide/plate.
8. Define experiment ID and click the settings icon next to "Automatic saving" to define the saving directory.
9. Press "Start" to initiate the imaging. Figure 1 shows a representative merged channel image of a full 24-well frame well.
10. Proceed to result analysis, the analysis script can be started directly after the imaging is started to process the samples as they are being acquired.

3.2.5 Image Analysis Using hpPCR Viewer

1. Launch "hpPCR viewer" from the desktop.
2. In the processor tab, define the folder path to the experiment images, enter panel ID, press save settings and analyze.
3. After the analysis is completed, a result folder is created in the saving directory with the data automatically exported in csv format. The sheet contains a sample list with corresponding target counts and quality metrics for each sample.
4. To further inspect the data of individual samples, click on the viewer tab in the software.
5. In this section 2D scatterplots for the samples (Fig. 2) can be viewed to inspect the data of individual samples. Here the user can inspect the cluster quality across different control channel C3–5.

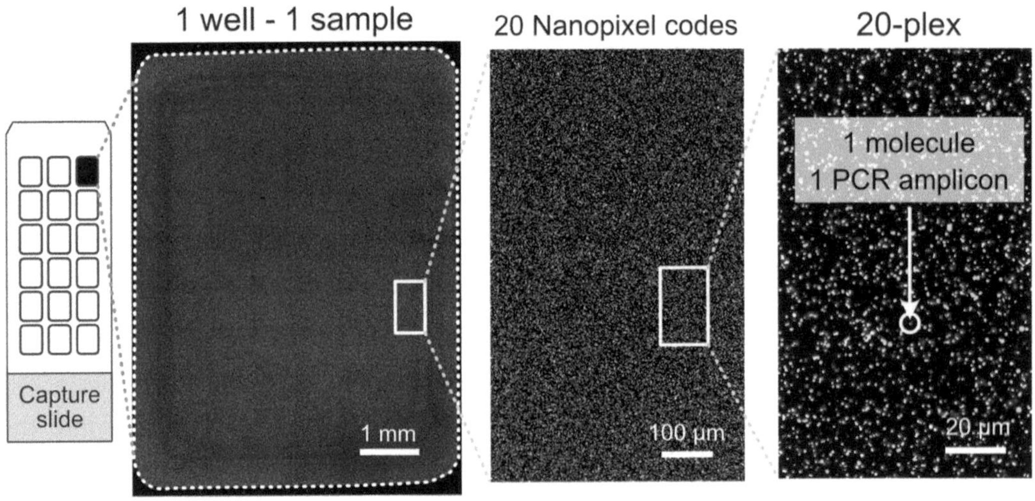

Fig. 1 Representative raw merged channel fluorescence microscopy image output of one well after hpPCR analysis of a control root sample. A total of approximately 600,000 counts were obtained in a full 6.8 × 5.4 mm well. Single RCA amplicons (the dots in the image) are specifically probed with several Nanopixel probes of each optical code. The number of amplicons of each color/spectral code are directly proportional to the pre-PCR input copies of each target sequence

6. To export plots for all samples of the experiment, select a sample, select control channels, and press generate experimental figures; a folder called experimental figures is created in the saving directory containing the plots.
7. Proceed to the readout section, select reference target, which will be used for data normalization.
8. The readout section can be used to inspect the target readout across samples. Mark samples to view from the dropdown list, then mark which targets to display. A column plot is then generated with data from the selected targets and samples (Fig. 3). Click save to save the plot in the experimental gure's subfolder.

3.3 Reverse Transcription and qPCR

Specific target sequences were selected for cross-validation with qPCR. These experiments were run on biological replicate tissue samples. The comparison between the qPCR and hpPCR results is shown in Figs. 4 and 5.

3.3.1 Protocol for cDNA Synthesis

1. Use 1000 ng of RNA from each sample for cDNA synthesis using the qScript Ultra SuperMix (Quantabio)
2. Thaw components, mix and centrifuge before use
3. Prepare the reaction mix consisting of nuclease-free water, template RNA, and 5X qScript Ultra SuperMix, adding up to final volume of 20 μL

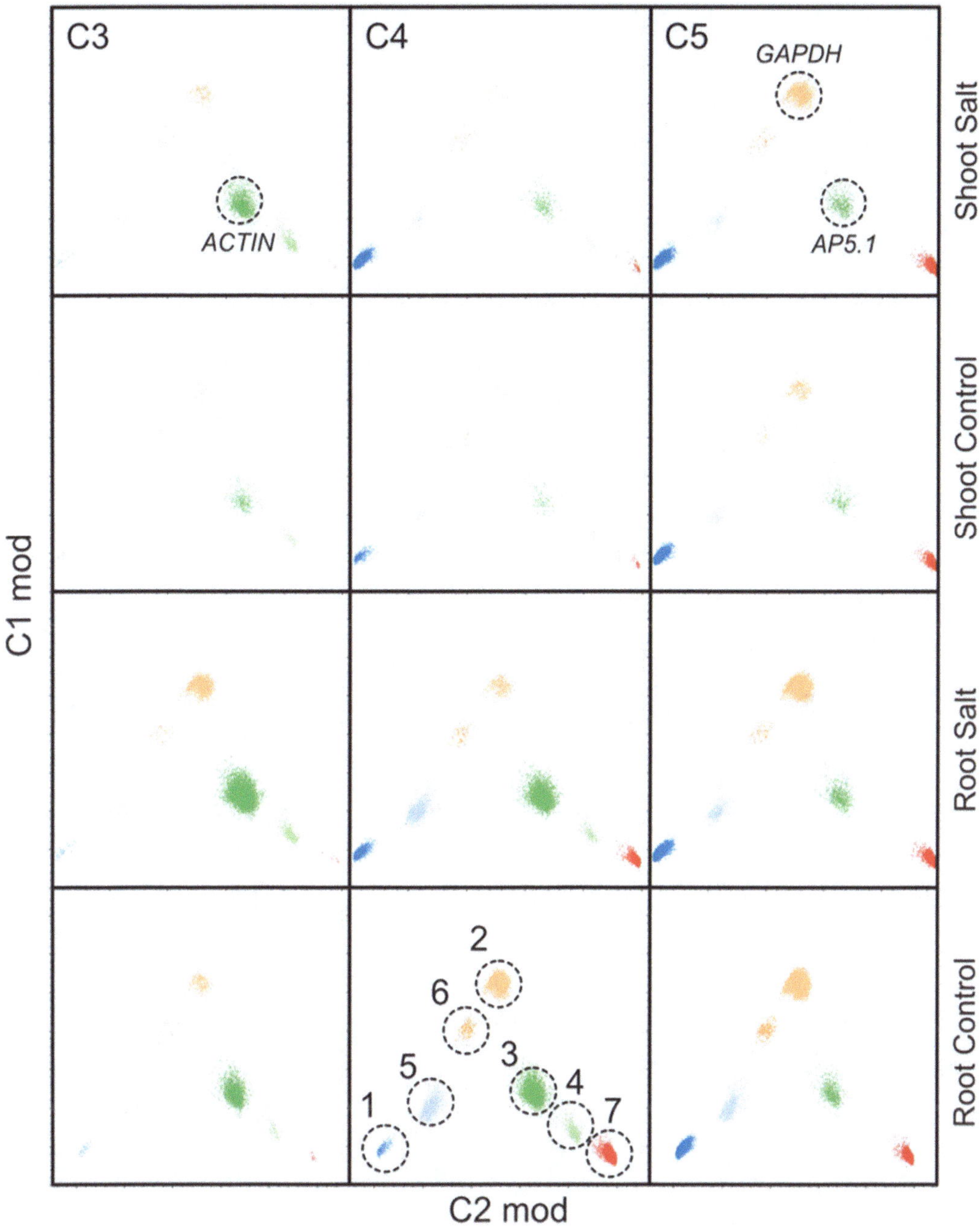

Fig. 2 Example scatter plots of channels C1 to C5 for control root/shoot and salt stress root/shoot samples. Each cluster of spots in the plots is an optical code generated by C1-C2 intensities normalized to the internal reference channel, comprising of multiple molecular counts within that spot. Unique combinations of cluster positions and control channels C3-C5 generates 21 (7 × 3) unique optical codes, of which 20 codes are used in the assay. C1-C2 code identifiers 1–7 refer to Nanopixel code according to Table 1. The code positions of the three internal control targets—*ACTIN*, *GAPDH,* and *AP5.1* are highlighted

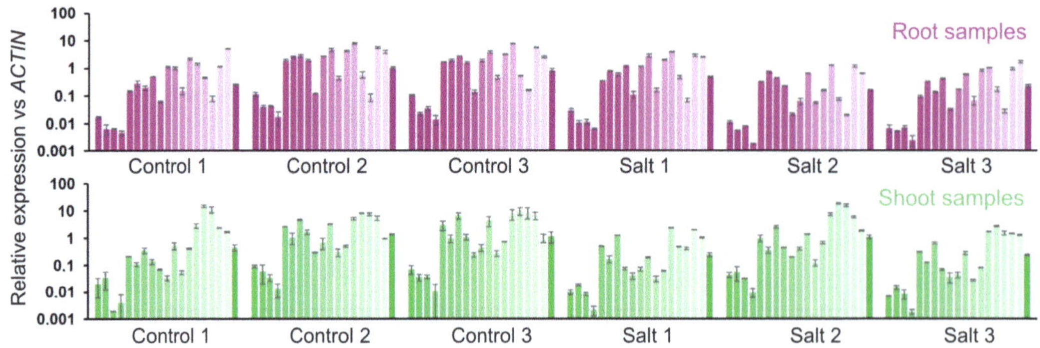

Fig. 3 Example of hpPCR relative gene expression output for the 19 target RNA sequences analyzed in three shoot and three root samples in both control and salt stress ("salt") conditions. Each bar represents 1 of the 19 targets, following the order listed in Table 1. *ACTIN* target counts were used as internal normalization control

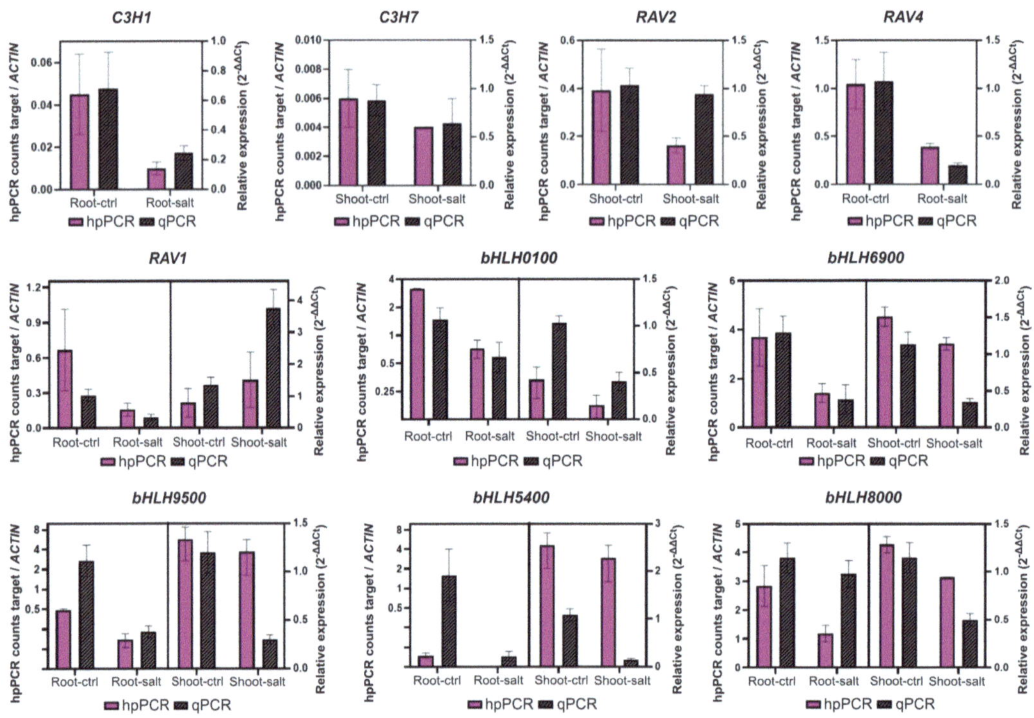

Fig. 4 Relative expression of each target gene, normalized to *ACTIN*, using hpPCR (in purple) and cross-validation using a 2-step RT-qPCR assay for selected targets (in stained black)

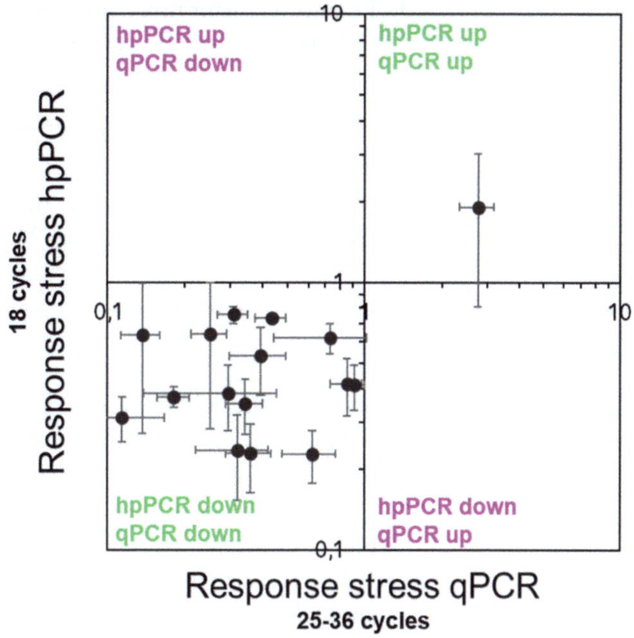

Fig. 5 Relative response of control vs. salt stress conditions for each target gene using hpPCR and 2-step RT-qPCR. Error bars refer to ±SD for qPCR (horizontal bars) and hpPCR (vertical bars)

4. Mix by gently vortexing, followed by brief centrifugation (use a tabletop centrifuge)
5. Incubate using a thermal cycler using the following protocol:
   (a) 2 min at 25 °C
   (b) 10 min at 55 °C
   (c) 1 min at 95 °C
   (d) Hold at 4 °C
6. Dilute the synthesized cDNA in 1:5 dilution with 80 μL nuclease-free water (stock dilution) (20 μL of cDNA +80 μL dH2O)
7. Transfer 20 μL of the stock dilution into new tube and further dilute with an additional 80 μL of nuclease free water (working dilution)
8. Store the synthesized cDNA samples at −20 °C

3.3.2 Protocol for qPCR

1. Defrost cDNA samples (working dilution) on ice and spin down using a table centrifuge
2. Add 4 μL of cDNA on the bottom of the 96-well plate (96-Well PCR Plates, Bio-Rad) using a multipipette (or a multichannel pipette)

3. Prepare a master mix consisting of 5 μL SYBR green and 0.5 μL of both forward and reverse primers (total volume 6 μL)
4. Add 6 μL of Master mix per well
5. Place plastic cover on the 96-well plate and spin for 1 min at 500 g
6. Run your protocol in the Bio-Rad CFX96 Real-Time PCR Detection System (*See* **Note 7**)

4 Notes

1. Seeds of the wheat variety BARI Gom-25, developed by the Bangladesh Agricultural Research Institute, that have previously been utilized in salt tolerance tests [12–15].
2. The incubation step at +38 °C requires a dark cover (preferably) to prevent evaporation and protect samples from potential light exposure if the oven interior is illuminated. Suitable covers include pipette box lids, petri dish lids, or press-seal with aluminum foil. The cover does not need to be in perfect contact with the top of the slide frame (if slides are used); its main purpose is to minimize evaporation.
3. Plant growth conditions and plant tissue extraction were performed according to a modified version of previously described procedures [14, 15]: seedlings were transferred to a hydroponic cultivation system after 4 days and had 6 days of hydroponic growth before being transferred to a new hydroponic solution with or without 100 mM NaCl, followed by tissue sample collection after 14 days.
4. Depending on the total number of samples analyzed, the hpPCR assay can be performed either on slides assembled with an SBS 24-well format frame or on 96-well plates. If using 96-well plates, skip the steps involving frame assembly, disassembly, and cover slip mounting.
5. For efficient liquid removal during hybridization and washing, use a multichannel pipette or dispenser. After discarding the liquid, ensure that the slide surface does not dry out by swiftly proceeding to the next step.
6. Slide frames can be reused with proper cleaning, e.g., rinsing under flowing tap water followed briefly by Milli-Q H_2O washing, then thoroughly drying them. If using 96-well plates, add mounting medium directly into the wells; the plate is ready for imaging. For slides, detailed mounting instructions are as follows: wear gloves, carefully place a coverslip on a dust-free section of the lab bench, and dispense 35 μL of mounting medium along the entire center line. Gently lower the slide so

the side where samples have been captured makes contact with the mounting medium, then allow it to spread evenly. Occasional trapped air bubbles can be removed by applying slight pressure. Carefully wipe off any excess mounting medium in case it overflows the edges. Avoid remounting coverslips if air bubbles form, as this often leads to poorer results.

7. RT-qPCR was performed on a Bio-Rad CFX96 system with SsoAdvanced Universal SYBR Green Supermix (Bio-Rad), using the wheat *ACTIN* gene as a reference gene. Data were analyzed via the 2–ΔΔCt method with three biological and three technical replicates. All primer pairs were experimentally optimized with analysis of melting curves to validate the absence of primer dimers and ensure detection specificity of all target genes.

Acknowledgments

We thank Tanvir Abedin, Emma Egonsson, Sameer Hassan, Richard Hembrom, and Frida Moberg for assistance with primer design, plant tissue isolation, and RT-qPCR. The authors gratefully acknowledge the financial support from the Swedish Research Council (2021-04265).

References

1. Agarwal PK, Gupta B, Huang B (2013) Bioengineering for salinity tolerance in plants: state of the art. Mol Biotechnol 54:102–123
2. Aleem M, Razzaq MK, Aleem M et al (2024) Genome-wide association study provides new insight into the underlying mechanism of drought tolerance during seed germination stage in soybean. Sci Rep 14:20765. https://doi.org/10.1038/s41598-024-71357-8
3. Arora NK (2019) Impact of climate change on agriculture production and its sustainable solutions. Environ Sustain 2:95–96. https://doi.org/10.1007/s42398-019-00078-w
4. Baillo EH, Kimotho RN, Zhang Z, Xu P (2019) Transcription factors associated with abiotic and biotic stress tolerance and their potential for crops improvement. Genes 10:771. https://doi.org/10.3390/genes10100771
5. Ballesta P, Mora F, Del Pozo A (2020) Association mapping of drought tolerance indices in wheat: QTL-rich regions on chromosome 4A. Sci Agric 77:e20180153
6. Giraldo P, Benavente E, Manzano-Agugliaro F, Gimenez E (2019) Worldwide research trends on wheat and barley: a bibliometric comparative analysis. Agronomy 9:352. https://doi.org/10.3390/agronomy9070352
7. Gupta B, Huang B (2014) Mechanism of salinity tolerance in plants: physiological, biochemical, and molecular characterization. https://doi.org/10.1155/2014/701596
8. Hassan S, Lethin J, Blomberg R, Mousavi H, Aronsson H (2019) In silico based screening of WRKY genes for identifying functional genes regulated by WRKY during salt stress. Comput Biol Chem 83:107131
9. Haoran R, Wenjing Y, Weikun J, Muhammad Owais S, Yuming L, Xianhan Q, Patrick C, Tao X, Nan M, Junping G, Xiaofeng Z (2024) Multi-omics analysis reveals key regulatory defense pathways and genes involved in salt tolerance of rose plants. Hortic Res 11:5. https://doi.org/10.1093/hr/uhae068
10. Hossain A, Skalicky M, Brestic M, Maitra S, Ashraful Alam M, Syed MA, Hossain J, Sarkar S, Saha S, Bhadra P, Shankar T (2021) Consequences and mitigation strategies of abiotic stresses in wheat (Triticum aestivum L.) under the changing climate. Agronomy 11:241

11. Majeed A, Muhammad Z (2019) Salinity: a major agricultural problem—causes, impacts on crop productivity and management strategies. In Plant abiotic stress tolerance: Agronomic, molecular and biotechnological approaches (pp. 83–99). Cham: Springer International Publishing
12. Lethin J, Byrt C, Berger B, Brien C, Jewell N, Roy S, Mousavi H, Sukumaran S, Olsson O, Aronsson H (2022) Improved salinity tolerance-associated variables observed in EMS mutagenized wheat lines. Int J Mol Sci 23:1386. https://doi.org/10.3390/ijms231911386
13. Lethin J, Shakil SSM, Hassan S, Sirijovski N, Töpel M, Olsson O, Aronsson H (2020) Development and characterization of an EMS-mutagenized wheat population and identification of salt-tolerant wheat lines. BMC Plant Biol 20:18. https://doi.org/10.1186/s12870-019-2137-8
14. Liu X, Sukumaran S, Viitanen E, Hassan S, Naik N, Aronsson H (2024) An accurate representation of the number of bZIP transcription factors in the Triticum aestivum (Wheat) genome and the regulation of functional genes during salt stress. Curr Issues Mol Biol 46: 4417–4436. https://doi.org/10.3390/cimb46050268
15. Sukumaran S, Lethin J, Liu X, Pelc J, Zeng P, Hassan S, Aronsson H (2023) Genome-wide analysis of MYB transcription factors in the wheat genome and their roles in salt stress response. Cells 12:1431. https://doi.org/10.3390/cells12101431
16. Seo PJ, Kim SG, Park CM (2008) Membrane-bound transcription factors in plants. Cell 3: 550–556
17. Singh KB, Foley RC, Onate-Sanchez L (2002) Transcription factors in plant defense and stress responses. Curr Opin Plant Biol 5:430–436
18. Yoon Y, Deok Seo DH, Shin H, Kim HJ, Kim CM, Jang G (2020) The role of stress-responsive transcription factors in modulating abiotic stress tolerance in plants. Agronomy 10: 788. https://doi.org/10.3390/agronomy10060788
19. Ren H, Yang W, Jing W, Owais Shahid M, Liu Y, Qiu X, Choisy P, Tao Xu , Nan Ma , Gao J, Zhou X (2024) Multi-omics analysis reveals key regulatory defense pathways and genes involved in salt tolerance of rose plants. Hortic Res, Vol 11, Issue 5, uhae068, https://doi.org/10.1093/hr/uhae068
20. Hassan S, Berk K, Aronsson H (2021) Evolution and identification of DREB transcription factors in the wheat genome: modeling, docking and simulation of DREB proteins associated with salt stress. J Biomol Struct Dyn. 23:1–14 https://doi.org/10.1080/07391102.2021.1894980

Index

Guillaume Pavlovic (ed.), *Digital PCR: Methods and Protocols*, Methods in Molecular Biology, vol. 2969, https://doi.org/10.1007/978-1-0716-4767-7,

MIX
Papier aus verantwortungsvollen Quellen
Paper from responsible sources
FSC® C105338

If you have any concerns about our products,
you can contact us on
ProductSafety@springernature.com

In case Publisher is established outside the EU,
the EU authorized representative is:
Springer Nature Customer Service Center GmbH
Europaplatz 3, 69115 Heidelberg, Germany

Printed by Libri Plureos GmbH
in Hamburg, Germany